ROWAN UNIVERSITY
CAMPBELL LIBRARY
201 MULLICA HILL RD.
GLASSBORO, NJ 08028-1701

Graduate Texts in Mathematics 226

Editorial Board
S. Axler   F.W. Gehring   K.A. Ribet

Graduate Texts in Mathematics

(continued after index)

Kehe Zhu

Spaces of Holomorphic Functions in the Unit Ball

 Springer

Kehe Zhu
Department of Mathematics
State University of New York at Albany
Albany, NY 12222
USA
kzhu@math.albany.edu

Editorial Board

S. Axler	F.W. Gehring	K.A. Ribet
Mathematics Department	Mathematics Department	Mathematics Department
San Francisco State	East Hall	University of California,
University	University of Michigan	Berkeley
San Francisco, CA 94132	Ann Arbor, MI 48109	Berkeley, CA 94720-3840
USA	USA	USA
axler@sfsu.edu	fgehring@math.lsa.umich.edu	ribet@math.berkeley.edu

Mathematics Subject Classification (2000): MSCM12198, SCM12198, SCM12007

Library of Congress Cataloging-in-Publication Data
Zhu, Kehe, 1961–
    Spaces of holomorphic functions in the unit ball / Kehe Zhu.
     p.   cm.
    Includes bibliographical references and index.
    ISBN 0-387-22036-4 (alk. paper)
    1. Holomorphic functions.   2. Unit ball.   I. Title.
  QA331.Z48   2004
  515'.98—dc22
                                              2004049191

ISBN 0-387-22036-4   Printed on acid-free paper.

© 2005 Springer Science+Business Media, Inc.
All rights reserved. This work may not be translated or copied in whole or in part without the written permission of the publisher (Springer Science+Business Media, Inc., 233 Spring Street, New York, NY 10013, USA), except for brief excerpts in connection with reviews or scholarly analysis. Use in connection with any form of information storage and retrieval, electronic adaptation, computer software, or by similar or dissimilar methodology now known or hereafter developed is forbidden.
The use in this publication of trade names, trademarks, service marks, and similar terms, even if they are not identified as such, is not to be taken as an expression of opinion as to whether or not they are subject to proprietary rights.

Printed in the United States of America.   (EB)

9 8 7 6 5 4 3 2 1

springeronline.com

3 3001 00910 309 3

To my family: Peijia, Peter, and Michael

Preface

There has been a flurry of activity in recent years in the loosely defined area of *holomorphic spaces*. This book discusses the most well-known and widely used spaces of holomorphic functions in the unit ball of $\mathbb{C}^n$: Bergman spaces, Hardy spaces, Besov spaces, Lipschitz spaces, BMOA, and the Bloch space.

The theme of the book is very simple. For each scale of spaces, I discuss integral representations, characterizations in terms of various derivatives, atomic decomposition, complex interpolation, and duality. Very few other properties are discussed.

I chose the unit ball as the setting because most results can be achieved there using straightforward formulas without much fuss. In fact, most of the results presented in the book are based on the explicit form of the Bergman and Cauchy-Szëgo kernels. The book can be read comfortably by anyone familiar with single variable complex analysis; no prerequisite on several complex variables is required.

Few of the results in the book are new, but most of the proofs are originally constructed and considerably simpler than the existing ones in the literature. There is some obvious overlap between this book and Walter Rudin's classic "*Function Theory in the Unit Ball of* $\mathbb{C}^n$". But the overlap is not substantial, and it is my hope that the two books will complement each other.

The book is essentially self-contained, with two exceptions worth mentioning. First, the existence of boundary values for functions in the Hardy spaces H^p is proved only for $p \geq 1$; a full proof can be found in Rudin's book. Second, the complex interpolation between the Hardy spaces H^1 and H^p (or BMOA) is not proved; a full proof requires more real variable techniques.

The exercises at the end of each chapter vary greatly in the level of difficulty. Some of them are simple applications of the main theorems, some are obvious generalizations or variations, while others are difficult results that complement the main text. In the latter case, at least one reference is provided for the reader.

I apologize in advance for any misrepresentation in the short sections entitled "Notes", for any omission of significant references, and for having not included one or several of your favorite theorems. I did not even try to compile a comprehensive bibliography.

The topics chosen for the book, and the way they are organized, reflect entirely my own taste, preference/prejudice, and research/teaching experience. Among the topics that I thought about seriously but eventually decided not to include are the so-called Arveson space, the so-called Q_p spaces, and general holomorphic Sobolev spaces. Of course, the Bergman spaces, the Bloch space, the holomorphic Besov spaces, and the holomorphic Lipschitz spaces can all be considered special cases of a more general family of holomorphic Sobolev spaces. It appears to me that the relatively elegant treatment of these special cases is more interesting and appealing than an otherwise more cumbersome presentation of an exhaustive class of functions.

During the preparation of the manuscript I received help and advice from Boo Rim Choe, Joe Cima, Richard Rochberg, and Jie Xiao. It is my pleasure to record my thanks to them here. I am particularly grateful to Ruhan Zhao, who read a preliminary version of the entire manuscript and caught numerous misprints and mistakes. My family—Peijia, Peter, and Michael—provided me with love, understanding, and blocks of uninterrupted time that is necessary for the completion of any mathematical project.

Albany, June 2004   *Kehe Zhu*

Contents

1

Preliminaries

In this chapter we set the stage and discuss the basic properties of the unit ball. Several results and techniques of this chapter will be used repeatedly in later chapters. These include the change of variables formula, the fractional differential and integral operators, the basic integral estimate of the kernel functions (Theorem 1.12), and the Marcinkiewicz interpolation theorem. Also, the radial derivative, the invariant Laplacian, the automorphism group, and the Bergman metric are essential concepts for the rest of the book.

1.1 Holomorphic Functions

Let $\mathbb{C}$ denote the set of complex numbers. Throughout the book we fix a positive integer n and let

$$\mathbb{C}^n = \mathbb{C} \times \cdots \times \mathbb{C}$$

denote the Euclidean space of complex dimension n. Addition, scalar multiplication, and conjugation are defined on $\mathbb{C}^n$ componentwise. For

$$z = (z_1, \cdots, z_n), \qquad w = (w_1, \cdots, w_n),$$

in $\mathbb{C}^n$, we define

$$\langle z, w \rangle = z_1 \overline{w}_1 + \cdots + z_n \overline{w}_n,$$

where $\overline{w}_k$ is the complex conjugate of w_k. We also write

$$|z| = \sqrt{\langle z, z \rangle} = \sqrt{|z_1|^2 + \cdots + |z_n|^2}.$$

The space $\mathbb{C}^n$ becomes an n-dimensional Hilbert space when endowed with the inner product above. The standard basis for $\mathbb{C}^n$ consists of the following vectors:

$$e_1 = (1, 0, 0, \cdots, 0), \quad e_2 = (0, 1, 0, \cdots, 0), \quad \cdots, \quad e_n = (0, 0, \cdots, 0, 1).$$

Via this basis we will identify the linear transformations of $\mathbb{C}^n$ with $n \times n$ matrices whose entries are complex numbers. Another vector in $\mathbb{C}^n$ that we often use is the zero vector,

$$0 = (0, 0, \cdots, 0).$$

The reader should have no problem accepting this slightly confusing notation.

The open unit ball in $\mathbb{C}^n$ is the set

$$\mathbb{B}_n = \{z \in \mathbb{C}^n : |z| < 1\}.$$

The boundary of $\mathbb{B}_n$ will be denoted by $\mathbb{S}_n$ and is called the unit sphere in $\mathbb{C}^n$. Thus

$$\mathbb{S}_n = \{z \in \mathbb{C}^n : |z| = 1\}.$$

Occasionally, we will also need the closed unit ball

$$\overline{\mathbb{B}}_n = \{z \in \mathbb{C}^n : |z| \leq 1\} = \mathbb{B}_n \cup \mathbb{S}_n.$$

The definition of holomorphic functions in several complex variables is more subtle than the one variable case, namely, several natural definitions exist and they all turn out to be equivalent. We will freely use these classical definitions but will not attempt to prove their mutual equivalence. Text books such as [61] or [89] all contain the necessary proofs.

Perhaps the most elementary definition of holomorphic functions in $\mathbb{B}_n$ is via complex partial derivatives. Thus a function $f : \mathbb{B}_n \to \mathbb{C}$ is said to be holomorphic in $\mathbb{B}_n$ if for every point $z \in \mathbb{B}_n$ and for every $k \in \{1, 2, \cdots, n\}$ the limit

$$\lim_{\lambda \to 0} \frac{f(z + \lambda e_k) - f(z)}{\lambda}$$

exists (and is finite), where $\lambda \in \mathbb{C}$. When f is holomorphic in $\mathbb{B}_n$, we use the notation

$$\frac{\partial f}{\partial z_k}(z)$$

to denote the above limit and call it the partial derivative of f with respect to z_k.

Equivalently, a function $f : \mathbb{B}_n \to \mathbb{C}$ is holomorphic if

$$f(z) = \sum_m a_m z^m, \qquad z \in \mathbb{B}_n.$$

Here the summation is over all multi-indexes

$$m = (m_1, \cdots, m_n),$$

where each m_k is a nonnegative integer and

$$z^m = z_1^{m_1} \cdots z_n^{m_n}.$$

The series above is called the Taylor series of f at the origin; it converges absolutely and uniformly on each of the sets

$$r\mathbb{B}_n = \{z \in \mathbb{C}^n : |z| \le r\}, \qquad 0 < r < 1.$$

If we let

$$f_k(z) = \sum_{|m|=k} a_m z^m$$

for each $k \ge 0$, where

$$|m| = m_1 + \cdots + m_n,$$

then the Taylor series of f can be rewritten as

$$f(z) = \sum_{k=0}^{\infty} f_k(z).$$

This is called the homogeneous expansion of f; each f_k is a homogeneous polynomial of degree k. Both the Taylor series and the homogeneous expansion of f are uniquely determined by f.

When a function $f : \mathbb{B}_n \to \mathbb{C}$ is holomorphic, all higher order partial derivatives exist and are still holomorphic. For a multi-index $m = (m_1, \cdots, m_n)$ we will employ the notation

$$\partial^m f = \frac{\partial^m f}{\partial z^m} = \frac{\partial^{|m|} f}{\partial z_1^{m_1} \cdots \partial z_n^{m_n}}.$$

Another common notation we adopt for a multi-index m is the following:

$$m! = m_1! \cdots m_n!.$$

In particular, we have the multi-nomial formula

$$(z_1 + \cdots + z_n)^N = \sum_{|m|=N} \frac{N!}{m!} z^m. \tag{1.1}$$

The space of all holomorphic functions in $\mathbb{B}_n$ will be denoted by $H(\mathbb{B}_n)$. We use $H^\infty(\mathbb{B}_n)$, or simply H^∞, to denote the space of all bounded holomorphic functions in $\mathbb{B}_n$. The ball algebra, denoted by $A(\mathbb{B}_n)$, consists of all functions in $H(\mathbb{B}_n)$ that are continuous up to the boundary $\mathbb{S}_n$.

1.2 The Automorphism Group

A mapping $F : \mathbb{B}_n \to \mathbb{C}^N$, where N is a positive integer, is given by N functions as follows:

$$F(z) = (f_1(z), \cdots, f_N(z)), \qquad z \in \mathbb{B}_n.$$

We say that F is a holomorphic mapping if each f_k is holomorphic in $\mathbb{B}_n$.

It is clear that any holomorphic mapping $F : \mathbb{B}_n \to \mathbb{C}^N$ has a Taylor type expansion

$$F(z) = \sum a_m z^m,$$

where $m = (m_1, \cdots, m_n)$ is a multi-index of nonnegative integers and each a_m belongs to $\mathbb{C}^N$. Similarly, F admits a homogeneous expansion

$$F(z) = \sum_{k=0}^{\infty} F_k(z),$$

where all N component functions of each F_k are homogeneous polynomials of degree k.

For a holomorphic mapping

$$F(z) = (f_1(z), \cdots, f_N(z)),$$

it will be convenient for us to write

$$F'(z) = \left(\frac{\partial f_i}{\partial z_j}(z) \right)_{N \times n} = \begin{pmatrix} \dfrac{\partial f_1}{\partial z_1} & \cdots & \dfrac{\partial f_1}{\partial z_n} \\ \cdots & \cdots & \cdots \\ \dfrac{\partial f_N}{\partial z_1} & \cdots & \dfrac{\partial f_N}{\partial z_n} \end{pmatrix}.$$

Thus the homogeneous expansion of F begins as follows:

$$F(z) = F(0) + F'(0)z + \cdots.$$

Here we think of the term $F'(0)z$ as the matrix $F'(0)$ times the column vector z in $\mathbb{C}^n$.

A mapping $F : \mathbb{B}_n \to \mathbb{B}_n$ is said to be bi-holomorphic if

(1) F is one-to-one and onto.
(2) F is holomorphic.
(3) F^{-1} is holomorphic.

The automorphism group of $\mathbb{B}_n$, denoted by $\mathrm{Aut}(\mathbb{B}_n)$, consists of all bi-holomorphic mappings of $\mathbb{B}_n$. It is clear that $\mathrm{Aut}(\mathbb{B}_n)$ is a group with composition being the group operation. Traditionally, bi-holomorphic mappings are also called automorphisms.

One class of automorphisms is easy to describe. Recall that $\mathbb{C}^n$ is a Hilbert space of complex dimension n. Thus every unitary mapping of $\mathbb{C}^n$ is an automorphism of $\mathbb{B}_n$. Relative to the basis $\{e_1, \cdots, e_n\}$, every $n \times n$ unitary matrix U is an automorphism. The following lemma shows that the unitary transformations are exactly the automorphisms that leave the origin of $\mathbb{C}^n$ fixed.

Lemma 1.1. *An automorphism φ of $\mathbb{B}_n$ is a unitary transformation of $\mathbb{C}^n$ if and only if $\varphi(0) = 0$.*

Proof. Assume that φ is an automorphism of $\mathbb{B}_n$ with $\varphi(0) = 0$. Fix any complex number λ with $|\lambda| = 1$ and consider the holomorphic mapping $F : \mathbb{B}_n \to \mathbb{B}_n$ defined by

$$F(z) = \varphi^{-1}(\overline{\lambda}\,\varphi(\lambda z)), \qquad z \in \mathbb{B}_n.$$

Clearly, $F(0) = 0$, and $F'(0)$ is the $n \times n$ identity matrix. If F is not the identity mapping of $\mathbb{B}_n$, then the homogeneous expansion of F can be written as

$$F(z) = z + \sum_{k=l}^{\infty} F_k(z),$$

where $l \geq 2$ and $F_l(z)$ is not zero. If we compose F with itself N times, then the resulting homogeneous expansion is

$$F \circ F \circ \cdots \circ F(z) = z + N F_l(z) + \cdots,$$

where the omitted terms consist of polynomials of degree greater than l. Letting $N \to \infty$ clearly leads to $F_l = 0$, which contradicts the earlier assumption that $F_l \neq 0$. This shows that $F(z) = z$, or $\varphi(\lambda z) = \lambda \varphi(z)$ for all $z \in \mathbb{B}_n$. This in turn implies that the homogeneous expansion of φ consists of the linear term alone, namely, φ is a linear transformation. Since φ maps $\mathbb{B}_n$ onto itself, we conclude that φ must be a unitary transformation. $\qquad\square$

Another class of automorphisms consists of symmetries of $\mathbb{B}_n$, which are also called involutive automorphisms or involutions. Thus for any point $a \in \mathbb{B}_n - \{0\}$ we define

$$\varphi_a(z) = \frac{a - P_a(z) - s_a Q_a(z)}{1 - \langle z, a \rangle}, \qquad z \in \mathbb{B}_n, \tag{1.2}$$

where $s_a = \sqrt{1 - |a|^2}$, P_a is the orthogonal projection from $\mathbb{C}^n$ onto the one dimensional subspace $[a]$ generated by a, and Q_a is the orthogonal projection from $\mathbb{C}^n$ onto $\mathbb{C}^n \ominus [a]$. It is clear that

$$P_a(z) = \frac{\langle z, a \rangle}{|a|^2} a, \qquad z \in \mathbb{C}^n, \tag{1.3}$$

and

$$Q_a(z) = z - \frac{\langle z, a \rangle}{|a|^2} a, \qquad z \in \mathbb{B}_n. \tag{1.4}$$

When $a = 0$, we simply define $\varphi_a(z) = -z$. It is obvious that each φ_a is a holomorphic mapping from $\mathbb{B}_n$ into $\mathbb{C}^n$.

Lemma 1.2. *For each $a \in \mathbb{B}_n$ the mapping φ_a satisfies*

$$1 - |\varphi_a(z)|^2 = \frac{(1 - |a|^2)(1 - |z|^2)}{|1 - \langle z, a \rangle|^2}, \qquad z \in \mathbb{B}_n, \tag{1.5}$$

and

$$\varphi_a \circ \varphi_a(z) = z, \qquad z \in \mathbb{B}_n. \tag{1.6}$$

In particular, each φ_a is an automorphism of $\mathbb{B}_n$ that interchanges the points 0 and a.

Proof. The case $a = 0$ is obvious. So we assume that $a \neq 0$.

Since $a - P_a(z)$ and $Q_a(z)$ are perpendicular in $\mathbb{C}^n$, we have

$$
\begin{aligned}
|a - P_a(z) - s_a Q_a(z)|^2 &= |a - P_a(z)|^2 + (1 - |a|^2)|Q_a(z)|^2 \\
&= |a|^2 - 2\operatorname{Re}\langle P_a(z), a \rangle + |P_a(z)|^2 \\
&\quad + (1 - |a|^2)(|z|^2 - |P_a(z)|^2).
\end{aligned}
$$

Manipulating the above expression using the facts that

$$
|a|^2 |P_a(z)|^2 = |\langle z, a \rangle|^2, \qquad \langle P_a(z), a \rangle = \langle z, a \rangle,
$$

we obtain

$$
|a - P_a(z) - s_a Q_a(z)|^2 = |1 - \langle z, a \rangle|^2 - (1 - |a|^2)(1 - |z|^2),
$$

which clearly leads to

$$
1 - |\varphi_a(z)|^2 = \frac{(1 - |a|^2)(1 - |z|^2)}{|1 - \langle z, a \rangle|^2}.
$$

In particular, we conclude that each φ_a is a holomorphic map from $\mathbb{B}_n$ into itself.

To prove the involutive property of φ_a, we first verify that

$$
1 - \langle \varphi_a(z), a \rangle = \frac{1 - |a|^2}{1 - \langle z, a \rangle},
$$

and

$$
P_a(\varphi_a(z)) = \frac{a}{|a|^2} \cdot \frac{|a|^2 - \langle z, a \rangle}{1 - \langle z, a \rangle}.
$$

Then a few lines of elementary calculations show that $\varphi_a \circ \varphi_a(z) = z$ for all $z \in \mathbb{B}_n$. This clearly implies that the mapping φ_a is invertible on $\mathbb{B}_n$ and its inverse is itself. In particular, its inverse is holomorphic, and so φ_a is an automorphism.

The properties that

$$
\varphi_a(0) = a, \qquad \varphi_a(a) = 0,
$$

follow easily from the definition of φ_a. $\qquad\square$

When identity (1.5) in the preceding lemma is polarized, the result is the following formula.

Lemma 1.3. *Suppose $a \in \mathbb{B}_n$. Then*

$$
1 - \langle \varphi_a(z), \varphi_a(w) \rangle = \frac{(1 - \langle a, a \rangle)(1 - \langle z, w \rangle)}{(1 - \langle z, a \rangle)(1 - \langle a, w \rangle)} \tag{1.7}
$$

for all z and w on the closed unit ball $\overline{\mathbb{B}}_n$.

The property $\varphi_a \circ \varphi_a(z) = z$ justifies the use of the term "involution" for φ_a. It turns out that the unitaries and the involutions generate the whole group $\mathrm{Aut}(\mathbb{B}_n)$.

Theorem 1.4. *Every automorphism φ of $\mathbb{B}_n$ is of the form*

$$\varphi = U\varphi_a = \varphi_b V,$$

where U and V are unitary transformations of $\mathbb{C}^n$, and φ_a and φ_b are involutions.

Proof. Suppose $\varphi \in \mathrm{Aut}(\mathbb{B}_n)$ and $a = \varphi^{-1}(0)$. Then the automorphism $\psi = \varphi \circ \varphi_a$ satisfies $\psi(0) = 0$. By Lemma 1.1, there exists a unitary transformation U of $\mathbb{C}^n$ such that $U = \varphi \circ \varphi_a$. Since φ_a is involutive, this gives $\varphi = U\varphi_a$. The other representation can be proved similarly. $\square$

Corollary 1.5. *Every φ in $\mathrm{Aut}(\mathbb{B}_n)$ extends to a homeomorphism of $\mathbb{S}_n$.*

Proof. It is obvious that every unitary transformation in $\mathrm{Aut}(\mathbb{B}_n)$ induces a homeomorphism of $\mathbb{S}_n$. By Lemma 1.2, every involution φ_a also extends to a homeomorphism on $\mathbb{S}_n$. $\square$

Given $\varphi \in \mathrm{Aut}(\mathbb{B}_n)$, we use $J_C\varphi(z)$ to denote the determinant of the complex $n \times n$ matrix $\varphi'(z)$ and call it the complex Jacobian of φ at z. If we identify $\mathbb{B}_n$ (in the natural way) with the unit ball in the $2n$-dimensional real Euclidean space $\mathbb{R}^{2n}$, then the mapping φ induces a real Jacobian determinant which we denote by $J_R\varphi(z)$. It is well known that

$$J_R\varphi(z) = |J_C\varphi(z)|^2 \tag{1.8}$$

for all $z \in \mathbb{B}_n$; see [61].

Lemma 1.6. *If we identify linear transformations of $\mathbb{C}^n$ with $n \times n$ matrices via the standard basis of $\mathbb{C}^n$, then for every $a \in \mathbb{B}_n - \{0\}$ we have*

$$\varphi_a'(0) = -(1 - |a|^2)P_a - \sqrt{1 - |a|^2}\, Q_a, \tag{1.9}$$

and

$$\varphi_a'(a) = -\frac{P_a}{1 - |a|^2} - \frac{Q_a}{\sqrt{1 - |a|^2}}. \tag{1.10}$$

Proof. For any $a \in \mathbb{B}_n$, $a \neq 0$, we can write

$$\varphi_a(z) = \left(a - P_a(z) - s_a Q_a(z)\right) \sum_{k=0}^{\infty} \langle z, a \rangle^k$$

$$= a + a\langle z, a \rangle - (P_a + s_a Q_a)(z) + O(|z|^2)$$

$$= a - s_a^2 P_a(z) - s_a Q_a(z) + O(|z|^2)$$

for $z \in \mathbb{B}_n$, where $s_a = \sqrt{1 - |a|^2}$. If we identify linear transformations of $\mathbb{C}^n$ with $n \times n$ matrices via the standard basis of $\mathbb{C}^n$, then the above shows that

$$\varphi_a'(0) = -s_a^2 P_a - s_a Q_a.$$

Similarly, a calculation using

$$\varphi_a(a + h) = \frac{-P_a(h) - s_a Q_a(h)}{s_a^2 - \langle h, a \rangle}$$

shows that

$$\varphi_a'(a) = -\frac{1}{s_a^2} P_a - \frac{1}{s_a} Q_a.$$

$\square$

Lemma 1.7. *For each $\varphi \in \mathrm{Aut}(\mathbb{B}_n)$ we have*

$$J_R \varphi(z) = \left(\frac{1 - |a|^2}{|1 - \langle z, a \rangle|^2} \right)^{n+1}, \tag{1.11}$$

where $a = \varphi^{-1}(0)$.

Proof. For any fixed a and z in $\mathbb{B}_n$ with $a \neq 0$, we let $w = \varphi_a(z)$ and consider the automorphism

$$U = \varphi_w \circ \varphi_a \circ \varphi_z.$$

Since $U(0) = 0$, Lemma 1.1 shows that U is a unitary. Rewrite $\varphi_a = \varphi_w \circ U \circ \varphi_z$ and apply the chain rule. We obtain

$$\varphi_a'(z) = \varphi_w'(0) U \varphi_z'(z),$$

and so

$$J_C \varphi_a(z) = \det(\varphi_w'(0)) \det(\varphi_z'(z)).$$

By (1.9), the linear transformation $\varphi_w'(0)$ has a one-dimensional eigenspace with eigenvalue $-(1 - |w|^2)$ and an $(n - 1)$-dimensional eigenspace with eigenvalue $-\sqrt{1 - |w|^2}$; so its determinant equals $(-1)^n (1 - |w|^2)^{(n+1)/2}$. This, together with a similar computation of the determinant of $\varphi_z'(z)$ using (1.10), shows that

$$J_R \varphi_a(z) = |J_C \varphi_a(z)|^2 = \left(\frac{1 - |w|^2}{1 - |z|^2} \right)^{n+1}.$$

An application of (1.5) then gives

$$J_R \varphi_a(z) = \left(\frac{1 - |a|^2}{|1 - \langle z, a \rangle|} \right)^{n+1}.$$

Every $\varphi \in \mathrm{Aut}(\mathbb{B}_n)$ can be written as $\varphi = U \varphi_a$, where $a = \varphi^{-1}(0)$. The general case follows from the special case obtained in the previous paragraph. $\square$

1.3 Lebesgue Spaces

Most spaces considered in the book will be defined in terms L^p integrals of the function or its derivatives. The measures we use in these integrals are based on the volume measure on the unit ball or the surface measure on the unit sphere.

We let dv denote the volume measure on $\mathbb{B}_n$, normalized so that $v(\mathbb{B}_n) = 1$. The surface measure on $\mathbb{S}_n$ will be denoted by $d\sigma$. Once again, we normalize σ so that $\sigma(\mathbb{S}_n) = 1$. The normalizing constants, namely, the actual volume of $\mathbb{B}_n$ and the actual surface area of $\mathbb{S}_n$, are not important to us, although their values will be determined as a by-product of the proof of Lemma 1.11 later in this section.

Lemma 1.8. *The measures v and σ are related by*

$$\int_{\mathbb{B}_n} f(z)\, dv(z) = 2n \int_0^1 r^{2n-1}\, dr \int_{\mathbb{S}_n} f(r\zeta)\, d\sigma(\zeta).$$

Proof. Let $dV = dx_1\, dy_1 \cdots dx_n\, dy_n$ be the actual Lebesgue measure in $\mathbb{C}^n$ (before normalization), where we identify each z_k with $x_k + iy_k$. Similarly, let dS be the surface measure on $\mathbb{S}_n$ before normalization. Then the Euclidean volume of the solid determined by dS in $\mathbb{S}_n$, $r > 0$, and $r + dr$, is given by

$$dV = \frac{dS}{S(1)}\left(V(r+dr) - V(r)\right).$$

Here, for $r > 0$, $V(r)$ is the actual volume of the ball

$$|z_1|^2 + \cdots + |z_n|^2 < r^2,$$

and $S(r)$ is the actual surface area of the sphere

$$|z_1|^2 + \cdots + |z_n|^2 = r^2.$$

From the change of variables $z_k = rw_k$, $1 \le k \le n$, we obtain

$$V(r) = \int_{|z_1|^2 + \cdots + |z_n|^2 < r^2} dV(z) = r^{2n} V(1).$$

It follows that

$$dV = \frac{V(1)}{S(1)}\left((r+dr)^{2n} - r^{2n}\right) dS.$$

Omitting powers of dr with exponents greater than 1, we get

$$dV = \frac{V(1)}{S(1)} 2n r^{2n-1}\, dr\, dS,$$

or

$$dv = 2n r^{2n-1}\, dr\, d\sigma.$$

$\square$

Lemma 1.8 will be referred to as integration in polar coordinates. The next lemma deals with integration on $\mathbb{S}_n$ of functions of fewer variables.

Lemma 1.9. *Suppose f is a function on $\mathbb{S}_n$ that depends only on $z_1, \cdots, z_k$, where $1 \le k < n$. Then f can be regarded as defined on $\mathbb{B}_k$ and*

$$\int_{\mathbb{S}_n} f \, d\sigma = \binom{n-1}{k} \int_{\mathbb{B}_k} (1 - |w|^2)^{n-k-1} f(w) \, dv_k(w),$$

where $\mathbb{B}_k$ is the open unit ball in $\mathbb{C}^k$ and dv_k is the normalized volume measure on $\mathbb{B}_k$.

Proof. For the purpose of this proof let P_k denote the orthogonal projection from $\mathbb{C}^n$ onto $\mathbb{C}^k$. Then

$$\int_{\mathbb{S}_n} f \, d\sigma = \int_{\mathbb{S}_n} f \circ P_k \, d\sigma.$$

By an approximation argument, it suffices for us to prove the result when f is continuous in $\mathbb{C}^k$ and has support in $r_0 \mathbb{B}_k$, where r_0 is some constant in $(0,1)$. Fix such an f and consider the integrals

$$I(r) = \int_{r\mathbb{B}_n} f \circ P_k \, dv, \qquad 0 < r < \infty.$$

We integrate in polar coordinates to get

$$I(r) = 2n \int_0^r t^{2n-1} \, dt \int_{\mathbb{S}_n} f \circ P_k(t\zeta) \, d\sigma(\zeta).$$

We then differentiate this to obtain

$$I'(1) = 2n \int_{\mathbb{S}_n} f \circ P_k \, d\sigma.$$

On the other hand, an application of Fubini's theorem shows that

$$I(r) = c \int_{\mathbb{B}_k} (r^2 - |w|^2)^{n-k} f(w) \, dv_k(w),$$

where $r > r_0$ and c is a certain constant depending on the normalization of dv and dv_k. Differentiation then gives

$$I'(1) = 2c(n-k) \int_{\mathbb{B}_k} (1 - |w|^2)^{n-k-1} f(w) \, dv_k(w).$$

Comparing this with the formula for $I'(1)$ in the previous paragraph, we obtain

$$\int_{\mathbb{S}_n} f \circ P_k \, d\sigma = c' \int_{\mathbb{B}_k} (1 - |w|^2)^{n-k-1} f(w) \, dv_k(w),$$

where c' is a constant independent of f. Thus the lemma is proved except for the multiplicative constant c'.

To determine the value of c', simply take $f = 1$ and compute the integral

$$\int_{\mathbb{B}_k} (1 - |w|^2)^{n-k-1} \, dv_k(w)$$

in polar coordinates.   □

Two special situations are worth mentioning. First, if $k = n - 1$, then

$$\int_{\mathbb{S}_n} f \, d\sigma = \int_{\mathbb{B}_k} f \, dv_k, \tag{1.12}$$

because the binomial coefficient in Lemma 1.9 becomes 1 in this case. On the other hand, if $k = 1$, $n > 1$, and f is a function of one complex variable, then for any $\eta \in \mathbb{S}_n$ we have

$$\int_{\mathbb{S}_n} f(\langle \zeta, \eta \rangle) \, d\sigma(\zeta) = (n-1) \int_{\mathbb{D}} (1 - |z|^2)^{n-2} f(z) \, dA(z). \tag{1.13}$$

This is because, by unitary invariance, we may assume that $\eta = e_1$, and hence $\langle \zeta, \eta \rangle = \zeta_1$.

We will also need to use the following formulas for integration on the unit sphere, the first of which is called integration by slices, and the second generalizes Lemma 1.9.

Lemma 1.10. *For $f \in L^1(\mathbb{S}_n, d\sigma)$ we have*

$$\int_{\mathbb{S}_n} f \, d\sigma = \int_{\mathbb{S}_n} d\sigma(\zeta) \frac{1}{2\pi} \int_0^{2\pi} f(e^{i\theta} \zeta) \, d\theta, \tag{1.14}$$

and if $1 < k < n$, then

$$\int_{\mathbb{S}_n} f \, d\sigma = c \int_{\mathbb{B}_k} (1 - |z|^2)^\alpha \, dv_k(z) \int_{\mathbb{S}_{n-k}} f(z, \sqrt{1 - |z|^2}\, \eta) \, d\sigma_{n-k}(\eta), \tag{1.15}$$

where $c = \dbinom{n-1}{k}$ and $\alpha = n - k - 1$.

Proof. It is obvious that

$$\int_{\mathbb{S}_n} f \, d\sigma = \int_{\mathbb{S}_n} f(e^{i\theta} \zeta) \, d\sigma(\zeta)$$

for all $0 \leq \theta \leq 2\pi$. Integrate with respect to $\theta \in [0, 2\pi]$ and apply Fubini's theorem. We then obtain (1.14), the formula of integration by slices.

If we write $\zeta = (\zeta', \zeta'')$, where $\zeta' \in \mathbb{C}^k$ and $\zeta'' \in \mathbb{C}^{n-k}$, then

$$\int_{\mathbb{S}_n} f(\zeta)\, d\sigma(\zeta) = \int_{\mathbb{S}_n} f(\zeta', \zeta'')\, d\sigma(\zeta).$$

By the unitary invariance of σ, we have

$$\int_{\mathbb{S}_n} f(\zeta)\, d\sigma(\zeta) = \int_{\mathbb{S}_n} f(\zeta', \sqrt{1 - |\zeta'|^2}\, \eta)\, d\sigma(\zeta),$$

where η is any fixed point on $\mathbb{S}_{n-k}$. Integrating over $\eta \in \mathbb{S}_{n-k}$ and applying Fubini's theorem, we obtain

$$\int_{\mathbb{S}_n} f(\zeta)\, d\sigma(\zeta) = \int_{\mathbb{S}_n} d\sigma(\zeta) \int_{\mathbb{S}_{n-k}} f(\zeta', \sqrt{1 - |\zeta'|^2}\, \eta)\, d\sigma_{n-k}(\eta).$$

The inner integral above defines a function that only depends on the first k variables. Therefore, we can apply Lemma 1.9 to get

$$\int_{\mathbb{S}_n} f\, d\sigma = c \int_{\mathbb{B}_k} (1 - |z|^2)^\alpha\, dv_k(z) \int_{\mathbb{S}_{n-k}} f(z, \sqrt{1 - |z|^2}\, \eta)\, d\sigma_{n-k}(\eta),$$

which completes the proof of the lemma. $\square$

One special case of (1.15) is especially useful, namely, if $k = n - 1$, we have

$$\int_{\mathbb{S}_n} f\, d\sigma = \int_{\mathbb{B}_{n-1}} dv_{n-1}(z) \frac{1}{2\pi} \int_0^{2\pi} f(z, \sqrt{1 - |z|^2}\, e^{i\theta})\, d\theta.$$

In the proof of Lemma 1.10 we used the obvious fact that both v and σ are invariant under unitary transformations. More specifically, if U is a unitary transformation of $\mathbb{C}^n$, then

$$\int_{\mathbb{B}_n} f(Uz)\, dv(z) = \int_{\mathbb{B}_n} f(z)\, dv(z) \tag{1.16}$$

and

$$\int_{\mathbb{S}_n} g(U\zeta)\, d\sigma(\zeta) = \int_{\mathbb{S}_n} g(\zeta)\, d\sigma(\zeta). \tag{1.17}$$

These equations are also referred to as the rotation invariance of v and σ, respectively.

We will also need a class of weighted volume measures on $\mathbb{B}_n$. Observe that if α is a real parameter, then integration in polar coordinates shows that the integral

$$\int_{\mathbb{B}_n} (1 - |z|^2)^\alpha\, dv(z)$$

is finite if and only if $\alpha > -1$. When $\alpha > -1$, we define a finite measure dv_α on $\mathbb{B}_n$ by

$$dv_\alpha(z) = c_\alpha (1 - |z|^2)^\alpha\, dv(z), \tag{1.18}$$

where c_α is a normalizing constant so that $v_\alpha(\mathbb{B}_n) = 1$. Using polar coordinates, we easily calculate that

$$c_\alpha = \frac{\Gamma(n + \alpha + 1)}{n!\,\Gamma(\alpha + 1)}. \tag{1.19}$$

When $\alpha \le -1$, we simply write

$$dv_\alpha(z) = (1 - |z|^2)^\alpha \, dv(z).$$

All the measures dv_α, $-\infty < \alpha < \infty$, are also unitarily invariant (or rotation invariant), that is,

$$\int_{\mathbb{B}_n} f(Uz)\, dv_\alpha(z) = \int_{\mathbb{B}_n} f(z)\, dv_\alpha(z) \tag{1.20}$$

for all $f \in L^1(\mathbb{B}_n, dv_\alpha)$ and all unitary transformations U of $\mathbb{C}^n$.

As a consequence of the rotation invariance under $Uz = ze^{i\theta}$, we easily check that if m and l are multi-indexes of nonnegative integers with $m \ne l$, then

$$\int_{\mathbb{S}_n} \zeta^m\, \overline{\zeta^l}\, d\sigma(\zeta) = 0, \qquad \int_{\mathbb{B}_n} z^m\, \overline{z^l}\, dv_\alpha(z) = 0, \tag{1.21}$$

where $\alpha > -1$. When $m = l$, we have the following formulas.

Lemma 1.11. *Suppose* $m = (m_1, \cdots, m_n)$ *is a multi-index of nonnegative integers and* $\alpha > -1$. *Then*

$$\int_{\mathbb{S}_n} |\zeta^m|^2\, d\sigma(\zeta) = \frac{(n-1)!\, m!}{(n - 1 + |m|)!}, \tag{1.22}$$

and

$$\int_{\mathbb{B}_n} |z^m|^2\, dv_\alpha(z) = \frac{m!\, \Gamma(n + \alpha + 1)}{\Gamma(n + |m| + \alpha + 1)}. \tag{1.23}$$

Proof. We identify $\mathbb{C}^n$ with $\mathbb{R}^{2n}$ using the real and imaginary parts of a complex number, and denote the usual Lebesgue measure on $\mathbb{C}^n$ by dV. If the Euclidean volume of $\mathbb{B}_n$ is c_n, then $c_n dv = dV$.

We evaluate the integral

$$I = \int_{\mathbb{C}^n} |z^m|^2 e^{-|z|^2}\, dV(z)$$

by two different methods. First, Fubini's theorem gives

$$I = \prod_{k=1}^n \int_{\mathbb{R}^2} (x^2 + y^2)^{m_k} e^{-(x^2 + y^2)}\, dx\, dy$$

$$= \pi^n \prod_{k=1}^n \int_0^\infty r^{m_k} e^{-r}\, dr$$

$$= \pi^n m!.$$

Then, integration in polar coordinates gives

$$I = 2nc_n \int_0^\infty r^{2|m|+2n-1} e^{-r^2}\, dr \int_{\mathbb{S}_n} |\zeta^m|^2\, d\sigma(\zeta)$$

$$= nc_n(|m|+n-1)! \int_{\mathbb{S}_n} |\zeta^m|^2\, d\sigma(\zeta).$$

Comparing the two answers, we obtain

$$\int_{\mathbb{S}_n} |\zeta^m|^2\, d\sigma(\zeta) = \frac{\pi^n m!}{nc_n(|m|+n-1)!}.$$

Choosing $m = (0, \cdots, 0)$ gives

$$c_n = \frac{\pi^n}{n!}.$$

It follows that

$$\int_{\mathbb{S}_n} |\zeta^m|^2\, d\sigma(\zeta) = \frac{(n-1)!\, m!}{(n-1+|m|)!}.$$

Another integration in polar coordinates gives

$$\int_{\mathbb{B}_n} |z^m|^2\, dv_\alpha(z) = 2nc_\alpha \int_0^1 r^{2|m|+2n-1}(1-r^2)^\alpha\, dr \int_{\mathbb{S}_n} |\zeta^m|^2\, d\sigma(\zeta)$$

$$= nc_\alpha \int_0^1 r^{|m|+n-1}(1-r)^\alpha\, dr \cdot \frac{(n-1)!\, m!}{(n-1+|m|)!}.$$

Identity (1.23) then follows from (1.22) and the fact that

$$\int_0^1 r^{n+|m|-1}(1-r)^\alpha\, dr = \frac{\Gamma(n+|m|)\Gamma(\alpha+1)}{\Gamma(n+|m|+\alpha+1)}.$$

This completes the proof of the lemma. □

As a by-product of the above proof we obtained the actual volume of $\mathbb{B}_n$ as $\pi^n/n!$. Therefore, the volume of the ball $r\mathbb{B}_n$ is

$$V(r) = \frac{\pi^n}{n!} r^{2n};$$

see the proof of Lemma 1.8. If we use $S(r)$ to denote the surface measure of the sphere $r\mathbb{S}_n$, then

$$V(r) = \int_0^r S(r)\, dr.$$

It follows that

$$S(r) = V'(r) = \frac{2\pi^n}{(n-1)!} r^{2n-1}.$$

In particular, the surface area of the unit sphere $\mathbb{S}_n$ is $(2\pi^n)/(n-1)!$.

As another consequence of Lemma 1.11 we obtain the following asymptotic estimates for certain important integrals on the ball and the sphere.

Theorem 1.12. *Suppose c is real and t > −1. Then the integrals*

$$I_c(z) = \int_{\mathbb{S}_n} \frac{d\sigma(\zeta)}{|1 - \langle z, \zeta \rangle|^{n+c}}, \qquad z \in \mathbb{B}_n,$$

and

$$J_{c,t}(z) = \int_{\mathbb{B}_n} \frac{(1 - |w|^2)^t \, dv(w)}{|1 - \langle z, w \rangle|^{n+1+t+c}}, \qquad z \in \mathbb{B}_n,$$

have the following asymptotic properties.

(1) If c < 0, then I_c and $J_{c,t}$ are both bounded in $\mathbb{B}_n$.
(2) If c = 0, then

$$I_c(z) \sim J_{c,t}(z) \sim \log \frac{1}{1 - |z|^2}$$

as $|z| \to 1^-$.
(3) If c > 0, then

$$I_c(z) \sim J_{c,t}(z) \sim (1 - |z|^2)^{-c}$$

as $|z| \to 1^-$.

Proof. Let $\lambda = (n + c)/2$. Then

$$\frac{1}{|1 - \langle z, \zeta \rangle|^{n+c}} = \left| \sum_{k=0}^{\infty} \frac{\Gamma(k + \lambda)}{k! \, \Gamma(\lambda)} \langle z, \zeta \rangle^k \right|^2.$$

For any fixed $z \in \mathbb{B}_n$, the functions $\langle z, \zeta \rangle^{k_1}$ and $\langle z, \zeta \rangle^{k_2}$ are orthogonal in $L^2(\mathbb{S}_n, d\sigma)$ whenever $k_1 \neq k_2$. It follows that

$$I_c(z) = \sum_{k=0}^{\infty} \left| \frac{\Gamma(k + \lambda)}{k! \, \Gamma(\lambda)} \right|^2 \int_{\mathbb{S}_n} |\langle z, \zeta \rangle|^{2k} \, d\sigma(\zeta).$$

If $z \neq 0$, then we can use the unit vector $\bar{z}/|z|$ in $\mathbb{C}^n$ as the first row to construct a unitary matrix U. Write $U\zeta = \zeta'$ and notice that the first coordinate of ζ' is

$$\zeta_1' = \langle \zeta, z \rangle / |z|.$$

By the unitary invariance of $d\sigma$, we have

$$\int_{\mathbb{S}_n} |\langle z, \zeta \rangle|^{2k} \, d\sigma(\zeta) = |z|^{2k} \int_{\mathbb{S}_n} |\zeta_1'|^{2k} \, d\sigma(\zeta').$$

This clearly holds for $z = 0$ as well. An application of Lemma 1.11 then gives

$$\int_{\mathbb{S}_n} |\langle z, \zeta \rangle|^{2k} \, d\sigma(\zeta) = \frac{(n-1)! \, k!}{(n-1+k)!} |z|^{2k}.$$

So,

$$I_c(z) = \sum_{k=0}^{\infty} \left| \frac{\Gamma(k+\lambda)}{k!\,\Gamma(\lambda)} \right|^2 \frac{(n-1)!\,k!}{(n-1+k)!} |z|^{2k}.$$

According to Stirling's formula, the coefficients of the series above are of order k^{c-1}. This proves the assertions about $I_c(z)$.

To prove the assertions about $J_{c,t}(z)$, we integrate in polar coordinates to obtain

$$J_{c,t}(z) = 2n \int_0^1 (1-r^2)^t I_{1+t+c}(rz) r^{2n-1}\,dr.$$

Combining this with the series for $I_c(z)$ in the previous paragraph, integrating term by term, and then applying Stirling's formula, we conclude that

$$J_{c,t}(z) \sim \sum_{k=0}^{\infty} k^{c-1} |z|^{2k}$$

as $|z| \to 1^-$. This completes the proof of the theorem. $\square$

The following change of variables formula will be very important for us later on.

Proposition 1.13. *Suppose α is real and f is in $L^1(\mathbb{B}_n, dv_\alpha)$. Then*

$$\int_{\mathbb{B}_n} f \circ \varphi(z)\,dv_\alpha(z) = \int_{\mathbb{B}_n} f(z) \frac{(1-|a|^2)^{n+1+\alpha}}{|1-\langle z,a\rangle|^{2(n+1+\alpha)}}\,dv_\alpha(z),$$

where φ is any automorphism of $\mathbb{B}_n$ and $a = \varphi(0)$.

Proof. By Theorem 1.4 there exists a unitary transformation U such that $\varphi = \varphi_a U$, where $a = \varphi(0)$. Since the measure dv_α is invariant under the action of unitary transformations, we may as well assume that $\varphi = \varphi_a$. In this case, we have $\varphi^{-1} = \varphi$ and its real Jacobian determinant at the point z is given by Lemma 1.7. Since

$$dv_\alpha(z) = c_\alpha (1-|z|^2)^\alpha\,dv(z),$$

where c_α is 1 for $\alpha \le -1$ and is given by (1.19) for $\alpha > -1$, a natural change of variables converts the integral

$$\int_{\mathbb{B}_n} f \circ \varphi(z)\,dv_\alpha(z)$$

to

$$c_\alpha \int_{\mathbb{B}_n} f(z)(1-|\varphi_a(z)|^2)^\alpha \left(\frac{1-|a|^2}{|1-\langle z,a\rangle|^2} \right)^{n+1}\,dv(z).$$

This along with (1.5) produces the desired result. $\square$

Two special weights are of particular interest to us. The first one is $\alpha = 0$. In this case, we have

$$\int_{\mathbb{B}_n} f \circ \varphi(z)\, dv(z) = \int_{\mathbb{B}_n} f(z) \frac{(1 - |a|^2)^{n+1}}{|1 - \langle z, a \rangle|^{2(n+1)}}\, dv(z). \tag{1.24}$$

The other weight is $\alpha = -(n+1)$. In this case we denote the resulting measure by

$$d\tau(z) = \frac{dv(z)}{(1 - |z|^2)^{n+1}}, \tag{1.25}$$

and call it the invariant measure on $\mathbb{B}_n$. The usage of the term "invariant measure" is justified by the following formula:

$$\int_{\mathbb{B}_n} f \circ \varphi(z)\, d\tau(z) = \int_{\mathbb{B}_n} f(z)\, d\tau(z). \tag{1.26}$$

In addition to the separable Lebesgue spaces $L^p(\mathbb{B}_n, dv_\alpha)$ and $L^p(\mathbb{S}_n, d\sigma)$, we will also encounter the spaces $L^\infty(\mathbb{B}_n)$ and $L^\infty(\mathbb{S}_n)$. We use $\mathbb{C}(\mathbb{B}_n)$ and $\mathbb{C}(\overline{\mathbb{B}}_n)$ to denote the spaces of all continuous functions on $\mathbb{B}_n$ and $\overline{\mathbb{B}}_n$, respectively. The space $\mathbb{C}_0(\mathbb{B}_n)$ consists of all functions in $\mathbb{C}(\overline{\mathbb{B}}_n)$ that vanish on the unit sphere.

1.4 Several Notions of Differentiation

In this section we discuss several different notions of differentiation on $\mathbb{B}_n$. The most basic one is of course the standard partial differentiation, namely, $\partial f / \partial z_k$.

A very important concept of differentiation on the unit ball is that of the radial derivative, which is based on the usual partial derivatives of a holomorphic function. Thus for a holomorphic function f in $\mathbb{B}_n$ we write

$$Rf(z) = \sum_{k=1}^{n} z_k \frac{\partial f}{\partial z_k}(z). \tag{1.27}$$

A simple verification shows that if

$$f(z) = \sum_{k=0}^{\infty} f_k(z)$$

is the homogeneous expansion of f, then

$$Rf(z) = \sum_{k=0}^{\infty} k f_k(z) = \sum_{k=1}^{\infty} k f_k(z). \tag{1.28}$$

This is called the radial derivative of f because

$$Rf(z) = \lim_{r \to 0} \frac{f(z + rz) - f(z)}{r} \tag{1.29}$$

whenever f is holomorphic. Here r is a real parameter, so that $z + rz$ is a radial variation of the point z.

For every holomorphic function f in $\mathbb{B}_n$, it is easy to see that

$$f(z) - f(0) = \int_0^1 \frac{Rf(tz)}{t}\,dt \tag{1.30}$$

for all $z \in \mathbb{B}_n$. This formula will come in handy when we need to recover a holomorphic function from its radial derivative.

With the help of homogeneous expansions we can introduce a class of fractional radial derivatives on the space $H(\mathbb{B}_n)$. Thus for each real parameter t we define an operator

$$R^t : H(\mathbb{B}_n) \to H(\mathbb{B}_n)$$

as follows:

$$R^t f(z) = \sum_{k=1}^{\infty} k^t f_k(z), \qquad f(z) = \sum_{k=0}^{\infty} f_k(z). \tag{1.31}$$

The operator R^t is clearly invertible on $H(\mathbb{B}_n)/\mathbb{C}$, with its inverse given by

$$R_t f(z) = R^{-t} f(z) = \sum_{k=1}^{\infty} k^{-t} f_k(z), \qquad f(z) = \sum_{k=0}^{\infty} f_k(z). \tag{1.32}$$

If we equip the space $H(\mathbb{B}_n)$ with the topology of uniform convergence on compact sets, then the operators R^t and R_t are continuous on $H(\mathbb{B}_n)$.

More generally, for any two real parameters α and t with the property that neither $n + \alpha$ nor $n + \alpha + t$ is a negative integer, we define an invertible operator

$$R^{\alpha,t} : H(\mathbb{B}_n) \to H(\mathbb{B}_n)$$

as follows. If

$$f(z) = \sum_{k=0}^{\infty} f_k(z)$$

is the homogeneous expansion of f, then

$$R^{\alpha,t} f(z) = \sum_{k=0}^{\infty} \frac{\Gamma(n + 1 + \alpha)\Gamma(n + 1 + k + \alpha + t)}{\Gamma(n + 1 + \alpha + t)\Gamma(n + 1 + k + \alpha)} f_k(z). \tag{1.33}$$

The inverse of $R^{\alpha,t}$, denoted by $R_{\alpha,t}$, is given by

$$R_{\alpha,t} f(z) = \sum_{k=0}^{\infty} \frac{\Gamma(n + 1 + \alpha + t)\Gamma(n + 1 + k + \alpha)}{\Gamma(n + 1 + \alpha)\Gamma(n + 1 + k + \alpha + t)} f_k(z). \tag{1.34}$$

The following result gives an alternative description of these operators.

Proposition 1.14. *Suppose neither $n + \alpha$ nor $n + \alpha + t$ is a negative integer. Then the operator $R^{\alpha,t}$ is the unique continuous linear operator on $H(\mathbb{B}_n)$ satisfying*

$$R^{\alpha,t}\left(\frac{1}{(1 - \langle z, w \rangle)^{n+1+\alpha}}\right) = \frac{1}{(1 - \langle z, w \rangle)^{n+1+\alpha+t}} \tag{1.35}$$

for all $w \in \mathbb{B}_n$. Similarly, the operator $R_{\alpha,t}$ is the unique continuous linear operator on $H(\mathbb{B}_n)$ satisfying

$$R_{\alpha,t}\left(\frac{1}{(1 - \langle z, w \rangle)^{n+1+\alpha+t}}\right) = \frac{1}{(1 - \langle z, w \rangle)^{n+1+\alpha}} \tag{1.36}$$

for all $w \in \mathbb{B}_n$.

Proof. The series

$$\frac{1}{(1 - \langle z, w \rangle)^{n+1+\alpha}} = \sum_{k=0}^{\infty} \frac{\Gamma(n+1+k+\alpha)}{k!\,\Gamma(n+1+\alpha)} \langle z, w \rangle^k$$

and

$$\frac{1}{(1 - \langle z, w \rangle)^{n+1+\alpha+t}} = \sum_{k=0}^{\infty} \frac{\Gamma(n+1+k+\alpha+t)}{k!\,\Gamma(n+1+\alpha+t)} \langle z, w \rangle^k$$

are actually homogeneous expansions. It is then obvious that the operators $R^{\alpha,t}$ and $R_{\alpha,t}$ have the desired mapping properties on kernel functions.

On the other hand, if $R^{\alpha,t}$ and $R_{\alpha,t}$ have the stated mapping properties on kernel functions, then applying the differential operators

$$\frac{\partial^m}{\partial \overline{w}^m}(0)$$

to these kernel equations shows that $R^{\alpha,t}$ and $R_{\alpha,t}$ have the desired effect on monomials and hence on general holomorphic functions. □

Proposition 1.15. *Suppose N is a positive integer and α is a real number such that $n + \alpha$ is not a negative integer. Then $R^{\alpha,N}$, as an operator acting on $H(\mathbb{B}_n)$, is a linear partial differential operator of order N with polynomial coefficients, that is,*

$$R^{\alpha,N} f(z) = \sum_{|m| \leq N} p_m(z) \frac{\partial^m f}{\partial z^m}(z),$$

where each p_m is a polynomial.

Proof. Fix $w \in \mathbb{B}_n$, replace the numerator of

$$\frac{1}{(1 - \langle z, w \rangle)^{n+1+\alpha+N}}$$

by

$$(1 - \langle z, w \rangle + \langle z, w \rangle)^N,$$

and apply the binomial formula. Then

$$\frac{1}{(1 - \langle z, w \rangle)^{n+1+\alpha+N}} = \sum_{k=0}^{N} \frac{N!}{k! \, (N-k)!} \frac{\langle z, w \rangle^k}{(1 - \langle z, w \rangle)^{n+1+\alpha+k}}.$$

For each k we apply the multi-nomial formula (1.1) to write

$$\langle z, w \rangle^k = \sum_{|m|=k} \frac{k!}{m!} z^m \overline{w}^m.$$

It is then clear that there exist constants c_{mk} such that

$$\frac{N!}{k! \, (N-k)!} \frac{\langle z, w \rangle^k}{(1 - \langle z, w \rangle)^{n+1+\alpha+k}} = \sum_{|m|=k} c_{mk} \, z^m \frac{\partial^k}{\partial z^m} \frac{1}{(1 - \langle z, w \rangle)^{n+1+\alpha}}.$$

Combining this with Proposition 1.14, we obtain

$$R^{\alpha, N} \frac{1}{(1 - \langle z, w \rangle)^{n+1+\alpha}} = \sum_{k=0}^{N} \sum_{|m|=k} c_{mk} \, z^m \frac{\partial^k}{\partial z^m} \frac{1}{(1 - \langle z, w \rangle)^{n+1+\alpha}}$$

for any fixed $w \in \mathbb{B}_n$. Differentiating with respect to $\overline{w}$ then leads to

$$R^{\alpha, N} = \sum_{k=0}^{N} \sum_{|m|=k} c_{mk} \, z^m \frac{\partial^k}{\partial z^m}.$$

This proves the proposition. $\qquad\qquad\qquad\qquad\qquad\qquad\qquad\qquad\qquad\qquad\quad$ $\square$

Let

$$\Delta = 4 \sum_{k=1}^{n} \frac{\partial^2}{\partial z_k \partial \overline{z}_k} = \sum_{k=1}^{n} \left(\frac{\partial^2}{\partial x_k^2} + \frac{\partial^2}{\partial y_k^2} \right)$$

be the ordinary Laplacian on $\mathbb{C}^n$. Here

$$\frac{\partial}{\partial z_k} = \frac{1}{2} \left(\frac{\partial}{\partial x_k} - i \frac{\partial}{\partial y_k} \right)$$

and

$$\frac{\partial}{\partial \overline{z}_k} = \frac{1}{2} \left(\frac{\partial}{\partial x_k} + i \frac{\partial}{\partial y_k} \right),$$

provided we use the identification $z_k = x_k + i y_k$ for $1 \le k \le n$. The complex-type derivatives are more convenient to use than the corresponding real ones.

Suppose f is a twice differentiable function in $\mathbb{B}_n$. We define

$$(\widetilde{\Delta} f)(z) = \Delta(f \circ \varphi_z)(0), \qquad z \in \mathbb{B}_n,$$

where φ_z is the involutive automorphism that interchanges the points 0 and z. The operator $\widetilde{\Delta}$ so defined is called the invariant Laplacian, because it has the following property with respect to the automorphism group.

Proposition 1.16. *Suppose f is twice differentiable in $\mathbb{B}_n$. Then*

$$\widetilde{\Delta}(f \circ \varphi) = (\widetilde{\Delta} f) \circ \varphi$$

for all $\varphi \in \text{Aut}(\mathbb{B}_n)$.

Proof. Fix $z \in \mathbb{B}_n$ and $\varphi \in \text{Aut}(\mathbb{B}_n)$. Let $a = \varphi(z)$. Then the automorphism

$$U = \varphi_a \circ \varphi \circ \varphi_z$$

fixes the origin and hence is a unitary by Lemma 1.1. It is easy to see that

$$\Delta(g \circ U)(0) = \Delta(g)(0)$$

for any twice differentiable function g. It follows that

$$\widetilde{\Delta}(f \circ \varphi)(z) = \Delta(f \circ \varphi \circ \varphi_z)(0) = \Delta(f \circ \varphi_a \circ U)(0)$$
$$= \Delta(f \circ \varphi_a)(0) = \widetilde{\Delta} f(a) = (\widetilde{\Delta} f) \circ \varphi(z).$$

$\square$

The invariant Laplacian can be described in terms of ordinary partial derivatives as follows.

Proposition 1.17. *If f is twice differentiable in $\mathbb{B}_n$, then*

$$(\widetilde{\Delta} f)(z) = 4(1 - |z|^2) \sum_{i,j=1}^{n} (\delta_{ij} - z_i \bar{z}_j) \frac{\partial^2 f}{\partial z_i \partial \bar{z}_j}(z)$$

for all $z \in \mathbb{B}_n$, where $\delta_{i,j}$ is the Kronecker delta.

Proof. Fix $z \in \mathbb{B}_n$ and write

$$\varphi_z(w) = (\varphi_1(w), \cdots, \varphi_n(w)), \qquad w \in \mathbb{B}_n.$$

By the chain rule,

$$(\widetilde{\Delta} f)(z) = \Delta(f \circ \varphi_z)(0) = 4 \sum_{i,j=1}^{n} \frac{\partial^2 f}{\partial z_i \partial \bar{z}_j}(z) \sum_{k=1}^{n} \frac{\partial \varphi_i}{\partial z_k}(0) \overline{\frac{\partial \varphi_j}{\partial z_k}(0)}.$$

The definition of φ_z shows that it admits the expansion

$$\varphi_z(w) = z - s_z w + \frac{s_z}{1 + s_z} \langle w, z \rangle z + \cdots,$$

where $s_z = \sqrt{1 - |z|^2}$ and the omitted terms have w-degree 2 or higher. It follows that

$$\frac{\partial \varphi_i}{\partial z_k}(0) = -s_z \delta_{ik} + \frac{s_z}{1 + s_z} \bar{z}_k z_i,$$

and, after some simplification,

$$\sum_{k=1}^{n} \frac{\partial \varphi_i}{\partial z_k}(0) \overline{\frac{\partial \varphi_j}{\partial z_k}(0)} = (1 - |z|^2)(\delta_{ij} - z_i \bar{z}_j).$$

This completes the proof.

$\square$

1.5 The Bergman Metric

The function

$$K(z, w) = \frac{1}{(1 - z\overline{w})^{n+1}}$$

is called the Bergman kernel of $\mathbb{B}_n$ and will be discussed in more detail in the next chapter. For now let us use it to define a Hermitian metric on $\mathbb{B}_n$.

We begin with the $n \times n$ complex matrix

$$B(z) = (b_{ij}(z)) = \frac{1}{n+1} \begin{pmatrix} \dfrac{\partial^2}{\partial \overline{z}_1 \partial z_1} \log K(z, z) & \cdots & \dfrac{\partial^2}{\partial \overline{z}_1 \partial z_n} \log K(z, z) \\ \cdots & \cdots & \cdots \\ \dfrac{\partial^2}{\partial \overline{z}_n \partial z_1} \log K(z, z) & \cdots & \dfrac{\partial^2}{\partial \overline{z}_n \partial z_n} \log K(z, z) \end{pmatrix}.$$

We will call this the Bergman matrix of $\mathbb{B}_n$. We also introduce an auxiliary matrix

$$A(z) = (z_i \overline{z}_j)_{n \times n} = \begin{pmatrix} z_1 \overline{z}_1 & \cdots & z_1 \overline{z}_n \\ \cdots & \cdots & \cdots \\ z_n \overline{z}_1 & \cdots & z_n \overline{z}_n \end{pmatrix}, \tag{1.37}$$

for each $z = (z_1, \cdots, z_n)$ in $\mathbb{C}^n$. If we identify linear transformations on $\mathbb{C}^n$ with $n \times n$ matrices via the standard basis of $\mathbb{C}^n$ (so that the adjoint of a linear transformation is just the conjugate transpose of the corresponding matrix), then it is easy to check that for $z \neq 0$,

$$A(z) = |z|^2 P_z, \tag{1.38}$$

where P_z is the orthogonal projection from $\mathbb{C}^n$ onto the one-dimensional subspace $[z]$ generated by z.

Proposition 1.18. *For $z \in \mathbb{B}_n$ the matrices $A(z)$ and $B(z)$ have the following properties:*

(a) $B(z) = [(1 - |z|^2)I + A(z)]/(1 - |z|^2)^2$, where I is the $n \times n$ identity matrix.
(b) $B(z)^{-1} = (1 - |z|^2)[I - A(z)]$.
(c) $B(z) = (\varphi_z'(z))^* \varphi_z'(z) = (\varphi_z'(z))^2$.
(d) $\det(B(z)) = K(z, z)$.
(e) $B(z) = P_z/(1 - |z|^2)^2 + Q_z/(1 - |z|^2)$ for $z \neq 0$.

It follows that for $n \geq 2$ and $z \neq 0$ the matrix $B(z)$ has two eigenvalues, namely, $(1 - |z|^2)^{-2}$ with eigenspace $[z]$, and $(1 - |z|^2)^{-2}$ with eigenspace $\mathbb{C}^n \ominus [z]$.

Proof. Since

$$\log K(z, z) = -(n + 1) \log(1 - |z|^2),$$

we have

$$\frac{\partial}{\partial z_j} \log K(z, z) = (n + 1) \frac{\overline{z}_j}{1 - |z|^2}$$

for $j = 1, \cdots, n$. It follows that

$$\frac{\partial^2}{\partial \bar{z}_j \partial z_j} \log K(z, z) = \frac{n+1}{(1 - |z|^2)^2}(1 - |z|^2 + |z_j|^2)$$

for $j = 1, \cdots, n$, and

$$\frac{\partial^2}{\partial \bar{z}_i \partial z_j} \log K(z, z) = (n+1)\frac{z_i \bar{z}_j}{(1 - |z|^2)^2}$$

for $i \neq j$. This shows that

$$B(z) = \frac{(1 - |z|^2)I + A(z)}{(1 - |z|^2)^2}.$$

A direct computation using rows and columns shows that

$$(I - A(z))((1 - |z|^2)I + A(z)) = (1 - |z|^2)I,$$

so that

$$B(z)^{-1} = (1 - |z|^2)(I - A(z)).$$

Recall from (1.10) that for $z \neq 0$,

$$\varphi'_z(z) = -\left(\frac{P_z}{1 - |z|^2} + \frac{Q_z}{\sqrt{1 - |z|^2}}\right),$$

where P_z is the orthogonal projection of $\mathbb{C}^n$ onto the one-dimensional subspace $[z]$ and Q_z is the orthogonal projection from $\mathbb{C}^n$ onto $\mathbb{C}^n \ominus [z]$. Here we identify linear transformations of $\mathbb{C}^n$ with $n \times n$ matrices via the standard basis of $\mathbb{C}^n$. Since

$$P_z^2 = P_z, \quad Q_z^2 = Q_z, \quad P_z Q_z = 0,$$

we easily obtain

$$(\varphi'_z(z))^2 = \frac{P_z}{(1 - |z|^2)^2} + \frac{Q_z}{1 - |z|^2}.$$

From $P_z + Q_z = I$ we then deduce that

$$(\varphi'_z(z))^2 = \frac{1}{(1 - |z|^2)^2}((1 - |z|^2)I + |z|^2 P_z).$$

Since $A(z) = |z|^2 P_z$, we have proved $B(z) = (\varphi'_z(z))^2$.

Since the real Jacobian determinant of the mapping φ_z is the modulus squared of the complex Jacobian determinant of φ_z, and the matrix $\varphi'_z(z)$ is self-adjoint, the identity $B(z) = (\varphi'_z(z))^2$ shows that

$$\det(B(z)) = |\det(\varphi'_z(z))|^2 = J_R \varphi_z(z).$$

This along with the formula for $J_R \varphi_z(z)$ in (1.11) shows that $\det(B(z)) = K(z, z)$.

□

A consequence of the above calculations is that the Bergman matrix $B(z)$ is positive and invertible. This is of course well known and is true in general. We also obtain the following representation for the square root of the Bergman matrix,

$$B(z)^{1/2} = \frac{P_z}{1 - |z|^2} + \frac{Q_z}{\sqrt{1 - |z|^2}}. \tag{1.39}$$

Proposition 1.19. *The Bergman matrix is invariant under automorphisms, that is,*

$$B(z) = (\varphi'(z))^* B(\varphi(z)) \varphi'(z)$$

for all $z \in \mathbb{B}_n$ and $\varphi \in \operatorname{Aut}(\mathbb{B}_n)$.

Proof. Without loss of generality we may assume that $\varphi = \varphi_a$ for some $a \in \mathbb{B}_n$. In this case, it follows from (1.5) and (1.11) that the Bergman kernel satisfies

$$K(z, z) = |J_C \varphi(z)|^2 K(\varphi(z), \varphi(z))$$

for all $z \in \mathbb{B}_n$ and $\varphi \in \operatorname{Aut}(\mathbb{B}_n)$. Thus

$$\log K(z, z) = \log |J_C \varphi(z)|^2 + \log K(\varphi(z), \varphi(z)).$$

By locally writing

$$\log |J_C \varphi(z)|^2 = \log J_C \varphi(z) + \log \overline{J_C \varphi(z)},$$

we see that

$$\frac{\partial^2}{\partial \bar{z}_i \partial z_j} \log |J_C \varphi(z)|^2 = 0$$

for all $z \in \mathbb{B}_n$.

Write

$$\varphi(z) = (\varphi_1(z), \cdots, \varphi_n(z)), \qquad z \in \mathbb{B}_n.$$

Then the chain rule gives

$$\frac{\partial}{\partial z_j} \log K(z, z) = \frac{\partial}{\partial z_j} \log |J_C \varphi(z)|^2 + \sum_{k=1}^n \frac{\partial}{\partial \varphi_k} \log K(\varphi(z), \varphi(z)) \frac{\partial \varphi_k}{\partial z_j}.$$

Another application of the chain rule produces

$$\frac{\partial^2}{\partial \bar{z}_i \partial z_j} \log K(z, z) = \sum_{k=1}^n \frac{\partial \varphi_k}{\partial z_j} \sum_{m=1}^n \frac{\partial^2}{\partial \varphi_k \partial \bar{\varphi}_m} \log K(\varphi(z), \varphi(z)) \overline{\left(\frac{\partial \varphi_m}{\partial z_i} \right)}$$

for all $i, j = 1, \cdots, n$. This proves the desired result. $\qquad\square$

For a smooth curve $\gamma : [0, 1] \to \mathbb{B}_n$ we define

$$l(\gamma) = \int_0^1 \left(\sum_{i,j=1}^n b_{ij}(\gamma(t)) \, \gamma_i'(t) \, \overline{\gamma_j'(t)} \right)^{1/2} dt$$

$$= \int_0^1 \langle B(\gamma(t))\gamma'(t), \gamma'(t) \rangle^{1/2} \, dt.$$

This definition clearly generalizes to the case of a piecewise smooth curve.

A metric $\beta : \mathbb{B}_n \times \mathbb{B}_n \to [0, \infty)$ can now be defined as follows. For any two points z and w in $\mathbb{B}_n$, let $\beta(z, w)$ be the infimum of the set consisting of all $l(\gamma)$, where γ is a piecewise smooth curve in $\mathbb{B}_n$ from z to w. That β is indeed a metric follows easily from the positivity of $B(z)$. We will call β the Bergman metric on $\mathbb{B}_n$.

Proposition 1.20. *The Bergman metric is invariant under automorphisms, that is,*

$$\beta(\varphi(z), \varphi(w)) = \beta(z, w)$$

for all $z, w \in \mathbb{B}_n$ and $\varphi \in \mathrm{Aut}(\mathbb{B}_n)$.

Proof. This follows easily from Proposition 1.19 and the definition of the Bergman metric. $\qquad\square$

Proposition 1.21. *If z and w are points in $\mathbb{B}_n$, then*

$$\beta(z, w) = \frac{1}{2} \log \frac{1 + |\varphi_z(w)|}{1 - |\varphi_z(w)|},$$

where φ_z is the involutive automorphism of $\mathbb{B}_n$ that interchanges 0 and z.

Proof. By invariance, we only need to prove the result for $w = 0$.

Fix a point $z \in \mathbb{B}_n$ and let $\gamma : [0, 1] \to \mathbb{B}_n$ be a piecewise smooth curve from 0 to z. Dividing the interval $[0, 1]$ into a finite number of subintervals if necessary, we may as well assume that γ is actually smooth and regular ($\gamma'(t)$ is non-vanishing) on $[0, 1]$. In this case, the function $\alpha(t) = |\gamma(t)|$ is smooth on $[0, 1]$.

Since $\alpha^2(t) = \langle \gamma(t), \gamma(t) \rangle$, differentiation gives

$$2\alpha(t)\alpha'(t) = 2\mathrm{Re}\,\langle \gamma'(t), \gamma(t) \rangle = 2\mathrm{Re}\,\langle P_{\gamma(t)}\gamma'(t), \gamma(t) \rangle,$$

where $P_{\gamma(t)}$ is the orthogonal projection from $\mathbb{C}^n$ onto the one-dimensional subspace spanned by $\gamma(t)$. It follows that

$$|\alpha'(t)| \leq |P_{\gamma(t)}\gamma'(t)|, \qquad t \in [0, 1].$$

On the other hand, according to part (e) of Proposition 1.18,

$$\langle B(\gamma(t))\gamma'(t), \gamma'(t) \rangle \geq \frac{|P_{\gamma(t)}\gamma'(t)|^2}{(1 - |\gamma(t)|^2)^2}.$$

So

$$l(\gamma) \geq \int_0^1 \frac{|P_{\gamma(t)}\gamma'(t)| \, dt}{1 - |\gamma(t)|^2} \geq \left| \int_0^1 \frac{\alpha'(t) \, dt}{1 - \alpha^2(t)} \right|.$$

By a change of variables,

$$\int_0^1 \frac{\alpha'(t) \, dt}{1 - \alpha^2(t)} = \int_0^{|z|} \frac{ds}{1 - s^2} = \frac{1}{2} \log \frac{1 + |z|}{1 - |z|}.$$

Therefore,

$$l(\gamma) \geq \frac{1}{2} \log \frac{1 + |z|}{1 - |z|}.$$

It is easy to check that equality holds if $\gamma(t) = tz, t \in [0, 1]$. This shows

$$\beta(0, z) = \frac{1}{2} \log \frac{1 + |z|}{1 - |z|},$$

and completes the proof of the proposition.   □

Corollary 1.22. *For z and w in $\mathbb{B}_n$ let*

$$\rho(z, w) = |\varphi_z(w)|.$$

Then ρ is a metric on $\mathbb{B}_n$. Moreover, ρ is invariant under automorphisms, that is,

$$\rho(\varphi(z), \varphi(w)) = \rho(z, w)$$

for all $z, w \in \mathbb{B}_n$ and $\varphi \in \mathrm{Aut}(\mathbb{B}_n)$.

Proof. A calculation shows that

$$\rho(z, w) = \tanh \beta(z, w) \tag{1.40}$$

for all $z, w \in \mathbb{B}_n$. The invariance of ρ, which can be checked directly, is thus a consequence of the invariance of β.

It remains to prove that ρ is a distance. It is obvious that ρ satisfies the positivity and symmetry conditions in the definition of distance. To prove the triangle inequality for ρ, consider the function

$$f(x) = \tanh(x + h) - \tanh(h) - \tanh(x), \qquad x \in [0, \infty),$$

where h is any positive constant. We have

$$f'(x) = \mathrm{sech}^2(x + h) - \mathrm{sech}^2(x), \qquad x \geq 0.$$

Since $\mathrm{sech}(x)$ is decreasing on $(0, \infty)$, we conclude that $f'(x) < 0$ for all $x > 0$, and so $f(x)$ is strictly decreasing on $(0, \infty)$. This together with $f(0) = 0$ shows that

$$\tanh(x + h) \leq \tanh(x) + \tanh(h)$$

for all x and h in $[0, \infty)$. The triangle inequality for ρ then follows from the monotonicity of $\tanh(x)$, the triangle inequality for β, and the above inequality.   □

The metric ρ will be called the pseudo-hyperbolic metric on $\mathbb{B}_n$. It is clear that ρ is bounded, while β is not.

For $z \in \mathbb{B}_n$ and $r > 0$ we let $D(z, r)$ denote the Bergman metric ball at z. Thus

$$D(z, r) = \{w \in \mathbb{B}_n : \beta(z, w) < r\}.$$

We now calculate the volume of the Bergman metric balls.

Lemma 1.23. *For any $z \in \mathbb{B}_n$ and $r > 0$ we have*

$$v(D(z, r)) = \frac{R^{2n}(1 - |z|^2)^{n+1}}{(1 - R^2|z|^2)^{n+1}}, \qquad (1.41)$$

where $R = \tanh(r)$. In particular, for any $r > 0$, there exist constants $c_r > 0$ and $C_r > 0$ such that

$$c_r(1 - |z|^2)^{n+1} \le v(D(z, r)) \le C_r(1 - |z|^2)^{n+1} \qquad (1.42)$$

for all $z \in \mathbb{B}_n$.

Proof. By Proposition 1.21, each $D(0, r)$ is actually a Euclidean ball of radius $R = \tanh(r)$ centered at the origin. Since the Bergman metric is invariant under automorphisms of $\mathbb{B}_n$, we have

$$D(z, r) = \varphi_z(D(0, r)).$$

Changing variables several times, we obtain

$$v(D(z, r)) = \int_{D(z,r)} dv(w) = (1 - |z|^2)^{n+1} \int_{|w|<R} \frac{dv(w)}{|1 - \langle z, w \rangle|^{2(n+1)}}$$

$$= (1 - |z|^2)^{n+1} \int_{\mathbb{B}_n} \frac{R^{2n} \, dv(w)}{|1 - \langle Rz, w \rangle|^{2(n+1)}} = \frac{R^{2n}(1 - |z|^2)^{n+1}}{(1 - R^2|z|^2)^{n+1}}.$$

This proves (1.41). The estimates in (1.42) clearly follow from (1.41). $\qquad \square$

Recall that for any real α we have

$$dv_\alpha(z) = c_\alpha(1 - |z|^2)^\alpha \, dv(z),$$

where $c_\alpha > 0$ is a constant. For more general α we have the following asymptotic estimate of $v_\alpha(D(z, r))$.

Lemma 1.24. *For any real α and positive r there exist constants $C > 0$ and $c > 0$ such that*

$$c(1 - |z|^2)^{n+1+\alpha} \le v_\alpha(D(z, r)) \le C(1 - |z|^2)^{n+1+\alpha}$$

for all $z \in \mathbb{B}_n$.

Proof. Let $R = \tanh(r)$ again and make a change of variables according to Proposition 1.13. We obtain

$$v_\alpha(D(z,r)) = c_\alpha \int_{D(z,r)} (1 - |z|^2)^\alpha \, dv(z)$$

$$= c_\alpha \int_{|w| < R} \frac{(1 - |z|^2)^{n+1+\alpha}}{|1 - \langle z, w \rangle|^{2(n+1+\alpha)}} (1 - |w|^2)^\alpha \, dv(w).$$

It is clear that there exist positive constants c and C such that

$$c \le \frac{c_\alpha(1 - |w|^2)^\alpha}{|1 - \langle z, w \rangle|^{2(n+1+\alpha)}} \le C$$

for all $z \in \mathbb{B}_n$ and $|w| < R$. $\qquad\square$

1.6 The Invariant Green's Formula

In this section we discuss Green's formula for the invariant Laplacian and the associated Green's function.

Theorem 1.25. *Suppose Ω is an open subset of $\mathbb{B}_n$, $\overline{\Omega} \subset \mathbb{B}_n$, whose boundary $\partial\Omega$ is sufficiently smooth. If u and v are twice differentiable in Ω and continuously differentiable on $\overline{\Omega}$, then*

$$\int_\Omega (u\widetilde{\Delta}v - v\widetilde{\Delta}u) \, d\widetilde{\tau} = \int_{\partial\Omega} \left(u\frac{\partial v}{\partial\widetilde{n}} - v\frac{\partial u}{\partial\widetilde{n}} \right) d\widetilde{\sigma},$$

where $d\widetilde{\tau}$ is the volume element on Ω in the Bergman metric, $\widetilde{\sigma}$ is the surface area element on $\partial\Omega$ determined by the Bergman metric, and $\partial/\partial\widetilde{n}$ is the outward normal derivative along $\partial\Omega$ with respect to the Bergman metric.

We will not prove this theorem in the book, because its proof is not complex analytic. See [41] or [102].

The volume element of Ω in the Bergman metric is simply the restriction to Ω of the Möbius invariant measure, that is,

$$d\widetilde{\tau}(z) = \frac{dv(z)}{(1 - |z|^2)^{n+1}}. \tag{1.43}$$

We will only apply Green's formula in the case when Ω is a shell in $\mathbb{B}_n$. On the surface $|z| = r$, $0 < r < 1$, the surface area element in the Bergman metric is given by

$$d\widetilde{\sigma}(r\zeta) = \frac{2nr^{2n-1} \, d\sigma(\zeta)}{(1 - r^2)^n}, \qquad \zeta \in \mathbb{S}_n. \tag{1.44}$$

Also, for any z on the surface $|z| = r$, the outward normal unit vector at z in the Bergman metric is given by

$$\widetilde{n} = \pm(1 - |z|^2)\frac{z}{|z|}, \tag{1.45}$$

and, consequently,

$$\frac{\partial u}{\partial \widetilde{n}} = \pm(1 - |z|^2)\frac{\partial u}{\partial |z|}. \tag{1.46}$$

Here $\pm$ depends on whether or not Ω is inside or outside of $|z| = r$.

The function

$$G(z) = \frac{1}{2n}\int_{|z|}^{1} \frac{(1-t^2)^{n-1}}{t^{2n-1}}\,dt, \qquad z \in \mathbb{B}_n, \tag{1.47}$$

is called Green's function for the invariant Laplacian $\widetilde{\Delta}$, or simply the invariant Green's function of $\mathbb{B}_n$. The following proposition tells us the rate of growth for G, both at the origin and near the boundary of $\mathbb{B}_n$.

Proposition 1.26. *The invariant Green's function G has the following asymptotic behavior.*

(a) As $|z| \to 0^+$, we have

$$G(z) \sim \log\frac{1}{|z|}$$

for $n = 1$, and

$$G(z) \sim \frac{1}{|z|^{2n-2}}$$

for $n > 1$.
(b) As $|z| \to 1^-$, we have

$$G(z) \sim (1 - |z|^2)^n$$

for $n \geq 1$.

Proof. This is elementary and we leave it as an exercise.   □

In particular, the singularity of $G(z)$ at the origin is integrable with respect to volume measure dv.

Theorem 1.27. *The invariant Green's function $G(z)$ has the following properties.*

(a) $\widetilde{\Delta}G = 0$ in $\mathbb{B}_n - \{0\}$.
(b) For every twice continuously differentiable function f with compact support in $\mathbb{B}_n$,

$$\int_{\mathbb{B}_n} G(z)\widetilde{\Delta}f(z)\,d\tau(z) = -f(0)$$

Proof. Rewrite

$$G(z) = \frac{1}{4n} \int_{|z|^2}^1 \frac{(1-t)^{n-1}\,dt}{t^n}, \qquad z \in \mathbb{B}_n.$$

By the chain rule,

$$\frac{\partial G}{\partial z_i} = -\frac{1}{4n} \cdot \frac{(1-|z|^2)^{n-1}}{|z|^{2n}} \cdot \bar{z}_i$$

for $1 \le i \le n$. Using

$$\frac{d}{dt} \frac{(1-t)^{n-1}}{t^n} = -\frac{(n-t)(1-t)^{n-2}}{t^{n+1}}$$

and the chain rule again, we obtain

$$\frac{\partial^2 G}{\partial z_i \partial \bar{z}_j} = -\frac{(1-|z|^2)^{n-2}}{4n|z|^{2n}} \left[(1-|z|^2)\delta_{ij} - \frac{(n-|z|^2)\bar{z}_i z_j}{|z|^2} \right].$$

Inserting this into the formula

$$\widetilde{\Delta} G(z) = 4(1-|z|^2) \sum_{i,j=1}^n (\delta_{ij} - z_i \bar{z}_j) \frac{\partial^2 G(z)}{\partial z_i \partial \bar{z}_j},$$

we easily check that $\widetilde{\Delta} G(z) = 0$ for all $z \in \mathbb{B}_n - \{0\}$.

If f is twice continuously differentiable and has compact support in $\mathbb{B}_n$, then

$$\int_{\mathbb{B}_n} G(z)\widetilde{\Delta} f(z)\,d\tau(z) = \lim_{r \to 0^+} \int_{r<|z|<1-r} G(z)\widetilde{\Delta} f(z)\,d\tau(z).$$

Fix a sufficiently small positive number r and apply Theorem 1.25 to the domain

$$\Omega = \{z \in \mathbb{B}_n : r < |z| < 1 - r\}.$$

We see that the integral

$$\int_{r<|z|<1-r} G(z)\widetilde{\Delta} f(z)\,d\tau(z)$$

is equal to

$$\int_{|z|=1-r} \left(G\frac{\partial f}{\partial \widetilde{n}} - f\frac{\partial G}{\partial \widetilde{n}} \right) d\widetilde{\sigma} - \int_{|z|=r} \left(G\frac{\partial f}{\partial \widetilde{n}} - f\frac{\partial G}{\partial \widetilde{n}} \right) d\widetilde{\sigma}, \qquad (1.48)$$

where $d\widetilde{\sigma}$ is the surface area element in the Bergman metric and

$$\frac{\partial f}{\partial \widetilde{n}} = (1-|z|^2)\frac{\partial f(z)}{\partial |z|}.$$

The first integral in (1.48) is zero, because f has compact support and r is sufficiently small. Since the singularity of G at the origin is either $-\log|z|$ (when $n=1$) or $1/|z|^{2n-2}$ (when $n>1$), and since

$$\frac{\partial G}{\partial \widetilde{n}} = -\frac{(1-|z|^2)^n}{2n|z|^{2n-1}},$$

it is easy to show that the second integral in (1.48) tends to $-f(0)$ as $r \to 0$. $\square$

1.7 Subharmonic Functions

In this section we collect a few results about subharmonic functions in the unit ball that will be needed later in the book. These results are real variable in nature, so we think of $\mathbb{B}_n$ as the open unit ball in the real Euclidean space $\mathbb{R}^{2n}$. We are going to make use of the following three properties of harmonic functions in several real variables without proof.

(a) Every harmonic function has the mean value property.
(b) The maximum principle holds for harmonic functions.
(c) Suppose B is any ball with boundary sphere S. If u is a continuous function on S, then u can be continuously extended to a harmonic function in B.

A function $f : \mathbb{B}_n \to [-\infty, \infty)$ is said to be upper semi-continuous if

$$\limsup_{z \to z_0} f(z) \leq f(z_0)$$

for every $z_0 \in \mathbb{B}_n$. An upper semi-continuous function $f : \mathbb{B}_n \to [-\infty, \infty)$ is said to be subharmonic if

$$f(a) \leq \int_{\mathbb{S}_n} f(a + r\zeta) \, d\sigma(\zeta) \tag{1.49}$$

for all $a \in \mathbb{B}_n$ and $0 \leq r < 1 - |a|$.

Theorem 1.28. *Suppose* $f : \mathbb{B}_n \to [-\infty, \infty)$ *is upper semi-continuous. Then the following conditions are equivalent.*

(a) f is subharmonic in $\mathbb{B}_n$.
(b) For every point a in $\mathbb{B}_n$ there exists some positive number $\epsilon < 1 - |a|$ such that

$$f(a) \leq \int_{\mathbb{S}_n} f(a + r\zeta) \, d\sigma(\zeta)$$

for all $0 \leq r < \epsilon$.
(c) If B is a ball whose boundary S is contained in $\mathbb{B}_n$, and if g is a harmonic function in B that is continuous up to S, then $f \leq g$ on S implies $f \leq g$ on B.

Proof. It is obvious that (a) implies (b).

Assume that f satisfies condition (c). For any $a \in \mathbb{B}_n$ and $0 < r < 1 - |a|$ we let B be the Euclidean ball centered at a with radius r. Clearly, the boundary sphere S of B is contained in $\mathbb{B}_n$. If f is continuous on $\mathbb{B}_n$, then f is continuous on S, and so there exists a continuous function g on the closure of B such that $g = f$ on S and g is harmonic in B. It follows that

$$f(a) \leq g(a) = \int_{\mathbb{S}_n} g(a + r\zeta) \, d\sigma(\zeta) = \int_{\mathbb{S}_n} f(a + r\zeta) \, d\sigma(\zeta).$$

If f is not necessarily continuous in $\mathbb{B}_n$, we then approximate f by a sequence of continuous functions $\{f_k\}$ with $f \leq f_{k+1} \leq f_k$ for all $k \geq 1$ (it is elementary to check that this is possible), the above estimate still holds for f because of the monotone convergence theorem. This proves that (c) implies (a).

To prove that (b) implies (c), we assume that there exists a ball B whose boundary sphere S is contained in $\mathbb{B}_n$, and that there exists a continuous function g on $\overline{B} = B \cup S$ such that g is harmonic in B, $f \le g$ on S, but $g(z) < f(z)$ for some $z \in B$. Let E be the set of points in $\overline{B}$ at which the upper semi-continuous function $h(z) = f(z) - g(z)$ attains its maximum value $M > 0$ in $\overline{B}$. Because $h \le 0$ on S, we must have $E \subset B$. Also, the semi-continuity of h implies that E is closed, so there exists some point $z_0 \in E$ such that no circular neighborhood of z_0 is entirely contained in E. We can then find a sequence $\{r_k\}$ tending to zero with the properties that each $|z - z_0| = r_k$ is contained in $\mathbb{B}_n$ but not entirely contained in E. The function h satisfies $h \le M$ on each $|z - z_0| = r_k$, with strict inequality on some nonempty open pieces of $|z - z_0| = r_k$. It follows that

$$\int_{\mathbb{S}_n} h(z_0 + r_k \zeta) \, d\sigma(\zeta) < M = f(z_0) - g(z_0)$$

for every k. Combining this with the mean value property for g, we obtain

$$\int_{\mathbb{S}_n} f(z_0 + r_k \zeta) \, d\sigma(\zeta) - g(z_0) < f(z_0) - g(z_0),$$

or

$$\int_{\mathbb{S}_n} f(z_0 + r_k \zeta) \, d\sigma(\zeta) < f(z_0)$$

for all k. In particular, condition (b) does not hold. This shows that (b) implies (c), and the proof of the theorem is complete. $\square$

We are going to use subharmonic functions mainly in the form of the following two corollaries.

Corollary 1.29. *If $\alpha > -1$ and f is subharmonic in $\mathbb{B}_n$, then*

$$f(a) \le \int_{\mathbb{B}_n} f(a + rz) \, dv_\alpha(z) \tag{1.50}$$

for all $a \in \mathbb{B}_n$ and $0 \le r < 1 - |a|$.

Proof. This follows from (1.49) and integration in polar coordinates. $\square$

Corollary 1.30. *If f is holomorphic in $\mathbb{B}_n$ and $0 < p < \infty$, then $\log |f|$ and $|f|^p$ are both subharmonic in $\mathbb{B}_n$.*

Proof. Fix a point $a \in \mathbb{B}_n$. If $f(a) = 0$, then obviously,

$$\log |f(a)| \le \int_{\mathbb{S}_n} \log |f(a + r\zeta)| \, d\sigma(\zeta),$$

and

$$|f(a)|^p \leq \int_{\mathbb{S}_n} |f(a + r\zeta)|^p \, d\sigma(\zeta),$$

where r is any radius satisfying $0 \leq r < 1 - |a|$.

If $f(a) \neq 0$, we can find a positive number $\epsilon < 1 - |a|$ such that f is nonvanishing in $|z - a| < \epsilon$. In particular, analytic branches of $\log f(z)$ and $f(z)^p$ can be defined on $|z - a| < \epsilon$. We then have

$$\log |f(a)| = \int_{\mathbb{S}_n} \log |f(a + r\zeta)| \, d\sigma(\zeta), \qquad 0 \leq r < \epsilon,$$

because $\log |f(z)|$ is harmonic in $|z - a| < \epsilon$. Similarly, the mean value property for holomorphic functions gives

$$f(a)^p = \int_{\mathbb{S}_n} f(a + r\zeta)^p \, d\sigma(\zeta), \qquad 0 \leq r < \epsilon.$$

Taking the modulus on both sides, we obtain

$$|f(a)|^p \leq \int_{\mathbb{S}_n} |f(a + r\zeta)|^p \, d\sigma(\zeta)$$

for all $0 \leq r < \epsilon$.

By condition (b) in Theorem 1.28, the functions $\log |f(z)|$ and $|f(z)|^p$ are subharmonic in $\mathbb{B}_n$. $\qquad\square$

1.8 Interpolation of Banach Spaces

Interpolation of Banach spaces is a powerful tool in analysis. In this section we introduce the notion of complex interpolation and present a version of the Marcinkiewicz interpolation theorem. Results in this section will be stated without proof, because they are just tools needed later, and because the techniques required for the proof are much different from those in the rest of the book. We refer the interested reader to the monograph [22] for details.

The following version of the Marcinkiewicz interpolation theorem will be used several times in the book.

Theorem 1.31. *Suppose μ is a positive measure on X, ν is a positive measure on Y, and T is an operator that associates to every ν-measurable function f on Y a nonnegative μ-measurable function Tf on X with the property that*

$$T(f + g) \leq Tf + Tg.$$

Let $1 \leq p_0 < p_1 \leq \infty$ and assume that there exists a positive constant C such that

$$\mu\{Tf > t\} \leq \left(\frac{C\|f\|_{p_0}}{t} \right)^{p_0}, \qquad f \in L^{p_0}(Y, \nu),$$

and in the case $p_1 < \infty$

$$\mu\{Tf > t\} \le \left(\frac{C\|f\|_{p_1}}{t}\right)^{p_1}, \qquad f \in L^{p_1}(Y, \nu),$$

and in the case $p_1 = \infty$

$$\|Tf\|_\infty \le C\|f\|_\infty, \qquad f \in L^\infty(Y, \nu).$$

Then for every $p_0 < p < p_1$ the operator T maps $L^p(Y, \nu)$ boundedly into $L^p(X, \mu)$.

Two Banach spaces X_0 and X_1 are called compatible if there exists a Hausdorff topological linear space X containing both of them. In this case, we can form two subspaces of X, $X_0 \cap X_1$ and $X_0 + X_1$, and they become Banach spaces with the following norms:

$$\|x\|_{X_0 \cap X_1} = \max(\|x\|_{X_0}, \|x\|_{X_1}),$$

and

$$\|x\|_{X_0 + X_1} = \inf\{\|x_0\|_{X_0} + \|x_1\|_{X_1} : x = x_0 + x_1, x_0 \in X_0, x_1 \in X_1\}.$$

Let $S = \{z \in \mathbb{C} : 0 < \mathrm{Re}\, z < 1\}$ denote the open strip and $\overline{S}$ its closure. If X_0 and X_1 are compatible Banach spaces, and if $\theta \in (0, 1)$, we can define a Banach space X_θ as follows. As a vector space, X_θ consists of vectors $x \in X_0 + X_1$ with the following property: there exists a function $f : \overline{S} \to X_0 + X_1$ such that

(a) f is bounded and continuous on $\overline{S}$.
(b) f is analytic in S.
(c) $f(\theta) = x$.
(d) $f(iy) \in X_0$ for every real y.
(e) $f(1 + iy) \in X_1$ for every real y.

For every f satisfying the above conditions we write

$$\|f\| = \max(\sup_{y \in \mathbb{R}} \|f(iy)\|_{X_0}, \sup_{y \in \mathbb{R}} \|f(1 + iy)\|_{X_1}).$$

The norm of $x \in X_\theta$ is then defined as the infimum of all such $\|f\|$.

To emphasize the dependence of X_θ on X_0 and X_1, we write

$$X_\theta = [X_0, X_1]_\theta,$$

and call it a complex interpolation space between X_0 and X_1. The construction of complex interpolation spaces is functorial in the following sense.

Theorem 1.32. *Suppose X_0 and X_1 are compatible, Y_0 and Y_1 are compatible, and $\theta \in (0, 1)$. If a linear operator $T : X_0 + X_1 \to Y_0 + Y_1$ maps X_0 boundedly into Y_0 (with norm M_0) and X_1 boundedly into Y_1 (with norm M_1), then T maps $[X_0, X_1]_\theta$ boundedly into $[Y_0, Y_1]_\theta$ (with norm not to exceed $M_0^{1-\theta} M_1^\theta$).*

Perhaps the most important example of complex interpolation spaces is the following result concerning L^p spaces.

Theorem 1.33. *If (X, μ) is a measure space and $1 \le p_0 < p_1 \le \infty$, then*

$$[L^{p_0}(X), L^{p_1}(X)]_\theta = L^p(X)$$

with equal norms, where $0 < \theta < 1$ and

$$\frac{1}{p} = \frac{1 - \theta}{p_0} + \frac{\theta}{p_1}.$$

Notes

Most of this chapter is classical. Sections 1.1 through 1.3 are adapted from [94]. Results about the invariant Laplacian are taken from [94] as well. The brief introduction to subharmonic functions follows the presentation in [61] and [94].

There are certainly many useful families of fractional differential operators on $H(\mathbb{B}_n)$. Because of Proposition 1.14, our choice of $R^{\alpha,t}$ here interacts particularly well with Bergman-type kernel functions.

The invariant Green's formula is more well known in differential geometry than in complex analysis. Theorem 1.25 can be found in [41] and [102], for example.

There are numerous versions of the Marcinkiewicz interpolation theorem. Our Theorem 1.31 can be found in [22] and [42]. The introductory material in Section 1.8 concerning complex interpolation (including Theorem 1.33) is from [22].

Exercises

1.1. Show that for any $z \in \mathbb{B}_n - \{0\}$ and $r > 0$ the Bergman metric ball $D(z, r)$ is an ellipsoid consisting of all points $w \in \mathbb{B}_n$ that satisfy

$$\frac{|P_z(w) - c|^2}{R^2 \sigma^2} + \frac{|Q_z(w)|^2}{R^2 \sigma} < 1,$$

where

$$R = \tanh(r), \qquad c = \frac{(1 - R^2)z}{1 - R^2 |z|^2}, \qquad \sigma = \frac{1 - |z|^2}{1 - R^2 |z|^2}.$$

1.2. Show that

$$J_R \varphi_a(0) = \lim_{r \to 0^+} \frac{v(D(a, r))}{v(D(0, r))} = (1 - |a|^2)^{n+1}$$

for any $a \in \mathbb{B}_n$.

1.3. Show that

$$\det(tI + A(z)) = t^{n-1}(t + |z|^2)$$

for all $z \in \mathbb{C}^n$ and $t \in \mathbb{C}$, where I is the $n \times n$ identity matrix and $A(z) = (z_i \bar{z}_j)$.

1.4. Find the eigenvalues and their associated eigenvectors of the matrix $A(z)$.

1.5. Suppose $\varphi : \mathbb{B}_n \to \mathbb{C}^n$ is holomorphic. Show that $J_R\varphi(z) = |J_C\varphi(z)|^2$ for all $z \in \mathbb{B}_n$.

1.6. Suppose f is twice differentiable in $\mathbb{B}_n$. If both $f(z)$ and its first order partial derivatives all tend to 0 as $|z| \to 1^-$, then

$$\lim_{r \to 1^-} \int_{|z| < r} G(z)\widetilde{\Delta}f(z)\, d\tau(z) = -f(0).$$

1.7. Prove Proposition 1.26.

1.8. If f is a function in $\mathbb{B}_n$ satisfying

$$\int_{\mathbb{B}_n} |f(z)|(1 - |z|^2)^n\, d\tau(z) < \infty,$$

we can define a function $G[f]$ in $\mathbb{B}_n$ by

$$G[f](z) = \int_{\mathbb{B}_n} G(\varphi_z(w))f(w)\, d\tau(w).$$

This function is called the invariant Green potential of f. Show that if f is twice continuously differentiable in $\mathbb{B}_n$ and continuous on $\overline{\mathbb{B}}_n$, then $G[f]$ is twice continuously differentiable in $\mathbb{B}_n$ and continuous on $\overline{\mathbb{B}}_n$. Furthermore, $\widetilde{\Delta}G[f] = f$ in $\mathbb{B}_n$ and $G[f] = 0$ on $\mathbb{S}_n$.

1.9. Show that

$$\int_{\mathbb{B}_n} G(z)\widetilde{\Delta}(1 - |z|^2)^\alpha\, d\tau(z) = -1$$

for all $\alpha > 1$. Note that the case $\alpha = n$ is especially interesting.

1.10. Show that $R^{\alpha,t}$ and $R_{\alpha,t}$ are continuous on $H(\mathbb{B}_n)$.

1.11. Show that if $f : \mathbb{B}_n \to [-\infty, \infty)$ is upper semi-continuous, then f is the decreasing limit of a sequence of continuous functions.

1.12. Show that if f is an upper semi-continuous function on a compact set K, then f attains its maximum.

1.13. Show that

$$R^{\alpha,t}R^{\alpha+t,s} = R^{\alpha,s+t}$$

whenever these operators are well defined.

1.14. If $\varphi \in \text{Aut}(\mathbb{B}_n)$ has an isolated fixed point in $\mathbb{B}_n$, then φ has a unique fixed point in $\mathbb{B}_n$. For this and the next seven problems see [125].

1.15. Show that each φ_a has a unique fixed point in $\mathbb{B}_n$.

1.16. Show that the fixed point of φ_a is the geodesic mid-point between 0 and a in the Bergman metric. Denote the fixed point of φ_a by m_a.

1.17. Show that $\varphi_{m_a} \circ \varphi_a = -\varphi_{m_a}$ for each $a \in \mathbb{B}_n - \{0\}$.

1.18. Show that

$$m_a = \frac{1 - \sqrt{1 - |a|^2}}{|a|^2} \, a,$$

where a is any point in $\mathbb{B}_n - \{0\}$.

1.19. Show that the geodesic mid-point in the Bergman metric between any two points a and b in $\mathbb{B}_n$ is given by

$$m_{a,b} = \varphi_a \left(\frac{1 - \sqrt{1 - |\varphi_a(b)|^2}}{|\varphi_a(b)|^2} \, \varphi_a(b) \right).$$

1.20. Suppose $a \in \mathbb{B}_n$ and f is a function in $\mathbb{B}_n$. Show that $f \circ \varphi_a = f$ if and only if $f = g \circ \varphi_{m_a}$, where g is an even function in $\mathbb{B}_n$; similarly, $f \circ \varphi_a = -f$ if and only if $f = g \circ \varphi_{m_a}$, where g is an odd function in $\mathbb{B}_n$.

1.21. If $J_a(z)$ is the complex Jacobian determinant of φ_a at $z \in \mathbb{B}_n$, show that

$$J_a(z) = (-1)^n \frac{(1 - |a|^2)^{(n+1)/2}}{(1 - \langle z, a \rangle)^{n+1}}.$$

1.22. Show that

$$\int_{\mathbb{C}^n} \frac{f(z) \, dv(z)}{|z|^{2n}} = n \int_{\mathbb{S}_n} d\sigma(\zeta) \int_{\mathbb{C}} \frac{f(\zeta w) \, dA(w)}{|w|^2}$$

whenever the integrals converge.

1.23. Suppose f and g are twice continuously differentiable functions with compact support in $\mathbb{B}_n$. Show that

$$\int_{\mathbb{B}_n} f \, \widetilde{\Delta} g \, d\tau = \int_{\mathbb{B}_n} g \, \widetilde{\Delta} f \, d\tau.$$

1.24. Suppose a and b are points in $\mathbb{B}_n$. Show that

$$|\varphi_b(\varphi_a(z))| = |\varphi_c(z)|, \qquad z \in \mathbb{B}_n,$$

where $c = \varphi_a(b)$.

1.25. Show that

$$2|1 - \langle z, w \rangle| \geq |1 - \langle z, w' \rangle|,$$

where $z \in \mathbb{B}_n$, $w \in \mathbb{B}_n - \{0\}$, and $w' = w/|w|$.

1.26. Show that

$$|\varphi_a(z)| \leq |a| + |z|$$

for all z and a in $\mathbb{B}_n$.

2

Bergman Spaces

In this chapter we study weighted Bergman spaces with standard radial weights. Main topics covered include integral representations, Bergman-type projections, characterizations in terms of various derivatives, and atomic decompositions. The integral representation formulas developed in this chapter, together with the fractional differential and fractional integral operators introduced in Chapter 1, will play a very important role in subsequent chapters.

2.1 Bergman Spaces

Recall that for $\alpha > -1$ the weighted Lebesgue measure dv_α is defined by

$$dv_\alpha(z) = c_\alpha(1 - |z|^2)^\alpha \, dv(z), \tag{2.1}$$

where

$$c_\alpha = \frac{\Gamma(n + \alpha + 1)}{n! \, \Gamma(\alpha + 1)} \tag{2.2}$$

is a normalizing constant so that dv_α is a probability measure on $\mathbb{B}_n$.

For $\alpha > -1$ and $p > 0$ the weighted Bergman space A_α^p consists of holomorphic functions f in $L^p(\mathbb{B}_n, dv_\alpha)$, that is,

$$A_\alpha^p = L^p(\mathbb{B}_n, dv_\alpha) \bigcap H(\mathbb{B}_n). \tag{2.3}$$

It is clear that A_α^p is a linear subspace of $L^p(\mathbb{B}_n, dv_\alpha)$.

When the weight $\alpha = 0$, we simply write A^p for A_α^p. These are the standard (unweighted) Bergman spaces.

We will use the notation

$$\|f\|_{p,\alpha} = \left[\int_{\mathbb{B}_n} |f(z)|^p \, dv_\alpha(z) \right]^{1/p} \tag{2.4}$$

for $f \in L^p(\mathbb{B}_n, dv_\alpha)$. Note that when $1 \leq p < \infty$ the space $L^p(\mathbb{B}_n, dv_\alpha)$ is a Banach space with the norm $\| \; \|_{p,\alpha}$. If $0 < p < 1$, the space $L^p(\mathbb{B}_n, dv_\alpha)$ is a complete metric space with the following distance:

$$\rho(f, g) = \|f - g\|_{p,\alpha}^p.$$

In the special case when $p = 2$, $L^2(\mathbb{B}_n, dv_\alpha)$ is a Hilbert space whose inner product will be denoted by $\langle \; , \; \rangle_\alpha$. Regardless of what p is, we are going to call $\| \; \|_{p,\alpha}$ the norm on $L^p(\mathbb{B}_n, dv_\alpha)$.

The following result shows how fast a function in A_α^p can grow near the boundary of $\mathbb{B}_n$.

Theorem 2.1. *Suppose $0 < p < \infty$ and $\alpha > -1$. Then*

$$|f(z)| \leq \frac{\|f\|_{p,\alpha}}{(1 - |z|^2)^{(n+1+\alpha)/p}}$$

for all $f \in A_\alpha^p$ and $z \in \mathbb{B}_n$.

Proof. If f is any holomorphic function in $\mathbb{B}_n$, then $|f|^p$ is subharmonic, so by Corollary 1.29,

$$|f(0)|^p \leq \int_{\mathbb{B}_n} |f(w)|^p \, dv_\alpha(w).$$

This proves the desired result when $z = 0$.

In general, for $f \in A_\alpha^p$ and $z \in \mathbb{B}_n$, we consider the function

$$F(w) = f \circ \varphi_z(w) \frac{(1 - |z|^2)^{(n+1+\alpha)/p}}{(1 - \langle w, z \rangle)^{2(n+1+\alpha)/p}}, \qquad w \in \mathbb{B}_n.$$

Changing variables according to Proposition 1.13, we see that

$$\|F\|_{p,\alpha} = \|f\|_{p,\alpha}.$$

The desired result then follows from $\|F\|_{p,\alpha} \geq |F(0)|$. $\square$

It is easy to show that the exponent of $(1 - |z|^2)$ in the preceding theorem is best possible. However, approximating a general function by polynomials (see Proposition 2.6), we obtain the following improved behavior of f near the boundary:

$$\lim_{|z| \to 1^-} (1 - |z|^2)^{(n+1+\alpha)/p} f(z) = 0$$

for all $f \in A_\alpha^p$.

The following result gives an integral representation for functions in A_α^1. The proofs of a large number of results in the book are based on this formula.

Theorem 2.2. *If $\alpha > -1$ and $f \in A_\alpha^1$, then*

$$f(z) = \int_{\mathbb{B}_n} \frac{f(w) \, dv_\alpha(w)}{(1 - \langle z, w \rangle)^{n+1+\alpha}} \tag{2.5}$$

for all $z \in \mathbb{B}_n$.

Proof. Let $f \in A_\alpha^1$. By the mean value property for holomorphic functions,

$$f(0) = \int_{\mathbb{S}_n} f(r\zeta) \, d\sigma(\zeta), \qquad 0 \le r < 1.$$

This together with integration in polar coordinates shows that

$$f(0) = \int_{\mathbb{B}_n} f(w) \, dv_\alpha(w).$$

Let $z \in \mathbb{B}_n$ and replace f by $f \circ \varphi_z$. Then

$$f(z) = \int_{\mathbb{B}_n} f \circ \varphi_z(w) \, dv_\alpha(w).$$

Making an obvious change of variables according to Proposition 1.13, we obtain

$$f(z) = \int_{\mathbb{B}_n} \frac{(1 - |z|^2)^{n+1+\alpha}}{|1 - \langle z, w \rangle|^{2(n+1+\alpha)}} f(w) \, dv_\alpha(w).$$

Fix $z \in \mathbb{B}_n$ and replace f by the function

$$f(w)(1 - \langle w, z \rangle)^{n+1+\alpha}.$$

We arrive at the desired reproducing formula. $\qquad\qquad\qquad\qquad\qquad\square$

For $1 \le p < \infty$ we have $A_\alpha^p \subset A_\alpha^1$, so the integral representation in Theorem 2.2 is also valid for functions in A_α^p when $1 \le p < \infty$. In particular, this integral representation holds for all $f \in H^\infty$.

Corollary 2.3. *Suppose $\alpha > -1$, $t > 0$, and f is holomorphic in $\mathbb{B}_n$. If neither $n + \alpha$ nor $n + \alpha + t$ is not a negative integer, then*

$$R^{\alpha,t} f(z) = \lim_{r \to 1^-} \int_{\mathbb{B}_n} \frac{f(rw) \, dv_\alpha(w)}{(1 - \langle z, w \rangle)^{n+1+\alpha+t}}, \tag{2.6}$$

and

$$R_{\alpha,t} f(z) = \lim_{r \to 1^-} \int_{\mathbb{B}_n} \frac{f(rw) \, dv_{\alpha+t}(w)}{(1 - \langle z, w \rangle)^{n+\alpha+1}}. \tag{2.7}$$

In particular, the limits above always exist.

Proof. For any fixed $r \in (0, 1)$ the dilation f_r, defined by $f_r(z) = f(rz)$, belongs to both A_α^1 and $A_{\alpha+t}^1$. So, according to Theorem 2.2,

$$f_r(z) = \int_{\mathbb{B}_n} \frac{f_r(w) \, dv_\alpha(w)}{(1 - \langle z, w \rangle)^{n+1+\alpha}}, \qquad z \in \mathbb{B}_n,$$

and

$$f_r(z) = \int_{\mathbb{B}_n} \frac{f_r(w)\, dv_{\alpha+t}(w)}{(1 - \langle z, w \rangle)^{n+1+\alpha+t}}, \qquad z \in \mathbb{B}_n.$$

The desired results then follow from Proposition 1.14 and the facts that

$$R^{\alpha,t} f(z) = \lim_{r \to 1^-} R^{\alpha,t} f_r(z)$$

and

$$R_{\alpha,t} f(z) = \lim_{r \to 1^-} R_{\alpha,t} f_r(z).$$

$\square$

Lemma 2.4. *Suppose $p > 0$, $\alpha > -1$, $0 < r < 1$, and $m = (m_1, \cdots, m_n)$ is a multi-index of nonnegative integers. Then there exists a positive constant C such that*

$$\left| \frac{\partial^m f}{\partial z^m}(z) \right| \le C \|f\|_{p,\alpha}$$

for all $f \in A_\alpha^p$ and all $z \in \mathbb{B}_n$ with $|z| \le r$.

Proof. Fix some $\delta \in (r, 1)$ and apply Theorem 2.2 in the special case $\alpha = 0$. We obtain

$$f(\delta z) = \int_{\mathbb{B}_n} \frac{f(\delta w)\, dv(w)}{(1 - \langle z, w \rangle)^{n+1}}, \qquad z \in \mathbb{B}_n.$$

Making a change of variables and replacing δz by z, we get

$$f(z) = \delta^2 \int_{|w| < \delta} \frac{f(w)\, dv(w)}{(\delta^2 - \langle z, w \rangle)^{n+1}}, \qquad |z| < \delta.$$

We can then differentiate under the integral sign and find a positive constant C such that

$$\left| \frac{\partial^m f}{\partial z^m}(z) \right| \le C \sup\{|f(w)| : |w| \le \delta\}$$

for all $|z| \le r$. This reduces the proof of the lemma to the case $|m| = 0$. When $|m| = 0$, the desired estimate clearly follows from Theorem 2.1. $\square$

Corollary 2.5. *For each $p > 0$ and $\alpha > -1$ the weighted Bergman space A_α^p is closed in $L^p(\mathbb{B}_n, dv_\alpha)$.*

Proof. Suppose $\{f_n\}$ is a sequence in A_α^p and

$$\lim_{n \to \infty} \|f_n - f\|_{p,\alpha} = 0$$

for some $f \in L^p(\mathbb{B}_n, dv_\alpha)$. Then some subsequence of $\{f_n(z)\}$ converges to $f(z)$ for almost all $z \in \mathbb{B}_n$. Also, $\{f_n\}$ is a Cauchy sequence in A_α^p, so by Lemma 2.4, the sequence $\{f_n(z)\}$ is uniformly Cauchy on each set $\{z \in \mathbb{B}_n : |z| < r\}$, and must converge to a holomorphic function there, where $0 < r < 1$. Since r is arbitrary, $\{f_n(z)\}$ converges to a holomorphic function $g(z)$ on $\mathbb{B}_n$. By the uniqueness of pointwise limits, we have $f(z) = g(z)$ for almost all $z \in \mathbb{B}_n$. This shows that f is holomorphic in $\mathbb{B}_n$ and hence belongs to A_α^p. $\square$

It follows that the weighted Bergman space A_α^p, with topology inherited from $L^p(\mathbb{B}_n, dv_\alpha)$, is a Banach space when $1 \leq p < \infty$, and is a complete metric space when $0 < p < 1$.

When $p = 2$, the space A_α^2 is a Hilbert space. Moreover, if

$$f(z) = \sum_m a_m z^m$$

is the Taylor expansion of f, then by (1.23),

$$\|f\|_\alpha^2 = \int_{\mathbb{B}_n} |f(z)|^2 \, dv_\alpha(z) = \sum_{m \geq 0} \frac{m!\,\Gamma(n+\alpha+1)}{\Gamma(n+|m|+\alpha+1)} |a_m|^2. \tag{2.8}$$

In particular, the functions

$$e_m(z) = \sqrt{\frac{\Gamma(n+|m|+\alpha+1)}{m!\,\Gamma(n+\alpha+1)}} \, z^m \tag{2.9}$$

form an orthonormal basis for A_α^2, where $m = (m_1, \cdots, m_n)$ runs over all multi-indexes of nonnegative integers.

Proposition 2.6. *Suppose $p > 0$ and $\alpha > -1$. Then the set of polynomials is dense in A_α^p.*

Proof. Writing the ball as

$$\mathbb{B}_n = \big\{z : |z| \leq 1 - \epsilon\big\} \bigcup \big\{z : 1 - \epsilon < |z| < 1\big\}$$

for a sufficiently small positive ϵ, we easily prove that

$$\lim_{r \to 1^-} \|f_r - f\|_{p,\alpha} = 0$$

for every $f \in A_\alpha^p$, where $f_r(z) = f(rz)$. For any fixed $r \in (0,1)$, the function f_r can be approximated uniformly by polynomials, and hence each f_r can be approximated in the norm topology of A_α^p by polynomials. $\square$

2.2 Bergman Type Projections

Since each point evaluation in $\mathbb{B}_n$ is a bounded linear functional on the Hilbert space A_α^2, where $\alpha > -1$, the classical Riesz representation theory in functional analysis shows that for each $w \in \mathbb{B}_n$ there exists a unique function K_w^α in A_α^2 such that

$$f(w) = \langle f, K_w^\alpha \rangle_\alpha = \int_{\mathbb{B}_n} f(z) \, \overline{K_w^\alpha(z)} \, dv_\alpha(z), \qquad f \in A_\alpha^2.$$

This will be called the reproducing formula for f in A_α^2. The function

$$K^\alpha(z, w) = K^\alpha_w(z), \qquad z, w \in \mathbb{B}_n,$$

is called the reproducing kernel of A^2_α. When $\alpha = 0$, the reproducing kernel

$$K(z, w) = K^0(z, w)$$

is also called the Bergman kernel.

Theorem 2.7. *For each $\alpha > -1$ the reproducing kernel of A^2_α is given by*

$$K^\alpha(z, w) = \frac{1}{(1 - \langle z, w \rangle)^{n+1+\alpha}}, \qquad z, w \in \mathbb{B}_n.$$

Proof. This follows from Theorem 2.2 and the uniqueness of the Riesz representation for a bounded linear functional on a Hilbert space. □

Since the function $K^\alpha(z, w)$ is bounded in z whenever w is fixed, we can consider integrals of the form

$$\int_{\mathbb{B}_n} f(z) K^\alpha(w, z) \, dv_s(z),$$

where $\alpha > -1$, $s \in \mathbb{R}$, $w \in \mathbb{B}_n$, and $f \in L^1(\mathbb{B}_n, dv_s)$. In particular, we will make use of the following integral operator

$$P_\alpha(f)(z) = \int_{\mathbb{B}_n} f(w) K^\alpha(z, w) \, dv_\alpha(w), \qquad f \in L^1(\mathbb{B}_n, dv_\alpha).$$

Lemma 2.8. *Suppose $\alpha > -1$. Then the restriction of P_α to $L^2(\mathbb{B}_n, dv_\alpha)$ is the orthogonal projection from $L^2(\mathbb{B}_n, dv_\alpha)$ onto A^2_α.*

Proof. Let P be the orthogonal projection from $L^2(\mathbb{B}_n, dv_\alpha)$ onto A^2_α. For $f \in L^2(\mathbb{B}_n, dv_\alpha)$ and $z \in \mathbb{B}_n$ the reproducing property of K^α and the self-adjointness of P give us

$$Pf(z) = \langle Pf, K^\alpha_z \rangle_\alpha = \langle f, PK^\alpha_z \rangle_\alpha.$$

Since $K^\alpha_z \in A^2_\alpha$, we have $PK^\alpha_z = K^\alpha_z$, and so

$$Pf(z) = \langle f, K^\alpha_z \rangle_\alpha = \int_{\mathbb{B}_n} f(w) K^\alpha(z, w) \, dv_\alpha(w).$$

This shows that P is the restriction of P_α to $L^2(\mathbb{B}_n, dv_\alpha)$. □

The above lemma shows that the operator P_α maps $L^2(\mathbb{B}_n, dv_\alpha)$ boundedly onto the Bergman space A^2_α. We also want to know how the operator P_α acts on other spaces such as $L^p(\mathbb{B}_n, dv_t)$. A useful tool for tackling such problems is the following Schur's test.

Theorem 2.9. *Suppose* (X, μ) *is a measure space,* $1 < p < \infty$, *and* $1/p + 1/q = 1$. *For a nonnegative kernel* $H(x, y)$ *consider the integral operator*

$$Tf(x) = \int_X H(x, y) f(y) \, d\mu(y).$$

If there exists a positive function h *on* X *and a positive constant* C *such that*

$$\int_X H(x, y) h(y)^q \, d\mu(y) \leq C h(x)^q$$

for almost all $x \in X$, *and*

$$\int_X H(x, y) h(x)^p \, d\mu(x) \leq C h(y)^p$$

for almost all $y \in X$, *then the operator* T *is bounded on* $L^p(X, \mu)$ *with* $\|T\| \leq C$.

Proof. Given a function $f \in L^p(X, \mu)$, Hölder's inequality gives

$$|Tf(x)| \leq \left[\int_X H(x, y) h(y)^q \, d\mu(y) \right]^{1/q} \left[\int_X H(x, y) h(y)^{-p} |f(y)|^p \, d\mu(y) \right]^{1/p}$$

for almost all $x \in X$. By the first inequality in the assumption,

$$|Tf(x)| \leq C^{1/q} h(x) \left[\int_X H(x, y) h(y)^{-p} |f(y)|^p \, d\mu(y) \right]^{1/p}$$

for almost all $x \in X$. By Fubini's theorem and the second inequality in the assumption, we obtain

$$\int_X |Tf(x)|^p \, d\mu(x) \leq C^p \int_X |f(y)|^p \, d\mu(y).$$

This proves Schur's test. $\qquad\square$

We now use Schur's test to describe the boundedness of a class of integral operators induced by Bergman type kernels on weighted Lebesgue spaces. Recall from Section 1.3 that

$$dv_t(z) = c_t (1 - |z|^2)^t \, dv(z), \qquad -\infty < t < \infty,$$

where c_t is a positive constant.

Theorem 2.10. *Fix two real parameters* a *and* b *and define two integral operators* T *and* S *by*

$$Tf(z) = (1 - |z|^2)^a \int_{\mathbb{B}_n} \frac{(1 - |w|^2)^b}{(1 - \langle z, w \rangle)^{n+1+a+b}} f(w) \, dv(w)$$

and

$$Sf(z) = (1 - |z|^2)^a \int_{\mathbb{B}_n} \frac{(1 - |w|^2)^b}{|1 - \langle z, w \rangle|^{n+1+a+b}} f(w) \, dv(w).$$

Then for $-\infty < t < \infty$ *and* $1 \leq p < \infty$ *the following conditions are equivalent:*

(a) T is bounded on $L^p(\mathbb{B}_n, dv_t)$.
(b) S is bounded on $L^p(\mathbb{B}_n, dv_t)$.
(c) $-pa < t + 1 < p(b+1)$.

Proof. It is obvious that (b) implies (a).

To prove (a) implies (c), let N be a large enough positive integer such that $N + b > -1$ and such that the function

$$f_N(z) = (1 - |z|^2)^N, \qquad z \in \mathbb{B}_n,$$

belongs to $L^p(\mathbb{B}_n, dv_t)$. The symmetry of $\mathbb{B}_n$ shows that

$$T f_N(z) = c_N (1 - |z|^2)^a, \qquad z \in \mathbb{B}_n,$$

where c_N is a positive constant. The boundedness of T on $L^p(\mathbb{B}_n, dv_t)$ then implies that the function $(1 - |z|^2)^a$ belongs to $L^p(\mathbb{B}_n, dv_t)$, which in turn implies that $pa + t > -1$, or $t + 1 > -pa$.

If $1 < p < \infty$ and $1/p + 1/q = 1$, the boundedness of T on $L^p(\mathbb{B}_n, dv_t)$ is equivalent to the boundedness of the adjoint of T on $L^q(\mathbb{B}_n, dv_t)$. It is easy to see that

$$T^* f(z) = (1 - |z|^2)^{b-t} \int_{\mathbb{B}_n} \frac{(1 - |w|^2)^{a+t}}{(1 - \langle z, w \rangle)^{n+1+a+b}} f(w) \, dv(w).$$

Combining this with the conclusion of the previous paragraph, we conclude that

$$t + 1 > -q(b - t),$$

which is equivalent to

$$t + 1 < p(b + 1).$$

Similarly, the boundedness of T on $L^1(\mathbb{B}_n, dv_t)$ implies the boundedness of T^* on $L^\infty(\mathbb{B}_n)$. Applying T^* to the constant function 1 then yields $b - t \geq 0$. To see that equality cannot occur here, consider the functions

$$f_z(w) = \frac{(1 - \langle z, w \rangle)^{n+1+a+b}}{|1 - \langle z, w \rangle|^{n+1+a+b}}, \qquad w \in \mathbb{B}_n.$$

Each f_z is a unit vector in $L^\infty(\mathbb{B}_n)$. If $b = t$, then

$$\|T^* f_z\|_\infty \geq |T^* f_z(z)| = c_t \int_{\mathbb{B}_n} \frac{(1 - |w|^2)^{a+t} \, dv(w)}{|1 - \langle z, w \rangle|^{n+1+a+t}}.$$

By part (2) of Theorem 1.12, the above integral tends to $+\infty$ as $|z| \to 1^-$. Thus the boundedness of T on $L^1(\mathbb{B}_n, dv_t)$ implies that $b - t > 0$, or $t + 1 < b + 1$. This completes the proof that (a) implies (c).

It remains to prove that (c) implies (b). The case $p = 1$ is a direct consequence of Fubini's theorem and part (3) of Theorem 1.12.

If $1 < p < \infty$ and $1/p + 1/q = 1$, then the condition $-pa < p(b+1)$ implies that the intervals

$$(A, B) = \left(-\frac{b+1}{q}, \frac{a}{q}\right)$$

and

$$(C, D) = \left(-\frac{a+1+t}{p}, \frac{b-t}{p}\right)$$

are both nonempty; the condition $-pa < t+1$ implies that $C < B$; and the condition $t+1 < p(b+1)$ implies that $A < D$. Therefore, the inequalities

$$-pa < t+1 < p(b+1)$$

imply that (A, B) and (C, D) have a nonempty intersection. Fix

$$s \in (A, B) \cap (C, D)$$

and let

$$h(z) = (1 - |z|^2)^s, \qquad z \in \mathbb{B}_n.$$

The boundedness of S on $L^p(\mathbb{B}_n, dv_t)$ then follows from Schur's test in conjunction with part (3) of Theorem 1.12.   □

The following theorem singles out several very important special cases of Theorem 2.10.

Theorem 2.11. *Suppose* $-1 < \alpha < \infty$, $-1 < t < \infty$, *and* $1 \le p < \infty$. *Then the operator* P_α *is a bounded projection from* $L^p(\mathbb{B}_n, dv_t)$ *onto* A_t^p *if and only if*

$$p(\alpha + 1) > t + 1.$$

In particular, P_α *is a bounded projection from* $L^p(\mathbb{B}_n, dv_\alpha)$ *onto* A_α^p *if and only if* $p > 1$, *and* P_α *is a bounded projection from* $L^1(\mathbb{B}_n, dv_t)$ *onto* A_t^1 *if and only if* $\alpha > t$.

In particular, we see that there exist a lot of bounded projections from the space $L^1(\mathbb{B}_n, dv_\alpha)$ onto A_α^1. This is in sharp contrast with the classical theory of Hardy spaces.

Theorem 2.12. *Suppose* $\alpha > -1$, $\beta > -1$, *and* $1 < p < \infty$. *Then*

$$(A_\alpha^p)^* = A_\beta^q$$

(with equivalent norms) under the integral pairing

$$\langle f, g \rangle_\gamma = \int_{\mathbb{B}_n} f(z)\,\overline{g(z)}\,dv_\gamma(z), \qquad f \in A_\alpha^p, g \in A_\beta^q,$$

where

$$\frac{1}{p} + \frac{1}{q} = 1, \qquad \gamma = \frac{\alpha}{p} + \frac{\beta}{q}.$$

Proof. If $g \in A_\beta^q$ and

$$F(f) = \langle f, g \rangle_\gamma = c_\gamma \int_{\mathbb{B}_n} (1 - |z|^2)^{\alpha/p} f(z) \overline{(1 - |z|^2)^{\beta/q} g(z)} \, dv(z), \qquad f \in A_\alpha^p,$$

it follows from Hölder's inequality that F is a bounded linear functional on A_α^p with $\|F\| \leq C \|g\|_{q,\beta}$, where C is a positive constant depending on c_α, c_β, and c_γ.

Conversely, if F is a bounded linear functional on A_α^p, then according to the Hahn-Banach extension theorem, F can be extended (without increasing its norm) to a bounded linear functional on $L^p(\mathbb{B}_n, dv_\alpha)$. By the usual duality of L^p spaces, there exists some $h \in L^q(\mathbb{B}_n, dv_\alpha)$ such that

$$F(f) = \int_{\mathbb{B}_n} f(z) \overline{h(z)} \, dv_\alpha(z), \qquad f \in L^p(\mathbb{B}_n, dv_\alpha).$$

Let

$$H(z) = \frac{c_\alpha}{c_\gamma} (1 - |z|^2)^{(\alpha - \beta)/q} h(z), \qquad z \in \mathbb{B}_n.$$

Then $H \in L^q(\mathbb{B}_n, dv_\beta)$ and

$$F(f) = \int_{\mathbb{B}_n} f(z) \overline{H(z)} \, dv_\gamma(z), \qquad f \in A_\alpha^p.$$

It is easy to check that the condition $\alpha > -1$ is equivalent to $q(\gamma + 1) > \beta + 1$, and the condition $\beta > -1$ is equivalent to $p(\gamma + 1) > \alpha + 1$. So, by Theorem 2.11, P_γ is a bounded projection from $L^p(\mathbb{B}_n, dv_\alpha)$ onto A_α^p, and P_γ is also a bounded projection from $L^q(\mathbb{B}_n, dv_\beta)$ onto A_β^q. Let $g = P_\gamma(H)$. Then $g \in A_\beta^q$ and

$$F(f) = \langle f, H \rangle_\gamma = \langle P_\gamma(f), H \rangle_\gamma = \langle f, P_\gamma(H) \rangle_\gamma = \langle f, g \rangle_\gamma$$

for all $f \in A_\alpha^p$. The proof is now complete. $\square$

A special case of the preceding theorem is when $\alpha = \beta$. In this case, we clearly have $\gamma = \alpha$ as well.

The dual of A_α^p for $0 < p \leq 1$ will be identified in the next chapter after we introduce the Bloch space.

2.3 Other Characterizations

In this section we characterize the weighted Bergman spaces in terms of various derivatives of a function. First recall that

$$Rf(z) = \sum_{k=1}^n z_k \frac{\partial f}{\partial z_k}(z)$$

is the radial derivative of f at z.

For a holomorphic function f in $\mathbb{B}_n$ we write

$$\nabla f(z) = \left(\frac{\partial f}{\partial z_1}(z), \cdots, \frac{\partial f}{\partial z_n}(z) \right) \tag{2.10}$$

and call $|\nabla f(z)|$ the (holomorphic) gradient of f at z. Similarly, we define

$$\widetilde{\nabla} f(z) = \nabla (f \circ \varphi_z)(0), \tag{2.11}$$

where φ_z is the biholomorphic mapping of $\mathbb{B}_n$ that interchanges 0 and z, and call $|\widetilde{\nabla} f(z)|$ the invariant gradient of f at z.

Lemma 2.13. *If f is holomorphic in $\mathbb{B}_n$, then*

$$|\widetilde{\nabla} f(z)|^2 = (1 - |z|^2)(|\nabla f(z)|^2 - |Rf(z)|^2)$$

for all $z \in \mathbb{B}_n$.

Proof. For any holomorphic function f in $\mathbb{B}_n$ we have

$$\widetilde{\Delta}(|f|^2)(0) = \Delta(|f|^2)(0) = 4|\nabla f(0)|^2 = 4|\widetilde{\nabla} f(0)|^2.$$

It follows that

$$4|\widetilde{\nabla} f(z)|^2 = 4|\widetilde{\nabla}(f \circ \varphi_z)(0)|^2 = \widetilde{\Delta}(|f \circ \varphi_z|^2)(0) = \widetilde{\Delta}(|f|^2)(z). \tag{2.12}$$

The desired result now follows from Proposition 1.17. $\qquad\square$

Since the invariant Laplacian is invariant under the action of the automorphism group, the relation (2.12) in the preceding proof shows that $|\widetilde{\nabla} f|$ is also Möbius invariant, namely,

$$|\widetilde{\nabla}(f \circ \varphi)(z)| = |(\widetilde{\nabla} f) \circ \varphi(z)| \tag{2.13}$$

for all f and all $\varphi \in \mathrm{Aut}(\mathbb{B}_n)$.

Lemma 2.14. *If f is holomorphic in $\mathbb{B}_n$, then*

$$(1 - |z|^2)|Rf(z)| \leq (1 - |z|^2)|\nabla f(z)| \leq |\widetilde{\nabla} f(z)|$$

for all $z \in \mathbb{B}_n$.

Proof. By the Cauchy-Schwarz inequality for $\mathbb{C}^n$,

$$|Rf(z)| \leq |z||\nabla f(z)| \leq |\nabla f(z)|.$$

This proves the first inequality. The second inequality follows from Lemma 2.13 and the fact that $|Rf(z)| \leq |z||\nabla f(z)|$. $\qquad\square$

The following lemma is critical for many problems concerning the spaces A_α^p when $0 < p < 1$.

Lemma 2.15. *Let $0 < p \leq 1$ and $\alpha > -1$. Then*

$$\int_{\mathbb{B}_n} |f(z)|(1 - |z|^2)^{(n+1+\alpha)/p - (n+1)} \, dv(z) \leq \frac{\|f\|_{p,\alpha}}{c_\alpha}$$

for all $f \in A_\alpha^p$, where c_α is the constant defined in (2.2).

Proof. Write

$$|f(z)| = |f(z)|^p |f(z)|^{1-p},$$

and estimate the second factor using Theorem 2.1. The desired result follows.   □

The exponent of $(1 - |z|^2)$ in the preceding lemma is the best possible; see Theorem 2.1. We should think of the above result as embedding the Bergman space A_α^p into a Bergman L^1 space.

Theorem 2.16. *Suppose $\alpha > -1$, $p > 0$, and f is holomorphic in $\mathbb{B}_n$. Then the following conditions are equivalent:*

(a) $f \in A_\alpha^p$.
(b) $|\widetilde{\nabla} f(z)|$ is in $L^p(\mathbb{B}_n, dv_\alpha)$.
(c) $(1 - |z|^2)|\nabla f(z)|$ is in $L^p(\mathbb{B}_n, dv_\alpha)$.
(d) $(1 - |z|^2)|Rf(z)|$ is in $L^p(\mathbb{B}_n, dv_\alpha)$.

Proof. Lemma 2.14 shows that (b) implies (c), and (c) implies (d).

To prove (a) implies (b), we fix $\beta > \alpha$ and observe that there exists a constant $C_1 > 0$ such that

$$|\nabla g(0)|^p \leq C_1 \int_{\mathbb{B}_n} |g(w)|^p \, dv_\beta(w)$$

for all holomorphic g in $\mathbb{B}_n$; see Lemma 2.4. Let $g = f \circ \varphi_z$, where $z \in \mathbb{B}_n$ and φ_z is the biholomorphic mapping of $\mathbb{B}_n$ that interchanges 0 and z, and make an obvious change of variables according to Proposition 1.13. We obtain

$$|\widetilde{\nabla} f(z)|^p \leq C_1 (1 - |z|^2)^{n+1+\beta} \int_{\mathbb{B}_n} \frac{|f(w)|^p \, dv_\beta(w)}{|1 - \langle z, w \rangle|^{2(n+1+\beta)}}.$$

An application of Fubini's theorem and part (3) of Theorem 1.12 then gives

$$\int_{\mathbb{B}_n} |\widetilde{\nabla} f(z)|^p \, dv_\alpha(z) \leq C_2 \int_{\mathbb{B}_n} |f(z)|^p \, dv_\alpha(z)$$

for some constant $C_2 > 0$ and all f holomorphic in $\mathbb{B}_n$. Actually, replacing f by $f - f(0)$, we have

$$\int_{\mathbb{B}_n} |\widetilde{\nabla} f(z)|^p \, dv_\alpha(z) \leq C_2 \int_{\mathbb{B}_n} |f(z) - f(0)|^p \, dv_\alpha(z).$$

To prove (d) implies (a), we assume that f is a holomorphic function in $\mathbb{B}_n$ such that the function $(1 - |z|^2)Rf(z)$ is in $L^p(\mathbb{B}_n, dv_\alpha)$. Let β be a sufficiently large positive constant. Then

$$Rf(z) = \int_{\mathbb{B}_n} \frac{Rf(w)\, dv_\beta(w)}{(1 - \langle z, w \rangle)^{n+1+\beta}}, \qquad z \in \mathbb{B}_n,$$

by Theorem 2.2. Since $Rf(0) = 0$, we have

$$Rf(z) = \int_{\mathbb{B}_n} Rf(w) \left(\frac{1}{(1 - \langle z, w \rangle)^{n+1+\beta}} - 1 \right) dv_\beta(w), \qquad z \in \mathbb{B}_n.$$

It follows that

$$f(z) - f(0) = \int_0^1 \frac{Rf(tz)}{t}\, dt = \int_{\mathbb{B}_n} Rf(w) L(z, w)\, dv_\beta(w),$$

where the kernel

$$L(z, w) = \int_0^1 \left(\frac{1}{(1 - t\langle z, w \rangle)^{n+1+\beta}} - 1 \right) \frac{dt}{t}$$

satisfies

$$|L(z, w)| \le \frac{C_3}{|1 - \langle z, w \rangle|^{n+\beta}}$$

for all z and w in $\mathbb{B}_n$; see Exercise 2.24. So

$$|f(z) - f(0)| \le C_4 \int_{\mathbb{B}_n} \frac{(1 - |w|^2)|Rf(w)|\, dv_{\beta-1}(w)}{|1 - \langle z, w \rangle|^{n+\beta}}.$$

If $1 \le p < \infty$ and β is large enough so that

$$0 < \alpha + 1 < p\beta,$$

then Theorem 2.10 shows that

$$\int_{\mathbb{B}_n} |f(z) - f(0)|^p\, dv_\alpha(z) \le C_5 \int_{\mathbb{B}_n} \left((1 - |z|^2)|Rf(z)| \right)^p dv_\alpha(z).$$

The case $0 < p < 1$ calls for a different proof. We start from the inequality

$$|f(z) - f(0)| \le C_6 \int_{\mathbb{B}_n} \left| \frac{Rf(w)}{(1 - \langle z, w \rangle)^{n+\beta}} \right| (1 - |w|^2)^\beta\, dv(w)$$

from the previous paragraph. Assume β is sufficiently large so we can set

$$\beta = \frac{n + 1 + \beta'}{p} - (n + 1),$$

with $\beta' > \alpha + p > -1$. By Lemma 2.15, we have

$$|f(z) - f(0)|^p \leq C_7 \int_{\mathbb{B}_n} \left| \frac{Rf(w)}{(1 - \langle z, w \rangle)^{n+\beta}} \right|^p dv_{\beta'}(w),$$

where $C_7 > 0$ is a constant independent of f. A use of Fubini's theorem and part (3) of Theorem 1.12 then gives

$$\int_{\mathbb{B}_n} |f(z) - f(0)|^p \, dv_\alpha(z) \leq C_8 \int_{\mathbb{B}_n} \left((1 - |z|^2) |Rf(z)| \right)^p dv_\alpha(z).$$

This completes the proof of the theorem. □

Note that the proof of the above theorem actually produces equivalent norms on A_α^p in terms of the radial derivative, the gradient, and the invariant gradient of f.

Theorem 2.17. *Suppose $\alpha > -1$, $p > 0$, N is a positive integer, and f is holomorphic in $\mathbb{B}_n$. Then $f \in A_\alpha^p$ if and only if the functions*

$$(1 - |z|^2)^N \frac{\partial^m f}{\partial z^m}(z), \qquad |m| = N,$$

all belong to $L^p(\mathbb{B}_n, dv_\alpha)$.

Proof. The case $N = 1$ follows from the equivalence of (a) and (c) in Theorem 2.16. We prove the case $N = 2$ here; the general case can then be proved using the same idea and induction.

So we assume $f \in A_\alpha^p$. By the equivalence of (a) and (c) in Theorem 2.16, each function $\partial f / \partial z_i$ is in $A_{\alpha+p}^p$, where $1 \leq i \leq n$. This in turn implies that each function

$$(1 - |z|^2) \frac{\partial^2 f}{\partial z_i \partial z_j}(z)$$

is in $L^p(\mathbb{B}_n, dv_{\alpha+p})$, or equivalently, each function

$$(1 - |z|^2)^2 \frac{\partial^2 f}{\partial z_i \partial z_j}(z)$$

is in $L^p(\mathbb{B}_n, dv_\alpha)$, where $1 \leq i, j \leq n$.

The arguments in the previous paragraph can be reversed. So the desired result is proved for $N = 2$. □

It should be clear that the integral

$$\int_{\mathbb{B}_n} |f(z)|^p \, dv_\alpha(z)$$

is comparable to the quantity

$$\sum_{|m| \leq N} \left| \frac{\partial^m f}{\partial z^m}(0) \right| + \sum_{|m| = N} \int_{\mathbb{B}_n} \left| (1 - |z|^2)^N \frac{\partial^m f}{\partial z^m}(z) \right|^p dv_\alpha(z)$$

whenever f is holomorphic in $\mathbb{B}_n$.

Lemma 2.18. *Suppose neither $n + s$ nor $n + s + t$ is a negative integer. If $\beta = s + N$ for some positive integer N, then there exists a one-variable polynomial h of degree N such that*

$$R^{s,t} \frac{1}{(1 - \langle z, w \rangle)^{n+1+\beta}} = \frac{h(\langle z, w \rangle)}{(1 - \langle z, w \rangle)^{n+1+\beta+t}}.$$

There also exists a polynomial $P(z, w)$ such that

$$R_{s,t} \frac{1}{1 - \langle z, w \rangle)^{n+1+\beta+t}} = \frac{P(z, w)}{(1 - \langle z, w \rangle)^{n+1+\beta}}.$$

Proof. Recall that

$$\frac{1}{(1 - \langle z, w \rangle)^{\lambda}} = \sum_{k=0}^{\infty} \frac{\Gamma(k + \lambda)}{k! \, \Gamma(\lambda)} \langle z, w \rangle^k \tag{2.14}$$

for any $\lambda \neq 0, -1, -2, \cdots$. It follows from the definition of $R^{s,t}$ that for $\beta = s + 1$ we can find a constant C such that

$$R^{s,t} \frac{1}{(1 - \langle z, w \rangle)^{n+1+\beta}} = C \sum_{k=0}^{\infty} \frac{k + n + 1 + s}{k!} \Gamma(n + 1 + k + s + t) \langle z, w \rangle^k,$$

which easily breaks into the sum of

$$C \langle z, w \rangle \sum_{k=0}^{\infty} \frac{\Gamma(n + 1 + k + \beta + t)}{k!} \langle z, w \rangle^k = \frac{C \Gamma(n + 1 + \beta + t)}{(1 - \langle z, w \rangle)^{n+1+\beta+t}}$$

and

$$C(n + 1 + s) \sum_{k=0}^{\infty} \frac{\Gamma(n + 1 + k + s + t)}{k!} \langle z, w \rangle^k = \frac{C(n + 1 + s) \Gamma(n + 1 + s + t)}{(1 - \langle z, w \rangle)^{n+1+s+t}}.$$

Adding these up proves the desired result for $R^{s,t}$ when $\beta = s + 1$.

In general, if $\beta = s + N$, then there exists a constant C_N such that

$$R^{s,t} \frac{1}{(1 - \langle z, w \rangle)^{n+1+\beta}} = C_N \sum_{k=0}^{\infty} \frac{p(k) \Gamma(n + 1 + k + s + t)}{k!} \langle z, w \rangle^k,$$

where $p(k)$ is a polynomial of k of degree N. We can write $p(k)$ as a linear combination of

$$1, \quad k, \quad k(k - 1), \quad \cdots, \quad k(k - 1) \cdots (k - N + 1).$$

And the proof for $R^{s,t}$ proceeds exactly the same as in the case $N = 1$.

To prove the result for $R_{s,t}$, we use Proposition 1.14 to write

$$R_{s,t} \frac{1}{(1 - \langle z, w \rangle)^{n+1+\beta+t}} = R_{s,t} R^{s+t, N} \frac{1}{(1 - \langle z, w \rangle)^{n+1+s+t}}.$$

We then use the commutativity of $R_{s,t}$ and $R^{s+t,N}$ to obtain

$$R_{s,t}\frac{1}{(1-\langle z,w\rangle)^{n+1+\beta+t}} = R^{s+t,N}R_{s,t}\frac{1}{(1-\langle z,w\rangle)^{n+1+s+t}}.$$

Use Proposition 1.14 again. We have

$$R_{s,t}\frac{1}{(1-\langle z,w\rangle)^{n+1+\beta+t}} = R^{s+t,N}\frac{1}{(1-\langle z,w\rangle)^{n+1+s}}.$$

The desired result for $R_{s,t}$ now follows from Proposition 1.15.   □

We can use the above lemma to compare the behavior of $R^{s,t}$ and $R^{\beta,t}$. In fact, in a very broad sense, $R^{s,t}f$ and $R^{\beta,t}f$ are comparable for any holomorphic function f. To see this, we write

$$h(\langle z,w\rangle) = \sum_{k=0}^{N} C_k(1-\langle z,w\rangle)^k,$$

where C_k are constants. Then

$$R^{s,t}\frac{1}{(1-\langle z,w\rangle)^{n+1+\beta}} = \sum_{k=0}^{N} C_k\frac{1}{(1-\langle z,w\rangle)^{n+1+\beta+t-k}},$$

which, according to Proposition 1.14, is the same as

$$R^{s,t}\frac{1}{(1-\langle z,w\rangle)^{n+1+\beta}} = \sum_{k=0}^{N} C_k R_{\beta+t-k,k}R^{\beta,t}\frac{1}{(1-\langle z,w\rangle)^{n+1+\beta}}.$$

Differentiating with respect to $\overline{w}$, we obtain

$$R^{s,t} = C_0 R^{\beta,t} + \sum_{k=1}^{N} C_k R_{\beta+t-k,k}R^{\beta,t}.$$

It is easy to see that $C_0 \neq 0$. Also, for each $1 \leq k \leq N$, the function $R_{\beta+t-k,k}R^{\beta,t}f$ is a k-th integral of $R^{\beta,t}f$, and so is more regular than $R^{\beta,t}f$. This shows that the behavior of $R^{s,t}f$ and $R^{\beta,t}f$ are often the same. Exercise 2.22 gives such an example.

Theorem 2.19. *Suppose $\alpha > -1$, $p > 0$, and $t > 0$. If neither $n+s$ nor $n+s+t$ is a negative integer, then there exist positive constants c and C such that*

$$c\int_{\mathbb{B}_n} |f(z)|^p \, dv_\alpha(z) \leq \int_{\mathbb{B}_n} |(1-|z|^2)^t R^{s,t}f(z)|^p \, dv_\alpha(z) \leq C\int_{\mathbb{B}_n} |f(z)|^p \, dv_\alpha(z)$$

for all holomorphic functions f in $\mathbb{B}_n$.

Proof. First assume that $f \in A_\alpha^p$. If $\beta = s + N$, where N is a sufficiently large positive integer, we have the integral representation

$$f(z) = \int_{\mathbb{B}_n} \frac{f(w)\, dv_\beta(w)}{(1 - \langle z, w \rangle)^{n+1+\beta}}, \qquad z \in \mathbb{B}_n.$$

Apply the operator $R^{s,t}$ inside the integral and use Lemma 2.18. We find a constant $C_1 > 0$ such that

$$|R^{s,t} f(z)| \le C_1 \int_{\mathbb{B}_n} \frac{|f(w)|\, dv_\beta(w)}{|1 - \langle z, w \rangle|^{n+1+\beta+t}}, \tag{2.15}$$

and so

$$(1 - |z|^2)^t |R^{s,t} f(z)| \le C_1 (1 - |z|^2)^t \int_{\mathbb{B}_n} \frac{|f(w)|\, dv_\beta(w)}{|1 - \langle z, w \rangle|^{n+1+\beta+t}}, \qquad z \in \mathbb{B}_n.$$

If $p \ge 1$ and N is large enough so that

$$\alpha + 1 < p(\beta + 1),$$

then it follows from Theorem 2.10 that

$$\int_{\mathbb{B}_n} \left| (1 - |z|^2)^t R^{s,t} f(z) \right|^p \, dv_\alpha(z) \le C_2 \int_{\mathbb{B}_n} |f(z)|^p \, dv_\alpha(z)$$

for some constant $C_2 > 0$ (independent of f).

If $0 < p < 1$, we write

$$\beta = \frac{n + 1 + \alpha'}{p} - (n + 1).$$

Here we assume that N is large enough so that $\alpha' > \alpha$. By (2.15) and Lemma 2.15, there exists a constant $C_3 > 0$ such that

$$|R^{s,t} f(z)|^p \le C_3 \int_{\mathbb{B}_n} \left| \frac{f(w)}{(1 - \langle w, z \rangle)^{n+1+\beta+t}} \right|^p \, dv_{\alpha'}(w)$$

$$= C_3 \int_{\mathbb{B}_n} \frac{|f(w)|^p \, dv_{\alpha'}(w)}{|1 - \langle z, w \rangle|^{n+1+\alpha'+pt}}.$$

An application of Fubini's theorem in combination of Theorem 1.12 shows that

$$\int_{\mathbb{B}_n} (1 - |z|^2)^{pt} |R^{s,t} f(z)|^p \, dv_\alpha(z) \le C_4 \int_{\mathbb{B}_n} |f(w)|^p \, dv_\alpha(w),$$

where C_4 is a positive constant independent of f.

Next assume that the function $(1 - |z|^2)^t R^{s,t} f(z)$ belongs to $L^p(\mathbb{B}_n, dv_\alpha)$. By the remark following Lemma 2.18 (also see Exercise 2.22), the function

$$g(z) = \frac{c_{\beta+t}}{c_\beta} (1 - |z|^2)^t R^{\beta,t} f(z)$$

also belongs to $L^p(\mathbb{B}_n, dv_\alpha)$. Using Corollary 2.3, Fubini's theorem, and Theorem 2.2, we check that $f = P_\beta g$. If $1 \leq p < \infty$, then Theorem 2.11 shows that f is in A_α^p. When $0 < p < 1$, we write $f = P_\beta g$ as

$$f(z) = C_5 \int_{\mathbb{B}_n} \frac{R^{\beta,t} f(w)}{(1 - \langle z, w \rangle)^{n+1+\beta}} (1 - |w|^2)^{\beta+t} \, dv(w), \qquad z \in \mathbb{B}_n,$$

where C_5 is a positive constant. We further assume that N is large enough so that

$$\beta + t = \frac{n + 1 + \alpha'}{p} - (n + 1)$$

for some $\alpha' > -1$. We can then apply Lemma 2.15 to obtain

$$|f(z)|^p \leq C_6 \int_{\mathbb{B}_n} \frac{|R^{\beta,t} f(w)|^p}{|1 - \langle w, z \rangle|^{(n+1+\beta)p}} \, dv_{\alpha'}(w)$$

for all $z \in \mathbb{B}_n$. Observe that

$$(n + 1 + \beta)p = n + 1 + \alpha' - pt = n + 1 + \alpha + (\alpha' - pt - \alpha),$$

and that we may assume that N is so large that

$$\alpha' - pt - \alpha > 0.$$

Using Fubini's theorem and Theorem 1.12, we deduce that

$$\int_{\mathbb{B}_n} |f(z)|^p \, dv_\alpha(z) \leq C_7 \int_{\mathbb{B}_n} (1 - |w|^2)^{pt} |R^{s,t} f(w)|^p \, dv_\alpha(w),$$

where C_7 is a positive constant. This completes the proof of the theorem. $\qquad\square$

When $p = 2$, both integrals in the preceding theorem can be evaluated using the Taylor expansions of f and $R^{s,t} f$. The desired result then follows easily from Stirling's formula for the gamma function.

2.4 Carleson Type Measures

In this section we are interested in measures μ on the unit ball with the property that $L^p(\mathbb{B}_n, \mu)$ contains the Bergman space A_α^p. Such measures will be termed Carleson type measures. It turns out that the requirements on μ are independent of p.

Recall that for $r > 0$ and $z \in \mathbb{B}_n$ the set

$$D(z, r) = \{w \in \mathbb{B}_n : \beta(z, w) < r\} \tag{2.16}$$

is a Bergman metric ball at z.

Lemma 2.20. *For each $r > 0$ there exists a positive constant C_r such that*

$$C_r^{-1} \le \frac{1 - |a|^2}{1 - |z|^2} \le C_r \tag{2.17}$$

and

$$C_r^{-1} \le \frac{1 - |a|^2}{|1 - \langle a, z \rangle|} \le C_r \tag{2.18}$$

for all a and z in $\mathbb{B}_n$ with $\beta(a, z) < r$. Moreover, if r is bounded above, then we may choose C_r to be independent of r.

Proof. Given any two points a and z in $\mathbb{B}_n$ with $\beta(a, z) < r$, we can write $z = \varphi_a(w)$ for some $w \in \mathbb{B}_n$ with $\beta(0, w) < r$. It follows from Lemma 1.2 that

$$1 - |z|^2 = \frac{(1 - |w|^2)(1 - |a|^2)}{|1 - \langle a, w \rangle|^2}.$$

Since $D(0, r)$ is actually a Euclidean ball centered at the origin with Euclidean radius less than 1, we can easily find a positive constant C such that

$$C^{-1} \le \frac{|1 - \langle a, w \rangle|^2}{1 - |w|^2} \le C$$

for all $a \in \mathbb{B}_n$ and $w \in D(0, r)$, so

$$C^{-1} \le \frac{1 - |a|^2}{1 - |z|^2} \le C$$

for all a and z with $\beta(a, z) < r$, and (2.17) is proved.

Since $z = \varphi_a(w)$ if and only if $w = \varphi_a(z)$, we can also apply Lemma 1.2 to obtain

$$1 - |w|^2 = \frac{(1 - |a|^2)(1 - |z|^2)}{|1 - \langle a, z \rangle|^2}.$$

By the previous paragraph, $1 - |w|^2$ is bounded by positive constants from both above and below, and $1 - |a|^2$ is comparable to $1 - |z|^2$, the estimates in (2.18) are now obvious. $\square$

Corollary 2.21. *Suppose $-\infty < \alpha < \infty$, $r_1 > 0$, $r_2 > 0$, and $r_3 > 0$. Then there exists a constant $C > 0$ such that*

$$C^{-1} \le \frac{v_\alpha(D(z, r_1))}{v_\alpha(D(w, r_2))} \le C$$

for all z and w in $\mathbb{B}_n$ with $\beta(z, w) \le r_3$.

Proof. This follows from Lemmas 1.24 and 2.20. $\square$

Many techniques in analysis involve covering lemmas, namely, ways to decompose the underlying domain into special nice pieces. We now present a useful decomposition of the open unit ball $\mathbb{B}_n$ into Bergman metric balls.

Lemma 2.22. *Given any positive number R and natural number M, there exists a natural number N such that every Bergman metric ball of radius r, where $r \leq R$, can be covered by N Bergman metric balls of radius r/M.*

Proof. Fix a Bergman metric ball $D(a, r)$ with $0 < r \leq R$. Let $\delta = r/M$ and let $\{D(a_k, \delta/2)\}$ be any finite covering of $D(a, r)$, where each $a_k \in D(a, r)$. Set $a_1' = a_1$ and let a_2' be the first of $\{a_2, a_3, \cdots\}$ such that $\beta(a_2', a_1') \geq \delta/2$, if such a term can be found. Let a_3' be the first term after a_2' whose Bergman distance is at least $\delta/2$ from both a_1' and a_2'. After this process stops, we obtain a covering $\{D(a_k', \delta)\}$ of $D(a, r)$ with $\beta(a_i', a_j') \geq \delta/2$ for $i \neq j$.

Since the sets $\{D(a_k', r/(4M))\}$ are disjoint and contained in $D(a, r + r/(4M))$, we have

$$\sum_k v\left(D\left(a_k', \frac{r}{4M}\right)\right) \leq v\left(D\left(a, r + \frac{r}{4M}\right)\right).$$

By Corollary 2.21, there exists a constant $C > 0$, independent of r but dependent on R and M, such that

$$v\left(D\left(a, r + \frac{r}{4M}\right)\right) \leq Cv\left(D\left(a_k', \frac{r}{4M}\right)\right)$$

for each k. It follows that $k \leq C$, so the natural number $N = [C] + 1$ has the desired properties. $\qquad\square$

Theorem 2.23. *There exists a positive integer N such that for any $0 < r \leq 1$ we can find a sequence $\{a_k\}$ in $\mathbb{B}_n$ with the following properties:*

(1) $\mathbb{B}_n = \cup_k D(a_k, r)$.
(2) The sets $D(a_k, r/4)$ are mutually disjoint.
(3) Each point $z \in \mathbb{B}_n$ belongs to at most N of the sets $D(a_k, 4r)$.

Proof. Fix any $r \in (0, 1]$. It is easy to find a sequence $\{a_k\}$ such that

$$\mathbb{B}_n = \bigcup_k D(a_k, r)$$

and that $\beta(a_i, a_j) \geq r/2$ for all $i \neq j$; see the first part of the proof of Lemma 2.22. Property (2) then follows from the triangle inequality.

By Lemma 2.22, there exists a positive integer N, independent of r, such that every Bergman metric ball of radius $4r$ can be covered by N Bergman metric balls of radius $r/4$. We show that property (3) must hold. In fact, if

$$z \in \bigcap_{i=1}^{N+1} D(a_{k_i}, 4r),$$

then $a_{k_i} \in D(z, 4r)$ for $1 \le i \le N + 1$. Let $D(z_i, r/4)$, $1 \le i \le N$, be a cover of $D(z, 4r)$. Then at least one of $D(z_i, r/4)$ must contain two of a_{k_j}, $1 \le j \le N + 1$. By the triangle inequality, these two points must have Bergman distance less than $r/2$, which contradicts the second assumption made on $\{a_k\}$ in the previous paragraph. This finishes the proof of the theorem. $\square$

Note that the radius $4r$ above is nothing special. We could have proved the result for any fixed radius greater than $r/4$. We are going to call r the separation constant for the sequence $\{a_k\}$, and we are going to call $\{a_k\}$ an r-lattice in the Bergman metric.

Lemma 2.24. *Suppose $r > 0$, $p > 0$, and $\alpha > -1$. Then there exists a constant $C > 0$ such that*

$$|f(z)|^p \le \frac{C}{(1 - |z|^2)^{n+1+\alpha}} \int_{D(z,r)} |f(w)|^p \, dv_\alpha(w)$$

for all $f \in H(\mathbb{B}_n)$ and all $z \in \mathbb{B}_n$.

Proof. Recall from Proposition 1.21 that $D(0, r)$ is a Euclidean ball centered at the origin with Euclidean radius $R = \tanh(r)$. So the subharmonicity of $|f|^p$ and Corollary 1.29 show that

$$|f(0)|^p \le \frac{1}{v_\alpha(D(0,r))} \int_{D(0,r)} |f(w)|^p \, dv_\alpha(w)$$

for all holomorphic f in $\mathbb{B}_n$. Replace f by $f \circ \varphi_z$ and make a change of variables according to Proposition 1.13. Then

$$|f(z)|^p \le \frac{1}{v_\alpha(D(0,r))} \int_{D(z,r)} |f(w)|^p \frac{(1 - |z|^2)^{n+1+\alpha}}{|1 - \langle z, w \rangle|^{2(n+1+\alpha)}} \, dv_\alpha(w).$$

The desired result then follows from Lemma 2.20. $\square$

Note that the above result can be restated as follows:

$$|f(z)|^p \le \frac{C}{v_\alpha(D(z,r))} \int_{D(z,r)} |f(w)|^p \, dv_\alpha(w), \qquad z \in \mathbb{B}_n, \qquad (2.19)$$

where f is holomorphic and C is a constant independent of f and z; see Lemma 1.24.

Theorem 2.25. *Suppose $p > 0$, $r > 0$, $\alpha > -1$, and μ is a positive Borel measure on $\mathbb{B}_n$. Then the following conditions are equivalent:*

(a) There exists a constant $C > 0$ such that

$$\int_{\mathbb{B}_n} |f(z)|^p \, d\mu(z) \le C \int_{\mathbb{B}_n} |f(z)|^p \, dv_\alpha(z)$$

for all holomorphic f in $\mathbb{B}_n$.

(b) There exists a constant $C > 0$ such that

$$\int_{\mathbb{B}_n} \frac{(1 - |a|^2)^{n+1+\alpha}}{|1 - \langle z, a \rangle|^{2(n+1+\alpha)}} \, d\mu(z) \leq C$$

for all $a \in \mathbb{B}_n$.
(c) There exists a constant $C > 0$ such that

$$\mu(D(a, r)) \leq C(1 - |a|^2)^{n+1+\alpha}$$

for all $a \in \mathbb{B}_n$.
(d) There exists a constant $C > 0$ such that

$$\mu(D(a_k, r)) \leq C(1 - |a_k|^2)^{n+1+\alpha}$$

for all $k \geq 1$, where $\{a_k\}$ is the sequence in Theorem 2.23.

Proof. It is easy to see that (a) implies (b). In fact, setting

$$f(z) = \left(\frac{(1 - |a|^2)^{n+\alpha+1}}{(1 - \langle z, a \rangle)^{2(n+1+\alpha)}} \right)^{1/p}$$

in (a) immediately yields (b).
    If (b) is true, then

$$\int_{D(a,r)} \frac{(1 - |a|^2)^{n+1}}{|1 - \langle z, a \rangle|^{2(n+1+\alpha)}} \, d\mu(z) \leq C$$

for all $a \in \mathbb{B}_n$. This along with Lemma 2.20 shows that (c) must be true.
    That (c) implies (d) is trivial.
    It remains to prove that (d) implies (a). So we assume that there exists a constant $C_1 > 0$ such that

$$\mu(D(a_k, r)) \leq C_1(1 - |a_k|^2)^{n+1+\alpha}$$

for all $k \geq 1$. If f is holomorphic in $\mathbb{B}_n$, then

$$\int_{\mathbb{B}_n} |f(z)|^p \, d\mu(z) \leq \sum_{k=1}^{\infty} \int_{D(a_k,r)} |f(z)|^p \, d\mu(z)$$

$$\leq \sum_{k=1}^{\infty} \mu(D(a_k, r)) \sup\{|f(z)|^p : z \in D(a_k, r)\}.$$

By Lemmas 2.24 and 2.20, there exists a constant $C_2 > 0$ such that

$$\sup\{|f(z)|^p : z \in D(a_k, r)\} \leq \frac{C_2}{(1 - |a_k|^2)^{n+1+\alpha}} \int_{D(a_k,2r)} |f(w)|^p \, dv_\alpha(w)$$

for all $k \geq 1$. It follows that

$$\int_{\mathbb{B}_n} |f(z)|^p \, d\mu(z) \leq C_1 C_2 \sum_{k=1}^{\infty} \int_{D(a_k, 2r)} |f(w)|^p \, dv_\alpha(w)$$

for all holomorphic f in $\mathbb{B}_n$. Since every point in $\mathbb{B}_n$ belongs to at most N of the sets $D(a_k, 2r)$, we must have

$$\int_{\mathbb{B}_n} |f(z)|^p \, d\mu(z) \leq C_1 C_2 N \int_{\mathbb{B}_n} |f(w)|^p \, dv_\alpha(w)$$

for all f holomorphic in $\mathbb{B}_n$. This completes the proof of the theorem. $\square$

Since $(1 - |a|^2)^{n+1+\alpha}$ is comparable to $v_\alpha(D(a, r))$, conditions (c) and (d) in Theorem 2.25 are equivalent to

$$\mu(D(a, r)) \leq C v_\alpha(D(a, r)), \qquad a \in \mathbb{B}_n,$$

and

$$\mu(D(a_k, r)) \leq C v_\alpha(D(a_k, r)), \qquad k \geq 1,$$

respectively.

We say that a sequence $\{f_k\}$ in A_α^p converges to 0 ultra-weakly if $\{\|f_k\|_{p,\alpha}\}$ is bounded and $f_k(z) \to 0$ for every $z \in \mathbb{B}_n$.

Theorem 2.26. *Suppose $p > 0$, $r > 0$, $\alpha > -1$, and μ is a positive Borel measure on $\mathbb{B}_n$. Then the following conditions are equivalent:*

(a) Whenever $\{f_k\}$ converges ultra-weakly to 0 in A_α^p, we have

$$\lim_{k \to \infty} \int_{\mathbb{B}_n} |f_k(z)|^p \, d\mu(z) = 0.$$

(b) The measure μ satisfies

$$\lim_{|a| \to 1^-} \int_{\mathbb{B}_n} \frac{(1 - |a|^2)^{n+1+\alpha}}{|1 - \langle z, a \rangle|^{2(n+1+\alpha)}} \, d\mu(z) = 0.$$

(c) The measure μ has the property that

$$\lim_{|a| \to 1^-} \frac{\mu(D(a, r))}{(1 - |a|^2)^{n+1+\alpha}} = 0.$$

(d) For the sequence $\{a_k\}$ from Theorem 2.23 we have

$$\lim_{k \to \infty} \frac{\mu(D(a_k, r))}{(1 - |a_k|^2)^{n+1+\alpha}} = 0.$$

Proof. The proof is similar to that of Theorem 2.25. We leave the details to the interested reader. $\square$

Once again, conditions (c) and (d) in Theorem 2.26 can be reformulated as

$$\lim_{|a|\to 1^-} \frac{\mu(D(a,r))}{v_\alpha(D(a,r))} = 0,$$

and

$$\lim_{k\to\infty} \frac{\mu(D(a_k,r))}{v_\alpha(D(a_k,r))} = 0,$$

respectively.

2.5 Atomic Decomposition

In this section we show that every function in the Bergman space A_α^p can be decomposed into a series of very nice functions (called atoms). These atoms are defined in terms of kernel functions and in some sense act as a basis for the space A_α^p.

Lemma 2.27. *For any $R > 0$ and any real b there exists a constant $C > 0$ such that*

$$\left| \frac{(1-\langle z,u\rangle)^b}{(1-\langle z,v\rangle)^b} - 1 \right| \le C\beta(u,v)$$

for all z, u, and v in $\mathbb{B}_n$ with $\beta(u,v) \le R$.

Proof. If u and v satisfy $\beta(u,v) \le R$, we can write $v = \varphi_u(w)$ with $|w| \le r$, where $r = \tanh R \in (0,1)$. Let $z' = \varphi_u(z)$. Then by Lemma 1.3,

$$1 - \langle z,u\rangle = \frac{1-|u|^2}{1-\langle z',u\rangle},$$

and

$$1 - \langle z,v\rangle = \frac{(1-|u|^2)(1-\langle z',w\rangle)}{(1-\langle z',u\rangle)(1-\langle u,w\rangle)}.$$

So

$$\frac{(1-\langle z,u\rangle)^b}{(1-\langle z,v\rangle)^b} - 1 = \frac{(1-\langle u,w\rangle)^b - (1-\langle z',w\rangle)^b}{(1-\langle z',w\rangle)^b}.$$

Since $|w| < r$, we have

$$1 - r < |1-\langle z',w\rangle| < 2.$$

Also, because $|\langle u,w\rangle| < r$ and $|\langle z',w\rangle| < r$, there exists a constant $C_1 > 0$, depending on r and b, such that

$$|(1-\langle u,w\rangle)^b - (1-\langle z',w\rangle)^b| \le C_1|\langle z',w\rangle - \langle u,w\rangle| \le 2C_1|w|.$$

On the relatively compact set $|w| < r$, the Bergman metric is equivalent to the Euclidean metric. Thus there exists a constant $C_2 > 0$ such that

$$|w| \le C_2\beta(0,w) = C_2\beta(0,\varphi_u(v)) = C_2\beta(u,v)$$

whenever $v = \varphi_u(w)$ with $\beta(u,v) < R$. This completes the proof of the lemma. □

Let $b = 1$ in the preceding lemma and apply the triangle inequality, then let $b = -1$ and apply the triangle inequality. We see that for any $R > 0$ there exists a constant $C > 0$ such that

$$C^{-1} \leq \frac{|1 - \langle z, u \rangle|}{|1 - \langle z, v \rangle|} \leq C \tag{2.20}$$

for all $z \in \mathbb{B}_n$ and all u and v with $\beta(u, v) \leq R$. This generalizes the estimates in Lemma 2.20.

In the remainder of this section we fix a sequence $\{a_k\}$ chosen according to Theorem 2.23. Let r be the separation constant for $\{a_k\}$.

Lemma 2.28. *For each $k \geq 1$ there exists a Borel set D_k satisfying the following conditions:*

(i) $D(a_k, r/4) \subset D_k \subset D(a_k, r)$ for every k.
(ii) $D_k \cap D_j = \emptyset$ for $k \neq j$.
(iii) $\mathbb{B}_n = \cup D_k$.

Proof. For any $k \geq 1$ let

$$E_k = D(a_k, r) - \bigcup_{j \neq k} D(a_j, r/4).$$

Then each E_k contains $D(a_k, r/4)$ and is contained in $D(a_k, r)$. Also, $\{E_k\}$ covers $\mathbb{B}_n$. In fact, if $z \in \mathbb{B}_n$, then $z \in D(a_k, r)$ for some k. If $z \in D(a_j, r/4)$ for some $j \neq k$, then $z \in E_j$; otherwise, $z \in E_k$.

Let $D_1 = E_1$ and inductively define

$$D_{k+1} = E_{k+1} - \bigcup_{i=1}^{k} D_i, \qquad k \geq 1.$$

Then $\{D_k\}$ is clearly a disjoint cover of $\mathbb{B}_n$. In fact, if $z \in \mathbb{B}_n$, then $z \in E_k$ for some k. If $k = 1$, then $z \in D_1$. If $k > 1$, then either we have $z \in D_i$ for some $1 \leq i < k$, or we have $z \in D_k$.

For each $k \geq 1$ we have

$$D_k \subset E_k \subset D(a_k, r).$$

It is clear that $D_1 = E_1$ contains $D(a_1, r/4)$. To see that each D_{k+1} contains $D(a_{k+1}, r/4)$, we fix $k \geq 1$ and fix $z \in D(a_{k+1}, r/4) \subset E_{k+1}$. Then $z \notin E_i$ for any $1 \leq i \leq k$, which implies that $z \notin D_i$ for any $1 \leq i \leq k$. This shows that

$$z \in E_{k+1} - \bigcup_{i=1}^{k} D_i = D_{k+1},$$

and the proof of the lemma is complete. □

We fix a real parameter $b > n$ and let $\beta = b - (n + 1)$. Define an operator T as follows.

$$Tf(z) = \int_{\mathbb{B}_n} \frac{(1 - |w|^2)^{b-(n+1)}}{|1 - \langle z, w \rangle|^b} \, f(w) \, dv(w), \qquad (2.21)$$

where $f \in L^1(\mathbb{B}_n, dv_\beta)$. We emphasize that the operator T depends on the parameter b.

We need to further partition the sets $\{D_k\}$ in Lemma 2.28. In fact, we partition the set D_1 and use automorphisms to carry the partition to other D_k's. To this end, we let η denote a positive radius that is much smaller than the separation constant r, in the sense that the quotient η/r is small. We fix a finite sequence $\{z_1, \cdots, z_J\}$ in $D(0, r)$, depending on η, such that $\{D(z_j, \eta)\}$ cover $D(0, r)$ and that $\{D(z_j, \eta/4)\}$ are disjoint. We then enlarge each set $D(z_j, \eta/4) \cap D(0, r)$ to a Borel set E_j in such a way that $E_j \subset D(z_j, \eta)$ and that

$$D(0, r) = \bigcup_{j=1}^{J} E_j$$

is a disjoint union; see the proof of Lemma 2.28 for how to achieve this.

For $k \geq 1$ and $1 \leq j \leq J$ we define $a_{kj} = \varphi_{a_k}(z_j)$ and

$$D_{kj} = D_k \cap \varphi_{a_k}(E_j).$$

It is clear that $a_{kj} \in D(a_k, r)$ for all $k \geq 1$ and $1 \leq j \leq J$. Since

$$D_k = \bigcup_{j=1}^{J} D_{kj}$$

is a disjoint union for every k, we obtain a disjoint decomposition

$$\mathbb{B}_n = \bigcup_{k=1}^{\infty} \bigcup_{j=1}^{J} D_{kj}$$

of $\mathbb{B}_n$.

We define an operator S on $H(\mathbb{B}_n)$ as follows.

$$Sf(z) = \sum_{k=1}^{\infty} \sum_{j=1}^{J} \frac{v_\beta(D_{kj}) f(a_{kj})}{(1 - \langle z, a_{kj} \rangle)^b}. \qquad (2.22)$$

We emphasize that the operator S depends on both the parameter b and the partition $\{D_{kj}\}$ (and hence also on the separation constants r and η).

The following lemma is the key to atomic decompositions for Bergman spaces, the Bloch space, and BMOA.

Lemma 2.29. *For any $p > 0$ and $\alpha > -1$ there exists a constant $C > 0$, independent of the separation constants r and η, such that*

$$|f(z) - Sf(z)| \leq C\sigma \sum_{k=1}^{\infty} \frac{(1 - |a_k|^2)^{(pb-n-1-\alpha)/p}}{|1 - \langle z, a_k \rangle|^b} \left[\int_{D(a_k, 2r)} |f(w)|^p \, dv_\alpha(w) \right]^{\frac{1}{p}}$$

for all $r \leq 1$, $z \in \mathbb{B}_n$, and $f \in H(\mathbb{B}_n)$, where

$$\sigma = \eta + \frac{\tanh(\eta)}{(\tanh(r))^{1-2n(1-1/p)}}.$$

Proof. Without loss of generality we may assume that $f \in A_\beta^1$. Then by Theorem 2.2,

$$f(z) = \int_{\mathbb{B}_n} \frac{f(w) \, dv_\beta(w)}{(1 - \langle z, w \rangle)^b}, \qquad z \in \mathbb{B}_n.$$

Since $\{D_{kj}\}$ is a partition of $\mathbb{B}_n$, we can write

$$f(z) - Sf(z) = \sum_{k=1}^{\infty} \sum_{j=1}^{J} \int_{D_{kj}} \left[\frac{f(w)}{(1 - \langle z, w \rangle)^b} - \frac{f(a_{kj})}{(1 - \langle z, a_{kj} \rangle)^b} \right] dv_\beta(w).$$

By the triangle inequality,

$$|f(z) - Sf(z)| \leq I(z) + H(z),$$

where

$$I(z) = \sum_{k=1}^{\infty} \sum_{j=1}^{J} \frac{1}{|1 - \langle z, a_{kj} \rangle|^b} \int_{D_{kj}} |f(w) - f(a_{kj})| \, dv_\beta(w),$$

and

$$H(z) = \sum_{k=1}^{\infty} \sum_{j=1}^{J} \frac{1}{|1 - \langle z, a_{kj} \rangle|^b} \int_{D_{kj}} \left| \frac{(1 - \langle z, a_{kj} \rangle)^b}{(1 - \langle z, w \rangle)^b} - 1 \right| |f(w)| \, dv_\beta(w).$$

We first estimate $I(z)$. For any $1 \leq k < \infty$ and $1 \leq j \leq J$ let

$$I_{kj} = \int_{D_{kj}} |f(w) - f(a_{kj})| \, dv_\beta(w).$$

By a change of variables,

$$I_{kj} = (1 - |a_{kj}|^2)^{n+1+\beta} \int_{E_{kj}} \frac{|f \circ \varphi_{a_{kj}}(w) - f \circ \varphi_{a_{kj}}(0)|}{|1 - \langle w, a_{kj} \rangle|^{2(n+1+\beta)}} \, dv_\beta(w),$$

where

$$
\begin{aligned}
E_{kj} &= \varphi_{a_{kj}}(D_{kj}) \subset \varphi_{a_{kj}} \circ \varphi_{a_k}(E_j) \\
&\subset \varphi_{a_{kj}} \circ \varphi_{a_k}(D(z_j, \eta)) \\
&= \varphi_{a_{kj}}(D(a_{kj}, \eta)) = D(0, \eta).
\end{aligned}
$$

For $w \in E_{kj}$ the quantities $(1 - |w|^2)^\beta$ and $|1 - \langle w, a_{kj} \rangle|$ are both bounded from below and from above. Also, since each $a_{kj} \in D(a_k, r)$, the quantities $1 - |a_{kj}|^2$ and $1 - |a_k|^2$ are comparable. Therefore, there exists a constant $C > 0$, independent of r and η, such that

$$
I_{kj} \leq C(1 - |a_k|^2)^{n+1+\beta} \int_{E_{kj}} |f \circ \varphi_{a_{kj}}(w) - f \circ \varphi_{a_{kj}}(0)| \, dv(w).
$$

In the rest of this proof, we let C denote a positive constant (independent of r, η, k, and j) whose exact value may change from one occurence to another.

Write $r' = \tanh(r)$, $\eta' = \tanh(\eta)$, and $R = \eta'/r'$. Since η is much smaller than r, we may as well assume that $R \leq \frac{1}{2}$. By Lemma 2.4, there exists a constant $C > 0$ such that

$$
|\nabla h(z)| \leq C \left(\int_{\mathbb{B}_n} |h(w)|^p \, dv(w) \right)^{1/p}, \qquad |z| \leq R,
$$

where h is any function in $H(\mathbb{B}_n)$. Consider $h(z) = g(r'z)$, where

$$
g(z) = f \circ \varphi_{a_{kj}}(z), \qquad z \in \mathbb{B}_n.
$$

After a change of variables, we obtain

$$
r'|\nabla g(r'z)| \leq C \left(\frac{1}{(r')^{2n}} \int_{D(0,r)} |g(w)|^p \, dv(w) \right)^{1/p}
$$

for all $|z| \leq R$. Equivalently,

$$
|\nabla g(z)| \leq \frac{C}{(r')^{1+(2n/p)}} \left(\int_{D(0,r)} |g(w)|^p \, dv(w) \right)^{1/p}
$$

for all $z \in D(0, \eta)$. For any $w \in E_{kj} \subset D(0, \eta)$, the identity

$$
g(w) - g(0) = \int_0^1 \left(\sum_{i=1}^n w_i \frac{\partial g}{\partial w_i}(tw) \right) dt
$$

leads to

$$
|g(w) - g(0)| \leq \eta' \sup\{|\nabla g(u)| : u \in D(0, \eta)\}.
$$

So, going back to I_{kj}, we have

$$
I_{kj} \leq C\eta'(1 - |a_k|^2)^{n+1+\beta} v(E_{kj}) \sup\{|\nabla g(u)| : u \in D(0, \eta)\}.
$$

Combining this with the earlier estimate on the complex gradient of g, we obtain

$$I_{kj} \leq \frac{C\eta'}{(r')^{1+(2n/p)}}(1-|a_k|^2)^{n+1+\beta}v(E_{kj})\left(\int_{D(0,r)}|g(w)|^p\,dv(w)\right)^{1/p}.$$

By a change of variables again,

$$\int_{D(0,r)}|g|^p\,dv = \int_{D(a_{kj},r)}|f(w)|^p\frac{(1-|a_{kj}|^2)^{n+1}\,dv(w)}{|1-\langle w,a_{kj}\rangle|^{2(n+1)}}.$$

By Lemma 2.27, the quantities $1-|a_{kj}|^2$ and $|1-\langle w,a_{kj}\rangle|$ are both comparable to $1-|a_k|^2$, where $w \in D(a_{kj},r)$. This along with the fact that $D(a_{kj},r) \subset D(a_k,2r)$ shows that

$$\int_{D(0,r)}|g|^p\,dv \leq \frac{C}{(1-|a_k|^2)^{n+1}}\int_{D(a_k,2r)}|f(w)|^p\,dv(w).$$

Since $1-|a_k|^2$ is comparable to $1-|w|^2$ for $w \in D(a_k,2r)$, we have

$$\int_{D(0,r)}|g|^p\,dv \leq \frac{C}{(1-|a_k|^2)^{n+1+\alpha}}\int_{D(a_k,2r)}|f(w)|^p\,dv_\alpha(w).$$

Combining this with the estimate in the previous paragraph, we obtain

$$I_{kj} \leq \frac{C\eta'}{(r')^{1+(2n/p)}}(1-|a_k|^2)^{(pb-n-1-\alpha)/p}v(E_{kj})\left(\int_{D(a_k,2r)}|f|^p\,dv_\alpha\right)^{1/p}.$$

Since

$$\sum_{j=1}^{J}v(E_{kj}) \leq Jv(D(0,\eta)) = J(\eta')^{2n}$$

and

$$v(D(0,r)) = \sum_{j=1}^{J}v(E_j) \geq \sum_{j=1}^{J}v(D(z_j,\eta/4)) \geq CJ(\eta')^{2n},$$

where the last inequality follows from Lemma 1.23, we have

$$\sum_{j=1}^{J}v(E_{kj}) \leq Cv(D(0,r)) = C(r')^{2n}.$$

Combining this with the estimate in the previous paragraph, we obtain

$$\sum_{j=1}^{J}I_{kj} \leq \frac{C\eta'}{(r')^{1-2n+(2n/p)}}(1-|a_k|^2)^{(pb-n-1-\alpha)/p}\left(\int_{D(a_k,2r)}|f|^p\,dv_\alpha\right)^{1/p}.$$

For each $k \geq 1$ and $1 \leq j \leq J$ it follows from Lemma 2.27 that $|1 - \langle z, a_{kj} \rangle|^b$ is comparable to $|1 - \langle z, a_k \rangle|^b$. Therefore,

$$I(z) \leq \frac{C\eta'}{(r')^{1-2n+(2n/p)}} \sum_{k=1}^{\infty} \frac{(1 - |a_k|^2)^{(pb-n-1-\alpha)/p}}{|1 - \langle z, a_k \rangle|^b} \left[\int_{D(a_k, 2r)} |f|^p \, dv_\alpha \right]^{1/p}.$$

To estimate $H(z)$, we let

$$H_{kj} = \int_{D_{kj}} \left| \frac{(1 - \langle z, a_{kj} \rangle)^b}{(1 - \langle z, w \rangle)^b} - 1 \right| |f(w)| \, dv_\beta(w)$$

for $k \geq 1$ and $1 \leq j \leq J$. By Lemma 2.27 and (2.17),

$$H_{kj} \leq C\eta (1 - |a_k|^2)^\beta \int_{D_{kj}} |f(w)| \, dv(w).$$

For every $w \in D_{kj}$ we use Lemma 2.24 to get

$$|f(w)| \leq C \left(\frac{1}{(1 - |a_k|^2)^{n+1+\alpha}} \int_{D(a_k, 2r)} |f(w)|^p \, dv_\alpha(w) \right)^{1/p}.$$

Then

$$H_{kj} \leq C\eta (1 - |a_k|^2)^{\beta - (n+1+\alpha)/p} v(D_{kj}) \left(\int_{D(a_k, 2r)} |f(w)|^p \, dv_\alpha(w) \right)^{1/p}.$$

Since

$$\sum_{j=1}^{J} v(D_{kj}) = v(D_k) \leq v(D(a_k, r)) \leq C(1 - |a_k|^2)^{n+1},$$

we deduce that

$$\sum_{j=1}^{J} H_{kj} \leq C\eta (1 - |a_k|^2)^{(pb-n-1-\alpha)/p} \left(\int_{D(a_k, 2r)} |f(w)|^p \, dv_\alpha(w) \right)^{1/p}.$$

According to Lemma 2.27, $|1 - \langle z, a_{kj} \rangle|^b$ is comparable to $|1 - \langle z, a_k \rangle|^b$. It follows that

$$H(z) \leq C\eta \sum_{k=1}^{\infty} \frac{(1 - |a_k|^2)^{(pb-n-1-\alpha)/p}}{|1 - \langle z, a_k \rangle|^b} \left(\int_{D(a_k, 2r)} |f(w)|^p \, dv_\alpha(w) \right)^{1/p}.$$

This completes the proof of the lemma.   □

We can now prove the main result of this section.

Theorem 2.30. *Suppose $p > 0$, $\alpha > -1$, and*

$$b > n \max\left(1, \frac{1}{p}\right) + \frac{\alpha + 1}{p}. \tag{2.23}$$

Then there exists a sequence $\{a_k\}$ in $\mathbb{B}_n$ such that A_α^p consists exactly of functions of the form

$$f(z) = \sum_{k=1}^{\infty} c_k \frac{(1 - |a_k|^2)^{(pb-n-1-\alpha)/p}}{(1 - \langle z, a_k \rangle)^b}, \qquad z \in \mathbb{B}_n, \tag{2.24}$$

where $\{c_k\}$ belongs to the sequence space l^p and the series converges in the norm topology of A_α^p.

Proof. First consider a function $f(z)$ defined by (2.24), where $\{a_k\}$ is an r-lattice in the Bergman metric whose existence is guaranteed by Theorem 2.23. We show that f must be in A_α^p. To this end, we write

$$f_k(z) = \frac{(1 - |a_k|^2)^{(pb-n-1-\alpha)/p}}{(1 - \langle z, a_k \rangle)^b}.$$

The assumption on b implies that $pb > n + 1 + \alpha$ for all $p > 0$. Thus $\{f_k\}$ is a bounded sequence in A_α^p by Theorem 1.12. Recall that the norm in $L^p(\mathbb{B}_n, dv_\alpha)$ is denoted by $\|\ \|_{p,\alpha}$.

If $0 < p \le 1$, then

$$\|f\|_{p,\alpha}^p \le \sum_{k=1}^{\infty} |c_k|^p \|f_k\|_{p,\alpha}^p.$$

Since $\{c_k\} \in l^p$ and $\{f_k\}$ is bounded in A_α^p, we see that f is in A_α^p.

When $p > 1$, we let $\{D_k\}$ denote the sets from Lemma 2.28 and consider the function

$$F(z) = \sum_{k=1}^{\infty} |c_k| v_\alpha(D_k)^{-1/p} \chi_k(z), \qquad z \in \mathbb{B}_n,$$

where χ_k is the characteristic function of D_k. It is clear that

$$\|F\|_{p,\alpha}^p = \sum_{k=1}^{\infty} |c_k|^p < \infty.$$

The assumption on b implies that

$$b > n + \frac{\alpha + 1}{p}, \quad \text{or,} \quad p(b - n) > \alpha + 1,$$

when $p > 1$. By Theorem 2.10, the operator T defined in (2.21) is bounded on $L^p(\mathbb{B}_n, dv_\alpha)$.

Since F is defined as a sum of nonnegative functions, we can apply T to the function F and integrate term by term. The result is

$$TF(z) = \sum_{k=1}^{\infty} |c_k| v_\alpha(D_k)^{-1/p} \int_{D_k} \frac{(1-|w|^2)^{b-n-1}}{|1-\langle z,w\rangle|^b}\, dv(w).$$

By Lemmas 1.24 and 2.20,

$$v_\alpha(D_k) \sim (1-|a_k|^2)^{n+1+\alpha}$$

and

$$1-|w|^2 \sim 1-|a_k|^2, \qquad w \in D_k.$$

Also, (2.20) tells us that $|1-\langle z,w\rangle|$ is comparable to $|1-\langle z,a_k\rangle|$ when $w \in D_k$. It follows that there exists a constant $\delta > 0$ such that

$$TF(z) \geq \delta \sum_{k=1}^{\infty} |c_k| \frac{(1-|a_k|^2)^{(pb-n-1-\alpha)/p}}{|1-\langle z,a_k\rangle|^b}$$

for all $z \in \mathbb{B}_n$. By the triangle inequality, we have

$$|f(z)| \leq \frac{1}{\delta} TF(z), \qquad z \in \mathbb{B}_n.$$

Since $F \in L^p(\mathbb{B}_n, dv_\alpha)$ and T is bounded on $L^p(\mathbb{B}_n, dv_\alpha)$ (see Theorem 2.10), we conclude that $f \in A_\alpha^p$ with

$$\int_{\mathbb{B}_n} |f(z)|^p \, dv_\alpha(z) \leq C \sum_k |c_k|^p$$

for some positive constant C independent of f.

The above proof, after some obvious minor adjustments, still works if $\{a_k\}$ is replaced by $\{a_{kj}\}$. In fact, if

$$f(z) = \sum_{k=1}^{\infty} \sum_{j=1}^{J} c_{kj} \frac{(1-|a_{kj}|^2)^{(pb-n-1-\alpha)/p}}{(1-\langle z,a_{kj}\rangle)^b},$$

then we can use the facts that $1-|a_{kj}|^2 \sim 1-|a_k|^2$ and $|1-\langle z,a_{kj}\rangle| \sim |1-\langle z,a_k\rangle|$ to obtain a constant $C > 0$ such that

$$|f(z)| \leq C \sum_{k=1}^{\infty} d_k \frac{(1-|a_k|^2)^{(pb-n-1-\alpha)/p}}{|1-\langle z,a_k\rangle|^b},$$

where

$$d_k = \sum_{j=1}^{J} |c_{kj}|.$$

By Hölder's inequality,

$$|d_k|^p \leq J^{p/q} \sum_{j=1}^{J} |c_{kj}|^p.$$

Therefore, the sequence $\{d_k\}$ is in l^p and it follows from the earlier proof that $f \in A_\alpha^p$.

We have now completed the first part of the proof of the theorem, that is, every function defined by (2.24) belongs to A_α^p, provided that we use a sequence $\{a_k\}$ guaranteed by Theorem 2.23 or an associated sequence $\{a_{kj}\}$ constructed before Lemma 2.29. Note that so far we have not made any assumption on the separation constants r and η.

To show that every function $f \in A_\alpha^p$ admits a representation given in (2.24), we fix an r-lattice $\{a_k\}$ in the Bergman metric and consider the (almost) η-lattice $\{a_{kj}\}$ and the corresponding finer partition $\{D_{kj}\}$ of $\mathbb{B}_n$ described before Lemma 2.29. Then by Lemma 2.29 and the first part of this proof, there exists a constant $C_1 > 0$ such that

$$\int_{\mathbb{B}_n} |f(z) - Sf(z)|^p \, dv_\alpha(z) \leq C_1 \sigma^p \sum_{k=1}^{\infty} \int_{D(a_k, 2r)} |f(z)|^p \, dv_\alpha(z),$$

where σ is the constant given in Lemma 2.29. Since each point of $\mathbb{B}_n$ belongs to at most of N of $D(a_k, 2r)$, we have

$$\int_{\mathbb{B}_n} |f(z) - Sf(z)|^p \, dv_\alpha(z) \leq C_1 N \sigma^p \int_{\mathbb{B}_n} |f(z)|^p \, dv_\alpha(z).$$

If η is small enough so that $C_1 N \sigma^p < 1$, then the operator $I - S$ on A_α^p has norm less than 1, where I is the identity operator. In this case, it follows from standard functional analysis that the operator S is invertible on A_α^p. Therefore, every $f \in A_\alpha^p$ admits a representation

$$f(z) = \sum_{k,j} c_{kj} \frac{(1 - |a_{kj}|^2)^{(pb-n-1-\alpha)/p}}{(1 - \langle z, a_{kj} \rangle)^b},$$

where

$$c_{kj} = \frac{v_\beta(D_{kj}) g(a_{kj})}{(1 - |a_{kj}|^2)^{(pb-n-1-\alpha)/p}}$$

and $g = S^{-1} f$. By Lemma 1.24,

$$v_\beta(D_{kj}) \leq v_\beta(D_k) \sim (1 - |a_k|^2)^{n+1+\beta} = (1 - |a_k|^2)^b.$$

Since $1 - |a_{kj}|^2$ is comparable to $1 - |a_k|^2$, we can find a constant $C_2 > 0$, independent of f, such that

$$\sum_{k,j} |c_{kj}|^p \leq C_2 \sum_{k,j} (1 - |a_k|^2)^{n+1+\alpha} |g(a_{kj})|^p.$$

Applying Lemma 2.24 to each $g(a_{kj})$, using the facts that $1 - |a_{kj}|^2$ is comparable to $1 - |a_k|^2$ and that $D(a_{kj}, r) \subset D(a_k, 2r)$, we obtain another constant $C_3 > 0$ such that

$$\sum_{kj} |c_{kj}|^p \leq C_3 J \sum_{k=1}^{\infty} \int_{D(a_k, 2r)} |g(z)|^p \, dv_\alpha(z).$$

Since every point of $\mathbb{B}_n$ belongs to at most N of the sets $D(a_k, 2r)$, we have

$$\sum_{kj} |c_{kj}|^p \leq C_3 J N \int_{\mathbb{B}_n} |g(z)|^p \, dv_\alpha(z).$$

This completes the proof of the theorem.   □

The proof of Theorem 2.30 tells us that if $\{c_k\} \in l^p$ and if f is given by the series representation (2.24), then

$$\int_{\mathbb{B}_n} |f(z)|^p \, dv_\alpha(z) \leq C \sum_k |c_k|^p$$

for some positive constant C independent of f. On the other hand, for any $f \in A_\alpha^p$, the proof of Theorem 2.30 tells us that we can choose a sequence $\{c_k\}$ to represent f as in (2.24) which also satisfies

$$\sum_k |c_k|^p \leq C \int_{\mathbb{B}_n} |f(z)|^p \, dv_\alpha(z),$$

where C is a positive constant independent of f. It follows that

$$\int_{\mathbb{B}_n} |f(z)|^p \, dv_\alpha(z) \sim \inf \left\{ \sum_k |c_k|^p : f \text{ satisfies (2.24)} \right\}. \qquad (2.25)$$

We state two special cases of the preceding theorem. The first case is when $p > 1$ and $b = n + 1 + \alpha$.

Corollary 2.31. *For any $\alpha > -1$ and $p > 1$ there exists a sequence $\{a_k\}$ in $\mathbb{B}_n$ such that A_α^p consists exactly of functions of the form*

$$f(z) = \sum_k c_k \frac{(1 - |a_k|^2)^{(n+1+\alpha)/q}}{(1 - \langle z, a_k \rangle)^{n+1+\alpha}}, \qquad (2.26)$$

where $1/p + 1/q = 1$ and $\{c_k\} \in l^p$.

The next case is for $p = 1$ and $b = 2(n + 1 + \alpha)$.

Corollary 2.32. *For any $\alpha > -1$ there exists a sequence $\{a_k\}$ in $\mathbb{B}_n$ such that A_α^1 consists exactly of functions of the form*

$$f(z) = \sum_k c_k \frac{(1 - |a_k|^2)^{n+1+\alpha}}{(1 - \langle z, a_k \rangle)^{2(n+1+\alpha)}}, \qquad (2.27)$$

where $\{c_k\}$ belongs to l^1.

Another consequence of the atomic decomposition is the following.

Corollary 2.33. *Suppose $\alpha > -1$ and $p > 0$. If r and q are positive numbers such that*

$$\frac{1}{p} = \frac{1}{q} + \frac{1}{r},$$

then every function $f \in A_\alpha^p$ admits a decomposition

$$f(z) = \sum_k g_k(z) h_k(z), \qquad z \in \mathbb{B}_n, \tag{2.28}$$

where each g_k is in A_α^q and each h_k is in A_α^r. Furthermore, if $0 < p \le 1$, then

$$\sum_k \|g_k\|_{q,\alpha} \|h_k\|_{r,\alpha} \le C\|f\|_{p,\alpha}, \tag{2.29}$$

where C is a positive constant independent of f.

Proof. If f is nonvanishing in $\mathbb{B}_n$, then we have the factorization $f = gh$, where $g = f^{p/q}$ is in A_α^q and $h = f^{p/r}$ is in A_α^r. In general, we use the atomic decomposition of f,

$$f(z) = \sum_k f_k(z),$$

where

$$f_k(z) = c_k \frac{(1 - |a_k|^2)^{b-(n+1+\alpha)/p}}{(1 - \langle z, a_k \rangle)^b}$$

is either identically zero (when $c_k = 0$) or nonvanishing on $\mathbb{B}_n$ (when $c_k \ne 0$). When $c_k \ne 0$, we simplify factor each $f_k = f_k^{p/q} f_k^{p/r}$. $\qquad\square$

2.6 Complex Interpolation

In this section we show that when $1 \le p < \infty$, the weighted Bergman spaces A_α^p interpolate just like the L^p spaces do. An extension to the case $p = \infty$ will be discussed in the next chapter.

Theorem 2.34. *Suppose $\alpha > -1$ and $1 \le p_0 < p_1 < \infty$. If*

$$\frac{1}{p} = \frac{1-\theta}{p_0} + \frac{\theta}{p_1}$$

for some $\theta \in (0,1)$, then

$$[A_\alpha^{p_0}, A_\alpha^{p_1}]_\theta = A_\alpha^p$$

with equivalent norms.

Proof. First assume that $f \in A_\alpha^p$. For any complex parameter ζ with $\operatorname{Re}\zeta \in [0, 1]$ we consider the function

$$h_\zeta(z) = \frac{f(z)}{|f(z)|}|f(z)|^{p\left(\frac{1-\zeta}{p_0} + \frac{\zeta}{p_1}\right)}, \qquad z \in \mathbb{B}_n.$$

Fix some $\beta > \alpha$ and let $f_\zeta = P_\beta(h_\zeta)$. It is clear that f_ζ is continuous in ζ when $0 \le \operatorname{Re}\zeta \le 1$ and analytic in ζ when $0 < \operatorname{Re}\zeta < 1$.

By Theorem 2.11, P_β is a bounded projection from $L^q(\mathbb{B}_n, dv_\alpha)$ onto A_α^q for any $1 \le q < \infty$. In particular, there exists a constant $C > 0$ such that

$$\|f_\zeta\|_{p_0,\alpha}^{p_0} \le C\|h_\zeta\|_{p_0,\alpha}^{p_0} = C\|f\|_{p,\alpha}^p$$

for all $\operatorname{Re}\zeta = 0$, and

$$\|f_\zeta\|_{p_1,\alpha}^{p_1} \le C\|h_\zeta\|_{p_1,\alpha}^{p_1} = C\|f\|_{p,\alpha}^p$$

for all $\operatorname{Re}\zeta = 1$. This shows that the function $f = f_\theta$ belongs to $[A_\alpha^{p_0}, A_\alpha^{p_1}]_\theta$ with $\|f\|_\theta \le C\|f\|_{p,\alpha}$.

Next assume that $f \in [A_\alpha^{p_0}, A_\alpha^{p_1}]_\theta$. Then f is holomorphic and

$$f \in [L^{p_0}(\mathbb{B}_n, dv_\alpha), L^{p_1}(\mathbb{B}_n, dv_\alpha)]_\theta = L^p(\mathbb{B}_n, dv_\alpha).$$

This shows $f \in A_\alpha^p$ and completes the proof of the theorem. □

Notes

The theory of Bergman spaces has been a central subject of study in complex analysis during the past few decades. This theory is especialy well developed for the unit disk in the complex plane; see [124], [51], and [32].

Most of the material in Sections 2.1 and 2.2 is from [39] and [94]. In particular, the idea of using Schur's test to prove the boundedness of Bergman type projections seems to have originated in [39]. Schur's test has become a standard (and very effective) tool for dealing with the boundedness of various operators, including Hankel operators on the Bergman space.

Theorem 2.1 in its present form first appeared in [115], although a less precise form can be found in [94]. Theorem 2.10 was proved in [130] in the case of the unit disk, but the proof there works in the higher dimensional case as well. Theorem 2.12 is due to Ruhan Zhao [120].

Carleson type measures for Bergman spaces was first introduced in [50], and has since been studied in numerous papers, including [65] and [122]. Theorem 2.23, which is a useful covering lemma, was essentially proved in [21] and [19].

The theorems in Section 2.3 are probably well known to the experts, although complete proofs are difficult to locate in the literature (except certain special cases). Lemma 2.15 is a key estimate for the study of Bergman spaces A_α^p when $0 < p \le 1$; it will be needed several times later in the book.

Atomic decomposition for Bergman spaces was due to Coifman and Rochberg [30]. Our proof here is modified from the original one in [30]. An alternative approach to the problem based on duality can be found in [67].

Complex interpolation of Bergman spaces is certainly well known to experts, although a precise reference is difficult to find. The one-dimensional case appeared in [124] as an exercise.

Exercises

2.1. Show that if $i \neq j$ and f is holomorphic in a neighborhood of the closed unit ball $\overline{\mathbb{B}}_n$, then

$$\overline{z}_i \frac{\partial f}{\partial z_j}(z) - \overline{z}_j \frac{\partial f}{\partial z_i}(z)$$

belongs to $L^2(\mathbb{B}_n, dv_\alpha) \ominus A_\alpha^2$.

2.2. Suppose $0 < p < \infty$, $\alpha > -1$, and f is holomorphic in $\mathbb{B}_n$. Show that

$$\int_{\mathbb{B}_n} |Rf(z)|^p \, dv_\alpha(z) < \infty$$

if and only if

$$\int_{\mathbb{B}_n} |\nabla f(z)|^p \, dv_\alpha(z) < \infty.$$

2.3. Show the every f in A_α^p can be approximated in norm by its Taylor polynomials if and only if $p > 1$. See [128].

2.4. Suppose α is real, $0 < p < \infty$, $\frac{p}{2} + \alpha > -1$, and f is holomorphic in $\mathbb{B}_n$. Show that

$$\int_{\mathbb{B}_n} (1 - |z|^2)^p |Rf(z)|^p \, dv_\alpha(z) < \infty$$

if and only if

$$\int_{\mathbb{B}_n} |\widetilde{\nabla} f(z)|^p \, dv_\alpha(z) < \infty.$$

2.5. Characterize Carleson type measures for A_α^p in terms of the Koranyi approach regions $Q_r(\zeta)$ (see Section 5.5 for definition).

2.6. Suppose $\alpha > -1$, $p > 0$, and $t > 0$. Show that a holomorphic function f is in A_α^p if and only if the function $(1 - |z|^2)^t R^t f(z)$ belongs to $L^p(\mathbb{B}_n, dv_\alpha)$.

2.7. Show that for any fixed $\alpha > -1$, the norm of P_α on $L^p(\mathbb{B}_n, dv_\alpha)$ is comparable to $\csc(\pi/p)$. See [137].

2.8. For real parameters a, b, and c define two operators $S = S_{a,b,c}$ and $T = T_{a,b,c}$ as follows:

$$Tf(z) = (1 - |z|^2)^a \int_{\mathbb{B}_n} \frac{(1 - |w|^2)^b f(w) \, dv(w)}{(1 - \langle z, w \rangle)^c},$$

and

$$Sf(z) = (1 - |z|^2)^a \int_{\mathbb{B}_n} \frac{(1 - |w|^2)^b f(w) \, dv(w)}{|1 - \langle z, w \rangle|^c}.$$

Suppose c is neither a negative integer nor 0. Then the following conditions are equivalent for any real t and $1 < p < \infty$:

(a) T is bounded on $L^p(\mathbb{B}_n, dv_t)$.
(b) S is bounded on $L^p(\mathbb{B}_n, dv_t)$.
(c) $-pa < t + 1 < p(b+1)$ and $c \leq n + 1 + a + b$.

See [63].

2.9. Fill in the details of the proof of Proposition 2.6. Prove Theorem 2.26.

2.10. If $\{a_k\}$ is a sequence from Theorem 2.23, show that there exist positive constants c and C such that

$$c \int_{\mathbb{B}_n} |f(z)|^p \, dv_\alpha(z) \leq \sum_k |f(a_k)|^p (1 - |a_k|^2)^{n+1+\alpha} \leq C \int_{\mathbb{B}_n} |f(z)|^p \, dv_\alpha(z)$$

for all $f \in A_\alpha^p$. Such a sequence $\{a_k\}$ is called a sampling sequence for A_α^p.

2.11. Suppose $\alpha > -1$ and $p > 0$. If $f \in A_\alpha^p$ and $f(0) = 0$, show that there exist functions $f_k \in A_\alpha^p$, $1 \leq k \leq n$, such that

$$f(z) = \sum_{k=1}^n z_k f_k(z), \qquad z \in \mathbb{B}_n.$$

2.12. Suppose $1 \leq p < \infty$ and $\alpha > -1$. Show that there exists a positive constant $C > 0$ such that

$$\int_{\mathbb{B}_n} |f(z)|^p \, dv_\alpha(z) \leq C \int_{\mathbb{B}_n} |u(z)|^p \, dv_\alpha(z)$$

for all holomorphic functions f in $\mathbb{B}_n$ with $f(0) = 0$, where $u(z) = \operatorname{Re} f(z)$.

2.13. For any $f \in L^1(\mathbb{B}_n, dv)$ define

$$Bf(z) = (1 - |z|^2)^{n+1} \int_{\mathbb{B}_n} \frac{f(w) \, dv(w)}{|1 - \langle z, w \rangle|^{2(n+1)}}, \qquad z \in \mathbb{B}_n.$$

The function Bf is called the Berezin transform of f. Show that the Berezin transform is bounded on $L^p(\mathbb{B}_n, dv)$ for $1 < p \leq \infty$.

2.14. Show that the Berezin transform B is one-to-one on $L^1(\mathbb{B}_n, dv)$.

2.15. Show that the Berezin transform B satisfies $Bf = f$ for all M-harmonic functions f in $L^1(\mathbb{B}_n, dv)$. Recall that f is M-harmonic if $\widetilde{\Delta} f = 0$.

2.16. If $1 \leq n \leq 11$, show that $Bf = f$ if and only if f is M-harmonic. Show that this fails when $n > 11$. See [3].

2.17. Find sharp growth estimates for the Taylor coefficients of functions in A_α^p.

2.18. If $\{f_k\}$ is an orthonormal basis for A_α^2, show that

$$K_\alpha(z, w) = \sum_{k=1}^\infty f_k(z)\overline{f_k(w)}$$

for z and w in $\mathbb{B}_n$.

2.19. Suppose $\alpha > -1, t > 0$, and $a > 0$. Show that there exists a function $F(z, w)$, holomorphic in z, conjugate holomorphic in w, and bounded in $\mathbb{B}_n \times \mathbb{B}_n$, such that

$$R_z^{\alpha,t}\left[\frac{1}{(1 - \langle z, w\rangle)^a}\right] = \frac{F(z, w)}{(1 - \langle z, w\rangle)^{a+t}}$$

for all z and w in $\mathbb{B}_n$. Furthermore, if t is a positive integer, then $F(z, w)$ is a polynomial in z and $\overline{w}$. See the proof of Theorem 3 in [136].

2.20. Suppose $-1 < \alpha < \beta$. Show that there exists a constant $C > 0$ such that

$$\int_{\mathbb{B}_n} \frac{\beta(z, w)(1 - |w|^2)^\alpha}{|1 - \langle z, w\rangle|^{n+1+\beta}}\, dv(w) \leq C(1 - |z|^2)^{\alpha - \beta}$$

for all $z \in \mathbb{B}_n$.

2.21. Suppose $1 \leq p < \infty$. Show that the operator T defined by

$$Tf(z) = \int_{\mathbb{B}_n} \frac{\beta(z, w)(1 - |w|^2)^\alpha}{|1 - \langle z, w\rangle|^{n+1+\alpha}} f(w)\, dv(w)$$

is bounded on $L^p(\mathbb{B}_n, dv_t)$ if and only if $p(\alpha + 1) > t + 1 > 0$.

2.22. Suppose $t > 0, \alpha > -1, 0 < p < \infty$, and $\beta = s + N$, where N is a positive integer. If $R^{s,t}$ is well defined and f is holomorphic in $\mathbb{B}_n$, show that $R^{s,t}f \in A_\alpha^p$ if and only if $R^{\beta,t}f \in A_\alpha^p$.

2.23. Show that
$$J_\varphi(z)K(\varphi(z), \varphi(w))\overline{J_\varphi(w)} = K(z, w),$$

where K is the (unweighted) Bergman kernel of $\mathbb{B}_n$ and $\varphi \in \mathrm{Aut}(\mathbb{B}_n)$.

2.24. Show that the kernel

$$L(z, w) = \int_0^1 \left[\frac{1}{(1 - t\langle z, w \rangle)^{n+1+\beta}} - 1 \right] \frac{dt}{t}$$

satisfies

$$|L(z, w)| \leq \frac{C}{|1 - \langle z, w \rangle|^{n+\beta}}, \qquad z, w \in \mathbb{B}_n,$$

where C is some positive constant independent of z and w.

2.25. If $1 < p < \infty$, show that ultra-weak convergence in A_α^p is the same as weak convergence, which is the same as weak-star convergence.

2.26. If $p = 1$, show that ultra-weak convergence in A_α^1 is the same as weak-star convergence but different from weak convergence.

2.27. Suppose $0 < p_1 < p_2 < \infty$ and

$$\frac{n + 1 + \alpha_1}{p_1} = \frac{n + 1 + \alpha_2}{p_2},$$

where $\alpha_1 > -1$ and $\alpha_2 > -1$. Show that $A_{\alpha_1}^{p_1} \subset A_{\alpha_2}^{p_2}$, and that the inclusion is continuous.

2.28. Suppose $0 < p < \infty$, $p \neq 2$, and $\alpha > -1$. Show that $\Phi : A_\alpha^p \to A_\alpha^p$ is a surjective linear isometry if and only if

$$\Phi(f)(z) = \lambda f \circ \varphi(z) \, (J_{\mathbb{C}}\varphi(z))^{2/p},$$

where $\varphi \in \mathrm{Aut}(\mathbb{B}_n)$ and λ is a unimodulus constant. See [59].

2.29. Suppose f is continuous on the closed unit ball. Show that

$$\lim_{\alpha \to 1^-} \int_{\mathbb{B}_n} |f(z)|^p \, dv_\alpha(z) = \int_{\mathbb{S}_n} |f(\zeta)|^p \, d\sigma(\zeta).$$

2.30. Suppose f is continuous on $\mathbb{B}_n$. Show that

$$\lim_{p \to \infty} \left[\int_{\mathbb{B}_n} |f(z)|^p \, dv_\alpha(z) \right]^{1/p} = \sup\{|f(z)| : z \in \mathbb{B}_n\}.$$

3

The Bloch Space

In this chapter we study the Bloch space $\mathcal{B}$ and the little Bloch space $\mathcal{B}_0$ in $\mathbb{B}_n$. We prove various characterizations of $\mathcal{B}$ and focus our attention on the following three properties.

The Bloch space can be thought of as the limit case of the Bergman spaces A_α^p as $p \to \infty$. In particular, $\mathcal{B}$ can be naturally identified with the dual space of A_α^1.

The Bloch space is intimately related to the Bergman metric; it consists exactly of those holomorphic functions that are Lipschitz from $\mathbb{B}_n$ with the Bergman metric to $\mathbb{C}$ with the Euclidean metric.

The Bloch space is prominent among Möbius invariant function spaces. In fact, it is the largest possible space of holomorphic functions whose (semi-)norm is invariant under the action of the automorphism group.

3.1 The Bloch space

In classical geometric function theory of the open unit disk $\mathbb{D}$ in the complex plane $\mathbb{C}$, the Bloch space is a central object of study and several outstanding problems remain unresolved. In the one dimensional case, the Bloch space consists of analytic functions f in $\mathbb{D}$ such that

$$\sup_{z \in \mathbb{D}} (1 - |z|^2)|f'(z)| < \infty.$$

This definition can be generalized to higher dimensions in several possible ways. The following two elementary generalizations are natural and will be shown to be equivalent:

$$\sup_{z \in \mathbb{B}_n} (1 - |z|^2)|\nabla f(z)| < \infty,$$

and

$$\sup_{z \in \mathbb{B}_n} (1 - |z|^2)|Rf(z)| < \infty,$$

where $|\nabla f(z)|$ is the holomorphic gradient of f at z and Rf is the radial derivative of f at z. However, these definitions are not invariant under the action of the automorphism group. Our approach is to use a Möbius invariant (although less straightforward) definition based on the invariant gradient or the invariant Laplacian, and then show that it is equivalent to the elementary definitions above.

For a holomorphic function f in $\mathbb{B}_n$ we define

$$Q_f(z) = \sup\left\{ \frac{|\langle \nabla f(z), \overline{w} \rangle|}{\sqrt{\langle B(z)w, w \rangle}} : w \in \mathbb{C}^n - \{0\}\right\}, \qquad z \in \mathbb{B}_n, \qquad (3.1)$$

where $B(z)$ is the Bergman matrix introduced in Section 1.5 and $\langle \, , \, \rangle$ is the natural inner product in $\mathbb{C}^n$. All vectors in $\mathbb{C}^n$ are considered column vectors. The following result will enable us to define a Möbius invariant semi-norm on the Bloch space in several equivalent ways.

Theorem 3.1. *For $z \in \mathbb{B}_n$ and f holomorphic in $\mathbb{B}_n$ the following quantities are all equal:*

(a) $Q_f(z)$.

(b) $\left\langle \overline{B(z)^{-1}} \nabla f(z), \nabla f(z) \right\rangle^{1/2}$.

(c) $\dfrac{1}{2}\left[\tilde{\Delta}(|f|^2)(z) \right]^{1/2}$.

(d) $|\tilde{\nabla} f(z)|$.

(e) $\left[(1 - |z|^2)(|\nabla f(z)|^2 - |Rf(z)|^2)\right]^{1/2}$.

Proof. It follows from Lemma 2.13 and (2.12) that the quantities in (c), (d), and (e) are the same.

We can replace w by $B(z)^{-1/2}w$ in the definition $Q_f(z)$ to obtain

$$Q_f(z) = \sup\left\{ \frac{|\langle \nabla f(z), \overline{B(z)^{-1/2}w} \rangle|}{|w|} : w \in \mathbb{B}_n - \{0\}\right\}$$

$$= \sup\left\{ \frac{|\langle \overline{B(z)^{-1/2}} \nabla f(z), \overline{w} \rangle|}{|w|} : w \in \mathbb{C}^n - \{0\}\right\}$$

$$= \left| \overline{B(z)^{-1/2}} \nabla f(z) \right|$$

$$= \left\langle \overline{B(z)^{-1}} \nabla f(z), \nabla f(z) \right\rangle^{1/2}.$$

This shows that the quantities in (a) and (b) are equal.

From (b) and the formula for $B(z)^{-1}$ in Proposition 1.18 we easily deduce that

$$Q_f(z) = \left[(1 - |z|^2)(|\nabla f(z)|^2 - |Rf(z)|^2)\right]^{1/2}.$$

So the quantities in (b) and (e) are equal. $\square$

We now define the Bloch space of $\mathbb{B}_n$, denoted by $\mathcal{B}$, as the space of holomorphic functions f in $\mathbb{B}_n$ such that

$$\|f\|_\mathcal{B} = \sup\{|\widetilde{\nabla} f(z)| : z \in \mathbb{B}_n\} < \infty. \tag{3.2}$$

This is only a semi-norm, with $\|f\|_\mathcal{B} = 0$ if and only if f is constant.

Proposition 3.2. *The semi-norm $\| \ \|_\mathcal{B}$ is complete and invariant under the action of* $\mathrm{Aut}(\mathbb{B}_n)$, *that is,*

$$\|f \circ \varphi\|_\mathcal{B} = \|f\|_\mathcal{B} \tag{3.3}$$

for all $f \in \mathcal{B}$ and all $\varphi \in \mathrm{Aut}(\mathbb{B}_n)$.

Proof. The Möbius invariance of the semi-norm $\|f\|_\mathcal{B}$ follows from that of the invariant gradient; see (2.13).

To show that $\| \ \|_\mathcal{B}$ is complete, assume that $\{f_k\}$ is a sequence of functions in $\mathcal{B}$ with the properties that each $f_k(0) = 0$ and that for any $\epsilon > 0$ there exists a natural number N such that

$$\|f_k - f_l\|_\mathcal{B} < \epsilon \qquad (k > N, l > N).$$

Since (see Lemma 2.14)

$$(1 - |z|^2)|\nabla f(z)| \le |\widetilde{\nabla} f(z)|,$$

it follows that for each $1 \le i \le n$ the sequence $\{\partial f_k / \partial z_i\}$ is uniformly Cauchy on every compact set in $\mathbb{B}_n$. This together with the assumption that each $f_k(0) = 0$ shows that there exists a holomorphic function f in $\mathbb{B}_n$ with $f(0) = 0$ and

$$\lim_{k \to \infty} f_k(z) = f(z), \quad \lim_{k \to \infty} \frac{\partial f_k}{\partial z_i}(z) = \frac{\partial f}{\partial z_i}(z),$$

uniformly on every compact set in $\mathbb{B}_n$, where $1 \le i \le n$.

If we write the norm $\| \ \|_\mathcal{B}$ using part (e) of Theorem 3.1, then

$$(1 - |z|^2)\left(|\nabla(f_k - f_l)(z)|^2 - |R(f_k - f_l)(z)|^2\right) < \epsilon^2, \quad k > N, l > N.$$

Let $l \to \infty$ and then take the supremum over $z \in \mathbb{B}_n$. We conclude that

$$\|f - f_k\|_\mathcal{B} \le \epsilon$$

for all $k > N$. This shows that $\mathcal{B}$ is complete in the semi-norm $\| \ \|_\mathcal{B}$. $\qquad \square$

Lemma 3.3. *Suppose β is a real constant and $g \in L^1(\mathbb{B}_n, dv)$. If*

$$f(z) = \int_{\mathbb{B}_n} \frac{g(w)\, dv(w)}{(1 - \langle z, w \rangle)^\beta}, \qquad z \in \mathbb{B}_n,$$

then

$$|\widetilde{\nabla} f(z)| \le \sqrt{2}\,|\beta|(1 - |z|^2)^{\frac{1}{2}} \int_{\mathbb{B}_n} \frac{|g(w)|\, dv(w)}{|1 - \langle z, w \rangle|^{\beta + \frac{1}{2}}}$$

for all $z \in \mathbb{B}_n$.

Proof. Fix $a \in \mathbb{B}_n$ and make the change of variables $w \mapsto \varphi_a(w)$ in

$$f \circ \varphi_a(z) = \int_{\mathbb{B}_n} \frac{g(w)\, dv(w)}{(1 - \langle \varphi_a(z), w \rangle)^\beta}.$$

We obtain

$$f \circ \varphi_a(z) = \int_{\mathbb{B}_n} \frac{g \circ \varphi_a(w)}{(1 - \langle \varphi_a(z), \varphi_a(w) \rangle)^\beta} J_a(w)\, dv(w),$$

where

$$J_a(w) = \frac{(1 - |a|^2)^{n+1}}{|1 - \langle w, a \rangle|^{2(n+1)}}$$

is the Jacobian determinant. Recall from Lemma 1.3 that

$$1 - \langle \varphi_a(z), \varphi_a(w) \rangle = \frac{(1 - |a|^2)(1 - \langle z, w \rangle)}{(1 - \langle z, a \rangle)(1 - \langle a, w \rangle)}.$$

So

$$f \circ \varphi_a(z) = \frac{(1 - \langle z, a \rangle)^\beta}{(1 - |a|^2)^\beta} \int_{\mathbb{B}_n} \frac{(1 - \langle a, w \rangle)^\beta}{(1 - \langle z, w \rangle)^\beta} g \circ \varphi_a(w) J_a(w)\, dv(w).$$

Differentiating in z at 0 using the product rule then produces

$$\tilde{\nabla} f(a) = \beta \int_{\mathbb{B}_n} \frac{(\overline{w} - \overline{a})(1 - \langle a, w \rangle)^\beta}{(1 - |a|^2)^\beta} g \circ \varphi_a(w) J_a(w)\, dv(w).$$

Make the change of variables $w \mapsto \varphi_a(w)$ again. We get

$$\tilde{\nabla} f(a) = \beta \int_{\mathbb{B}_n} \frac{(\overline{\varphi_a(w)} - \overline{a}) g(w)\, dv(w)}{(1 - \langle a, w \rangle)^\beta},$$

so

$$|\tilde{\nabla} f(a)| \le |\beta| \int_{\mathbb{B}_n} \frac{|\varphi_a(w) - a||g(w)|\, dv(w)}{|1 - \langle a, w \rangle|^\beta}.$$

It is easy to check that

$$|\varphi_a(w) - a|^2 = \frac{(1 - |a|^2)(|w|^2 - |\langle w, a \rangle|^2)}{|1 - \langle a, w \rangle|^2} \le \frac{(1 - |a|^2)(1 - |\langle w, a \rangle|^2)}{|1 - \langle a, w \rangle|^2},$$

the desired result then follows from the obvious estimate

$$1 - |\langle w, a \rangle|^2 = (1 + |\langle w, a \rangle|)(1 - |\langle w, a \rangle|) \le 2|1 - \langle a, w \rangle|.$$

$\square$

Note that Lemma 3.3 is interesting only when $n > 1$.

The Bloch space $\mathcal{B}$ becomes a Banach space with the following norm:

$$\|f\| = |f(0)| + \|f\|_{\mathcal{B}}, \qquad f \in \mathcal{B}.$$

The Bloch semi-norm $\| \ \|_{\mathcal{B}}$ is Möbius invariant, but it is usually inconvenient for us to verify that a certain function belongs to the Bloch space by using the definition. The following theorem gives us several conditions that are equivalent to but more easily verifiable than the definition.

Theorem 3.4. *Suppose $\alpha > -1$ and f is holomorphic in $\mathbb{B}_n$. Then the following conditions are equivalent:*

(a) f is in $\mathcal{B}$.

(b) $(1 - |z|^2)|\nabla f(z)|$ is bounded in $\mathbb{B}_n$.

(c) $(1 - |z|^2)|Rf(z)|$ is bounded in $\mathbb{B}_n$.

(d) $f = P_\alpha g$ for some $g \in L^\infty(\mathbb{B}_n)$.

Proof. By Lemma 2.14, condition (a) implies (b), and condition (b) implies (c).

To show that (c) implies (d), suppose $(1 - |z|^2)Rf(z)$ is bounded in $\mathbb{B}_n$. Consider the function

$$g(z) = \frac{c_{\alpha+1}}{c_\alpha}(1 - |z|^2) \int_{\mathbb{B}_n} \frac{f(w)\, dv_\alpha(w)}{(1 - \langle z, w\rangle)^{n+2+\alpha}}, \qquad z \in \mathbb{B}_n.$$

We can rewrite

$$g(z) = \frac{c_{\alpha+1}}{c_\alpha}(1 - |z|^2) \int_{\mathbb{B}_n} \frac{1 - \langle z, w\rangle + \langle z, w\rangle}{(1 - \langle z, w\rangle)^{n+2+\alpha}}\, f(w)\, dv_\alpha(w)$$

and break the integral into two. The result is that

$$g(z) = \frac{c_{\alpha+1}}{c_\alpha}\left[(1 - |z|^2)f(z) + \frac{(1 - |z|^2)Rf(z)}{n + 1 + \alpha}\right]. \tag{3.4}$$

Since the boundedness of $(1 - |z|^2)Rf(z)$ in $\mathbb{B}_n$ together with

$$f(z) - f(0) = \int_0^1 \frac{Rf(tz)}{t}\, dt$$

shows that f grows at most as fast as $-\log(1 - |z|^2)$, it follows that g is bounded in $\mathbb{B}_n$. By Fubini's theorem and the reproducing property of P_α and $P_{\alpha+1}$, we easily check that $f = P_\alpha g$. This proves that (c) implies (d).

Finally, we assume that $f = P_\alpha g$ for some $\alpha > -1$ and $g \in L^\infty(\mathbb{B}_n)$. By Lemma 3.3, there exists a positive constant C such that

$$|\widetilde{\nabla} f(z)| \leq C\|g\|_\infty (1 - |z|^2)^{1/2} \int_{\mathbb{B}_n} \frac{(1 - |w|^2)^\alpha\, dv(w)}{|1 - \langle z, w\rangle|^{n+1+\alpha+\frac{1}{2}}}$$

for all $z \in \mathbb{B}_n$. We deduce from Theorem 1.12 that $|\widetilde{\nabla} f(z)|$ is bounded in $\mathbb{B}_n$. So (d) implies (a) and the proof of the theorem is complete. $\qquad\square$

The Bloch space can also be described in terms of higher order derivatives, and more generally, in terms of fractional radial derivatives.

Theorem 3.5. *Suppose N is a positive integer, $t > 0$, and f is holomorphic in $\mathbb{B}_n$. If α is a real parameter such that neither $n + \alpha$ nor $n + \alpha + t$ is a negative integer, then the following conditions are equivalent:*

(1) $f \in \mathcal{B}$.
(2) The function $(1 - |z|^2)^t R^{\alpha,t} f(z)$ is bounded in $\mathbb{B}_n$.
(3) The functions

$$(1 - |z|^2)^N \frac{\partial^N f}{\partial z^m}(z), \qquad |m| = N,$$

are bounded in $\mathbb{B}_n$.

Proof. If $f \in \mathcal{B}$, then by Theorem 3.4 there exists a function $g \in L^\infty(\mathbb{B}_n)$ such that

$$f(z) = \int_{\mathbb{B}_n} \frac{g(w)\, dv_\beta(w)}{(1 - \langle z, w \rangle)^{n+1+\beta}}.$$

Here $\beta = \alpha + K$ and K is a large enough positive integer so that $\beta > -1$. By Lemma 2.18, there exists a one-variable polynomial h such that

$$R^{\alpha,t} f(z) = c_\alpha \int_{\mathbb{B}_n} \frac{h(\langle z, w \rangle) g(w)\, dv_\beta(w)}{(1 - \langle z, w \rangle)^{n+1+\beta+t}}.$$

An application of Theorem 1.12 then shows that the function

$$(1 - |z|^2)^t R^{\alpha,t} f(z)$$

is bounded in $\mathbb{B}_n$. A similar argument (using differentiation under the integral sign) shows that the functions

$$(1 - |z|^2)^N \frac{\partial^m f}{\partial z^m}(z), \qquad |m| = N,$$

are all bounded in $\mathbb{B}_n$.

Next assume that the function $(1 - |z|^2)^t R^{\alpha,t} f(z)$ is bounded in $\mathbb{B}_n$. By the remark following Lemma 2.18, the function

$$g(z) = \frac{c_{\beta+t}}{c_\beta}(1 - |z|^2)^t R^{\beta,t} f(z)$$

is also bounded in $\mathbb{B}_n$, where $\beta = \alpha + K$ is as in the previous paragraph. A use of Fubini's theorem and Theorem 2.2 reveals that $f = P_\beta g$, so $f \in \mathcal{B}$ in view of Theorem 3.4. This proves that conditions (1) and (2) are equivalent.

Finally, if the functions

$$(1 - |z|^2)^N \frac{\partial^N f}{\partial z^m}(z), \qquad |m| = N,$$

are all bounded in $\mathbb{B}_n$, then successive integration shows that the functions

$$(1 - |z|^2)\frac{\partial f}{\partial z_k}(z), \qquad 1 \le k \le n,$$

are all bounded in $\mathbb{B}_n$, and so f belongs to the Bloch space. Therefore, conditions (1) and (3) are equivalent. □

The Bloch space can also be characterized in terms of the Bergman metric. In fact, they are related to each other in a very precise way.

Theorem 3.6. *If f is holomorphic in $\mathbb{B}_n$, then*

$$\|f\|_{\mathcal{B}} = \sup\left\{\frac{|f(z) - f(w)|}{\beta(z, w)} : z, w \in \mathbb{B}_n, z \ne w\right\},$$

where β is the Bergman metric on $\mathbb{B}_n$.

Proof. First assume that $f \in \mathcal{B}$. Fix any two points z and w in $\mathbb{B}_n$ and let

$$\gamma = \gamma(t), \qquad 0 \le t \le 1,$$

be a geodesic from w to z in the Bergman metric. Then

$$f(z) - f(w) = \int_0^1 \left(\sum_{k=1}^n \gamma_k'(t)\frac{\partial f}{\partial z_k}(\gamma(t))\right) dt.$$

From the definition of Q_f we see that

$$\left|\sum_{k=1}^n \gamma_k'(t)\frac{\partial f}{\partial z_k}(\gamma(t))\right| \le Q_f(\gamma(t))\sqrt{\langle B(\gamma(t))\gamma'(t), \gamma'(t)\rangle}.$$

It follows that

$$|f(z) - f(w)| \le \|f\|_{\mathcal{B}} \int_0^1 \sqrt{\langle B(\gamma(t))\gamma'(t), \gamma'(t)\rangle} \, dt = \|f\|_{\mathcal{B}} \, \beta(z, w).$$

This shows that

$$\sup\left\{\frac{|f(z) - f(w)|}{\beta(z, w)} : z, w \in \mathbb{B}_n, z \ne w\right\} \le \|f\|_{\mathcal{B}}$$

for all holomorphic functions f in $\mathbb{B}_n$.

Next assume that

$$C = \sup\left\{\frac{|f(z) - f(w)|}{\beta(z, w)} : z, w \in \mathbb{B}_n, z \ne w\right\} < \infty.$$

In particular,

$$|f(z) - f(0)| \leq C\beta(z,0) = \frac{C}{2} \log \frac{1+|z|}{1-|z|},$$

and so

$$\frac{|f(z) - f(0)|}{|z|} \leq \frac{C}{2|z|} \log \frac{1+|z|}{1-|z|}$$

for all $z \in \mathbb{B}_n - \{0\}$. If u is any unit vector in $\mathbb{C}^n$, then taking the directional derivative of f at 0 in the u-direction yields

$$\left| \sum_{k=1}^{n} u_k \frac{\partial f}{\partial z_k}(0) \right| \leq C \lim_{|z| \to 0^+} \frac{1}{2|z|} \log \frac{1+|z|}{1-|z|} = C.$$

This shows that $Q_f(0) \leq C$. But the Möbius invariance of the Bergman metric implies that

$$C = \sup \left\{ \frac{|f \circ \varphi(z) - f \circ \varphi(w)|}{\beta(z,w)} : z, w \in \mathbb{B}_n, z \neq w \right\}$$

for any $\varphi \in \mathrm{Aut}(\mathbb{B}_n)$. We conclude that

$$Q_f(z) = Q_{f \circ \varphi_z}(0) \leq C$$

for all $z \in \mathbb{B}_n$. This shows that $f \in \mathcal{B}$ with $\|f\|_{\mathcal{B}} \leq C$.   □

Corollary 3.7. *A holomorphic function f in $\mathbb{B}_n$ belongs to the Bloch space $\mathcal{B}$ if and only if there exists a constant $C > 0$ such that*

$$|f(z) - f(w)| \leq C\beta(z,w)$$

for all z and w in $\mathbb{B}_n$.

A consequence of Corollary 3.7 is that every function f in $\mathcal{B}$ grows at most logarithmically near the boundary of $\mathbb{B}_n$, that is, there must exist a constant $C > 0$ such that

$$|f(z) - f(0)| \leq C \log \frac{1}{1-|z|^2} \tag{3.5}$$

for all $z \in \mathbb{B}_n$. In particular, $\mathcal{B} \subset A_\alpha^p$ for all $p > 0$ and $\alpha > -1$. The logarithmic growth rate is actually achieved by the following functions in $\mathcal{B}$,

$$f_\lambda(z) = \log(1 - \langle z, \lambda \rangle), \qquad z \in \mathbb{B}_n,$$

where λ is any point from $\mathbb{S}_n$.

Corollary 3.8. *Suppose $\alpha > -1$, $p > 0$, and f is holomorphic in $\mathbb{B}_n$. Then $f \in \mathcal{B}$ if and only if there exists a constant $C > 0$ such that*

$$\int_{\mathbb{B}_n} |f \circ \varphi_a(z) - f(a)|^p \, dv_\alpha(z) \leq C, \qquad a \in \mathbb{B}_n, \tag{3.6}$$

or equivalently,

$$\int_{\mathbb{B}_n} |f(z) - f(a)|^p \frac{(1 - |a|^2)^{n+1+\alpha}}{|1 - \langle z, a \rangle|^{2(n+1+\alpha)}} \, dv_\alpha(z) \le C, \qquad a \in \mathbb{B}_n. \tag{3.7}$$

Proof. That the two integrals are equal follows from a change of variables; see Proposition 1.13.

If f is in the Bloch space, then there exists a constant $C > 0$ such that

$$|f(z) - f(w)| \le C\beta(z, w)$$

for all z and w in $\mathbb{B}_n$. So

$$\int_{\mathbb{B}_n} |f \circ \varphi_a(z) - f(a)|^p \, dv_\alpha(z) \le C^p \int_{\mathbb{B}_n} \beta(\varphi_a(z), a)^p \, dv_\alpha(z)$$

$$= C^p \int_{\mathbb{B}_n} \beta(z, 0)^p \, dv_\alpha(z).$$

Since

$$\int_{\mathbb{B}_n} \beta(z, 0)^p \, dv_\alpha(z) < \infty,$$

we see that $f \in \mathcal{B}$ implies that

$$\sup_{a \in \mathbb{B}_n} \int_{\mathbb{B}_n} |f \circ \varphi_a(z) - f(a)|^p \, dv_\alpha(z) < \infty.$$

To prove the reverse implication, recall from Lemma 2.4 that there exists a constant $C > 0$ such that

$$|\nabla g(0)|^p \le C \int_{\mathbb{B}_n} |g(z) - g(0)|^p \, dv_\alpha(z)$$

for all holomorphic g in $\mathbb{B}_n$. Replacing g by $f \circ \varphi_a$, we obtain

$$|\widetilde{\nabla} f(a)|^p \le C \int_{\mathbb{B}_n} |f \circ \varphi_a(z) - f(a)|^p \, dv_\alpha(z)$$

for all $a \in \mathbb{B}_n$, and the desired result follows. $\square$

Not only can the Bloch norm be defined using the Bergman metric, the following result shows that the Bergman metric can also be recovered from the Bloch semi-norm.

Theorem 3.9. *For any z and w in $\mathbb{B}_n$ we have*

$$\beta(z, w) = \sup\{|f(z) - f(w)| : \|f\|_\mathcal{B} \le 1\}.$$

Proof. According to Theorem 3.6,

$$|f(z) - f(w)| \le \|f\|_{\mathcal{B}} \, \beta(z, w)$$

for all $f \in \mathcal{B}$ and all points z and w in $\mathbb{B}_n$. It follows that

$$\sup\{|f(z) - f(w)| : \|f\|_{\mathcal{B}} \le 1\} \le \beta(z, w)$$

for all z and w in $\mathbb{B}_n$.

To prove the reverse inequality, we fix any $z \in \mathbb{B}_n - \{0\}$ and consider the following function in $\mathbb{B}_n$.

$$h(w) = \frac{1}{2} \log \frac{|z| + \langle w, z \rangle}{|z| - \langle w, z \rangle}, \qquad w \in \mathbb{B}_n.$$

Since

$$\frac{\partial h}{\partial w_k}(w) = \frac{\bar{z}_k |z|}{|z|^2 - \langle w, z \rangle^2}, \qquad 1 \le k \le n,$$

a calculation using part (e) of Theorem 3.1 shows that

$$|\widetilde{\nabla} h(w)|^2 = \frac{|z|^2 (1 - |w|^2)(|z|^2 - |\langle w, z \rangle|^2)}{\left| |z|^2 - \langle w, z \rangle^2 \right|^2}$$

for all $w \in \mathbb{B}_n$, or

$$|\widetilde{\nabla} h(w)|^2 = \frac{(1 - |w|^2)(1 - |\langle w, z' \rangle|^2)}{\left| 1 - \langle w, z' \rangle^2 \right|^2}$$

for all $w \in \mathbb{B}_n$, where $z' = z/|z|$. By the triangle inequality, we have

$$1 - |w|^2 \le |1 - \langle w, z' \rangle^2|, \quad 1 - |\langle w, z' \rangle|^2 \le |1 - \langle w, z' \rangle^2|.$$

So $|\widetilde{\nabla} h(w)|^2 \le 1$ for every $w \in \mathbb{B}_n$, or $\|h\|_{\mathcal{B}} \le 1$. It follows that

$$\beta(z, 0) = \frac{1}{2} \log \frac{1 + |z|}{1 - |z|} = |h(z) - h(0)|$$

$$\le \sup\{|f(z) - f(0)| : \|f\|_{\mathcal{B}} \le 1\}.$$

By Möbius invariance, we must have

$$\beta(z, w) \le \sup\{|f(z) - f(w)| : \|f\|_{\mathcal{B}} \le 1\}$$

for all z and w in $\mathbb{B}_n$. $\qquad\square$

3.2 The Little Bloch Space

The Bloch space is not separable. In this section we discuss a separable subspace of the Bloch space, the little Bloch space.

First recall that $C(\overline{\mathbb{B}}_n)$ is the space of all continuous functions on the closed unit ball, and $C_0(\mathbb{B}_n)$ is the closed subspace of $C(\overline{\mathbb{B}}_n)$ consisting of those functions that vanish on the boundary $\mathbb{S}_n$.

The little Bloch space will be denoted by $\mathcal{B}_0$. It consists of functions $f \in \mathcal{B}$ such that

$$\lim_{|z| \to 1^-} |\widetilde{\nabla} f(z)| = 0.$$

Since $|\widetilde{\nabla} f(z)|$ is clearly continuous in $\mathbb{B}_n$, the above condition simply says that the function $|\widetilde{\nabla} f(z)|$ belongs to $C_0(\mathbb{B}_n)$. Theorem 3.1 immediately gives several equivalent definitions of $\mathcal{B}_0$. In particular, it follows from part (e) of Theorem 3.1 that if f is holomorphic in a neighborhood of $\overline{\mathbb{B}}_n$, then f belongs to $\mathcal{B}_0$.

Proposition 3.10. $\mathcal{B}_0$ *is a closed subspace of* $\mathcal{B}$ *and the set of polynomials is dense in* $\mathcal{B}_0$.

Proof. It is obvious that $\mathcal{B}_0$ is closed in $\mathcal{B}$. Given $f \in \mathcal{B}_0$, we have $\|f - f_r\|_{\mathcal{B}} \to 0$ as $r \to 1^-$, where $f_r(z) = f(rz)$. Since each f_r can be uniformly approximated by polynomials, and the sup-norm norm in $\mathbb{B}_n$ dominates the Bloch norm (see Exercise 3.5), we conclude that every $f \in \mathcal{B}_0$ can be approximated in the Bloch norm by polynomials. □

The following is the little oh version of Theorem 3.4.

Theorem 3.11. *Suppose* $\alpha > -1$ *and* f *is holomorphic in* $\mathbb{B}_n$. *Then the following conditions are equivalent:*

(a) $f \in \mathcal{B}_0$.
(b) The function $(1 - |z|^2)|\nabla f(z)|$ *belongs to* $C_0(\mathbb{B}_n)$.
(c) The function $(1 - |z|^2)Rf(z)$ *belongs to* $C_0(\mathbb{B}_n)$.
(d) There exists a function $g \in C_0(\mathbb{B}_n)$ *such that* $f = P_\alpha g$.

Proof. By Lemma 2.14 we have that (a) implies (b), and (b) implies (c).

Recall from the proof of Theorem 3.4 that $f = P_\alpha g$ whenever $f \in \mathcal{B}$, where

$$g(z) = \frac{c_{\alpha+1}}{c_\alpha} \left[(1 - |z|^2)f(z) + \frac{(1 - |z|^2)Rf(z)}{n + 1 + \alpha} \right].$$

Every function in $\mathcal{B}$ grows at most logarithmically near the boundary $\mathbb{S}_n$, so the function $(1 - |z|^2)f(z)$ is in $C_0(\mathbb{B}_n)$ for every $f \in \mathcal{B}$. Thus condition (c) implies that $g \in C_0(\mathbb{B}_n)$, that is, (c) implies (d).

If (d) holds, then $f = P_\alpha g$ for some $g \in C_0(\mathbb{B}_n) \subset C(\overline{\mathbb{B}}_n)$. By the Stone-Weierstrass approximation theorem, every function in $C(\overline{\mathbb{B}}_n)$ can be approximated uniformly on $\mathbb{B}_n$ by finite linear combinations of functions of the form $h(z) = z^m \bar{z}^{m'}$. It is easy to check that $P_\alpha h$ is a polynomial, and hence is in the little Bloch space. Since P_α maps $L^\infty(\mathbb{B}_n)$ boundedly into the Bloch space, and the little Bloch space is closed in $\mathcal{B}$, we conclude that $f = P_\alpha g$ belongs to $\mathcal{B}_0$. ☐

The following theorem shows that the space $C_0(\mathbb{B}_n)$ can often be replaced by the space $C(\overline{\mathbb{B}}_n)$ in the study of the little Bloch space.

Theorem 3.12. *Suppose $\alpha > -1$ and f is holomorphic in $\mathbb{B}_n$. Then the following conditions are equivalent:*

(a) f belongs to the little Bloch space.
(b) $|\widetilde{\nabla} f(z)|$ belongs to $C(\overline{\mathbb{B}}_n)$.
(c) $(1 - |z|^2)|\nabla f(z)|$ belongs to $C(\overline{\mathbb{B}}_n)$.
(d) $(1 - |z|^2)Rf(z)$ belongs to $C(\overline{\mathbb{B}}_n)$.
(e) $f = P_\alpha g$ for some $g \in C(\overline{\mathbb{B}}_n)$.

Proof. It is trivial that (a) implies (b). It follows from Lemma 2.14 that (b) implies (c), and (c) implies (d).

That (d) implies (e) follows from the same construction used in the proof of Theorems 3.4 and 3.11.

The last part of the proof of Theorem 3.11 actually shows that (e) implies (a). ☐

Recall that the ball algebra $A(\mathbb{B}_n)$ is the space of all holomorphic functions in $\mathbb{B}_n$ that are continuous up to the boundary. The above result shows that the ball algebra is contained in the little Bloch space.

In terms of higher order derivatives and fractional derivatives, we have the following analog of Theorem 3.5 from the last section.

Theorem 3.13. *Suppose N is a positive integer, α is real, and t is positive. If neither $n + \alpha$ nor $n + \alpha + t$ is a negative integer, then the following conditions are equivalent for a holomorphic function f in $\mathbb{B}_n$:*

(a) $f \in \mathcal{B}_0$.
(b) $(1 - |z|^2)^t R^{\alpha,t} f(z)$ is in $C_0(\mathbb{B}_n)$.
(c) $(1 - |z|^2)^t R^{\alpha,t} f(z)$ is in $C(\overline{\mathbb{B}}_n)$.
(d) $(1 - |z|^2)^N \dfrac{\partial f^N}{\partial z^m}(z)$ is in $C_0(\mathbb{B}_n)$ for every multi-index m with $|m| = N$.
(e) $(1 - |z|^2)^N \dfrac{\partial f^N}{\partial z^m}(z)$ is in $C(\overline{\mathbb{B}}_n)$ for every multi-index m with $|m| = N$.

Proof. The proof is similar to that of Theorem 3.5. We omit the details. ☐

Approximating a function f by polynomials, we find that every function f in the little Bloch space satisfies

$$\lim_{|z|\to 1^-} f(z)/\log\frac{1}{1-|z|^2} = 0. \tag{3.8}$$

The next result strengthens Theorem 3.9 of the previous section.

Theorem 3.14. *For any z and w in $\mathbb{B}_n$ we have*

$$\beta(z,w) = \sup\{|f(z) - f(w)| : \|f\|_{\mathcal{B}} \le 1, f \in \mathcal{B}_0\}.$$

Proof. The proof is the same as that of Theorem 3.9, except that we use the functions

$$h_r(w) = \frac{1}{2}\log\frac{|z| + r\langle w, z\rangle}{|z| - r\langle w, z\rangle}, \qquad w \in \mathbb{B}_n,$$

where $z \in \mathbb{B}_n - \{0\}$ is fixed and $r \in (0,1)$, instead of

$$h(w) = \frac{1}{2}\log\frac{|z| + \langle w, z\rangle}{|z| - \langle w, z\rangle}.$$

Each function h_r is in the little Bloch space, and

$$\begin{aligned}
|\widetilde{\nabla} h_r(w)|^2 &= (1 - |w|^2)\big[r^2|\nabla h(rw)|^2 - |Rh(rw)|^2\big] \\
&\le (1 - |w|^2)\big[|\nabla h(rw)|^2 - |Rh(rw)|^2\big] \\
&= |\widetilde{\nabla} h(rw)|^2.
\end{aligned}$$

So $\|h_r\|_{\mathcal{B}} \le \|h\|_{\mathcal{B}} \le 1$ (see the proof of Theorem 3.9), and

$$|h_r(z) - h_r(0)| \le \sup\{|f(z) - f(0)| : f \in \mathcal{B}_0, \|f\|_{\mathcal{B}} \le 1\}.$$

Letting $r \to 1^-$ then yields

$$\beta(z,0) \le \sup\{|f(z) - f(0)| : f \in \mathcal{B}_0, \|f\|_{\mathcal{B}} \le 1\}.$$

This together with Möbius invariance gives

$$\beta(z,w) \le \sup\{|f(z) - f(w)| : f \in \mathcal{B}_0, \|f\|_{\mathcal{B}} \le 1\}.$$

The reversed inequality follows from Theorem 3.9. $\square$

We mention a trick that is often useful for constructing non-trivial functions in the Bloch space or the little Bloch space of $\mathbb{B}_n$. If n' is any integer with $1 \le n' \le n$, and if f is a function in the Bloch (or little Bloch) space of the unit ball of $\mathbb{C}^{n'}$, then the function

$$\tilde{f}(z_1, \cdots, z_{n'}, \cdots, z_n) = f(z_1, \cdots, z_{n'})$$

belongs to the Bloch (or little Bloch space) of $\mathbb{B}_n$. This clearly follows from condition (b) or (c) in Theorems 3.4 and 3.11. In particular, functions in the Bloch (or little Bloch) space of the unit disk can be lifted to functions in the Bloch (or little Bloch) space of $\mathbb{B}_n$. The following is a classical way of constructing non-trivial Bloch functions in the unit disk $\mathbb{D}$ using lacunary series.

Theorem 3.15. *Suppose $\{n_k\}$ is a series of positive integers satisfying $n_{k+1} \geq \lambda n_k$ for all $k \geq 1$, where λ is a constant greater than 1. Consider an analytic function f in $\mathbb{D}$ whose Taylor series is of the form*

$$f(z) = \sum_{k=1}^{\infty} a_k z^{n_k}, \qquad z \in \mathbb{D}. \tag{3.9}$$

Then f belongs to the Bloch space of $\mathbb{D}$ if and only if $\{a_k\}$ is bounded; and f belongs to the little Bloch space of $\mathbb{D}$ if and only if $a_k \to 0$ as $k \to \infty$.

Proof. Recall that f is in the Bloch space of $\mathbb{D}$ if and only if

$$\sup\{(1 - |z|^2)|f'(z)| : z \in \mathbb{D}\} < \infty;$$

and f is in the little Bloch space of $\mathbb{D}$ if and only if

$$\lim_{|z| \to 1^-} (1 - |z|^2) f'(z) = 0.$$

When

$$f(z) = \sum_{k=0}^{\infty} a_k z^k$$

is in the Bloch space of $\mathbb{D}$, we easily check that

$$a_{k+1} = (k + 2) \int_{\mathbb{D}} (1 - |z|^2) f'(z) \bar{z}^k \, dA(z)$$

for all $k \geq 0$, which implies that the sequence $\{a_k\}$ is bounded. Similarly, when f is in the little Bloch space of $\mathbb{D}$, an elementary argument shows that the Taylor coefficients of f must converge to 0.

If $\{a_k\}$ satisfies $|a_k| \leq M$ for all $k \geq 1$ and $\{n_k\}$ satisfies $n_{k+1} \geq \lambda n_k$ for all $k \geq 1$, then the lacunary series (3.9) defines a function in the Bloch space of $\mathbb{D}$. In fact, the constant $C = \lambda/(\lambda - 1)$ satisfies $1 < C < \infty$ and

$$n_{k+1} \leq C(n_{k+1} - n_k), \qquad k \geq 1.$$

It follows that

$$n_{k+1}|z|^{n_{k+1}-1} \leq C(n_{k+1} - n_k)|z|^{n_{k+1}-1} \leq C(|z|^{n_k} + \cdots + |z|^{n_{k+1}-1})$$

for all $k \geq 1$. It is easy to see that

$$n_1 |z|^{n_1 - 1} \leq C(1 + |z| + \cdots + |z|^{n_1 - 1}).$$

Thus

$$|f'(z)| \leq M \sum_{k=1}^{\infty} n_k |z|^{n_k - 1} \leq MC \sum_{l=0}^{\infty} |z|^l = \frac{MC}{1 - |z|}$$

for all $z \in \mathbb{D}$, and f is in the Bloch space of $\mathbb{D}$.

Similarly, if f is defined by a lacunary series whose coefficients tend to 0, then f belongs to the little Bloch space of $\mathbb{D}$. $\qquad\square$

In particular, we now know how to construct unbounded holomorphic functions in the little Bloch space of $\mathbb{B}_n$. We mention this point here because in the next section we are going to consider limits of the form

$$\lim_{r \to 1^-} \int_{\mathbb{B}_n} f(rz)\overline{g(rz)}\, dv_\alpha(z),$$

where f is in the Bloch space $\mathbb{B}_n$ and g belongs to A_α^1. The limit is necessary because f might be unbounded, even when f is in the little Bloch space.

3.3 Duality

In this section we think of the Bloch space as a Banach space and will use norms, but not semi-norms, on it.

Theorem 3.16. *Suppose $\alpha > -1$. The Banach dual of $\mathcal{B}_0$ can be identified with A_α^1 (with equivalent norms) under the integral pairing*

$$\langle f, g \rangle_\alpha = \lim_{r \to 1^-} \int_{\mathbb{B}_n} f(z)\overline{g(rz)}\, dv_\alpha(z), \qquad f \in \mathcal{B}_0, g \in A_\alpha^1.$$

In particular, the limit above always exists.

This theorem is a special case of Theorem 7.5 in Chapter 7. Also, the next theorem is a special case of Theorem 7.6 in Chapter 7. Since the proofs for these special cases are not any easier, we will not include them here. Be assured that no circular arguments exist in the book.

The space A_α^p is not a Banach space when $0 < p < 1$. However, we can still consider its dual space. In fact, we define the dual space of A_α^p for $0 < p < 1$ in exactly the same way as we do for $p \geq 1$. Thus the dual space of A_α^p consists of all linear functionals $F : A_\alpha^p \to \mathbb{C}$ such that

$$|F(f)| \leq C\|f\|_{p,\alpha}, \qquad f \in A_\alpha^p,$$

where C is a positive constant depending on F.

Theorem 3.17. *Suppose* $0 < p \leq 1$, $\alpha > -1$, *and*

$$s = \frac{n+1+\alpha}{p} - (n+1).$$

Then we can identity the dual space of A_α^p with $\mathcal{B}$ (with equivalent norms) under the integral pairing

$$\langle f, g \rangle_s = \lim_{r \to 1^-} \int_{\mathbb{B}_n} f(rz)\,\overline{g(z)}\,dv_s(z), \qquad f \in A_\alpha^p, g \in \mathcal{B}.$$

In particular, the limit above always exists.

A special case is worth mentioning here. When $p = 1$, we have $s = \alpha$, and so the dual space of A_α^1 can be identified with $\mathcal{B}$ under the natural integral pairing $\langle\ ,\ \rangle_\alpha$.

3.4 Maximality

Let X be a linear space of holomorphic functions in $\mathbb{B}_n$ equipped with a semi-norm (including the case of a norm) $\|\ \|$. A functional $F : X \to \mathbb{C}$ is continuous (or bounded) if there exists a constant $C > 0$ such that $|F(f)| \leq C\|f\|$ for all $f \in X$.

In this book we use the term *Möbius invariant Banach space* to denote a semi-normed linear space X of holomorphic functions in $\mathbb{B}_n$ with the property that

$$\|f \circ \varphi\| = \|f\|, \qquad f \in X, \varphi \in \mathrm{Aut}(\mathbb{B}_n).$$

By considering its completion if necessary, we will always assume that X is already complete in the semi-norm. We further assume that the mapping

$$(\theta_1, \cdots, \theta_n) \mapsto f(z_1 e^{i\theta_1}, \cdots, z_n e^{i\theta_n})$$

is continuous from $[0, 2\pi]^n$ to X.

Lemma 3.18. *Suppose $(X, \|\ \|)$ is a Möbius invariant Banach space. If X contains a nonconstant function, then X contains all the polynomials.*

Proof. Let f be a nonconstant function in X. If

$$f(z) = \sum_m a_m z^m$$

is the Taylor expansion of f, then there exists some nonzero multi-index m such that $a_m \neq 0$. Fix such an $m = (m_1, \cdots, m_n)$ and consider the function

$$F(z) = \frac{1}{(2\pi)^n} \int_0^{2\pi} \cdots \int_0^{2\pi} f(z_1 e^{i\theta_1}, \cdots, z_n e^{i\theta_n}) e^{-i(m_1\theta_1 + \cdots + m_n\theta_n)}\, d\theta_1 \cdots d\theta_n.$$

Since X is Möbius invariant, it follows easily that the above integral converges in the norm topology of X and $\|F\|_X \leq \|f\|_X$. But an easy computation with the Taylor expansion of f shows that $F(z) = a_m z^m$, so X contains the monomial z^m.

Composing z^m with all possible unitary transformations and using the Möbius invariance of X, we conclude that X contains all homogeneous polynomials of degree $|m|$. In particular, for each $\varphi \in \text{Aut}(\mathbb{B}_n)$, the function $z_1^{|m|} \circ \varphi$ belongs to X.

Let $\varphi = \varphi_a$, where $a = (\lambda, 0, \cdots, 0)$ with $|\lambda| < 1$. We have

$$z_1^{|m|} \circ \varphi = \left(\frac{\lambda - z_1}{1 - \bar{\lambda} z_1} \right)^{|m|}. \tag{3.10}$$

It is clear that for each nonnegative integer l (the special case $l = 1$ is enough for our purpose) we can find some λ such that the Taylor coefficient of z_1^l of the function in (3.10) is nonzero. By the argument used in the first paragraph of this proof, the function z_1^l is in X for each $l \geq 0$. Combining this with the remarks in the previous paragraph, we conclude that X contains all polynomials. $\square$

We now show that the Bloch space is maximal among Möbius invariant Banach spaces.

Theorem 3.19. *Suppose $(X, \| \|)$ is a Möbius invariant Banach space in $\mathbb{B}_n$. If X possesses a nonzero bounded linear functional L, then $X \subset \mathcal{B}$ and there exists a constant $C > 0$ such that $\|f\|_{\mathcal{B}} \leq C\|f\|$ for all $f \in X$. If L further satisfies $L(1) \neq 0$, then $X \subset H^\infty(\mathbb{B}_n)$ and there exists a constant $C > 0$ such that $\|f\|_\infty \leq C\|f\|$ for all $f \in X$.*

Proof. Let L be a nonzero bounded linear functional on X with

$$|L(f)| \leq C\|f\|, \qquad f \in X.$$

We first consider the case where $L(1) \neq 0$. For any $f \in X$, it is easy to see that

$$f(0)L(1) = \frac{1}{2\pi} \int_0^{2\pi} L(f(e^{it}z)) \, dt.$$

It follows from the boundedness of L on X that

$$|f(0)||L(1)| \leq C\|f\|.$$

Replace f by $f \circ \varphi_z$ and use the Möbius invariance of X. We obtain

$$|f(z)||L(1)| \leq C\|f\|$$

for all $z \in \mathbb{B}_n$. This shows that $f \in H^\infty(\mathbb{B}_n)$ and

$$\|f\|_\infty \leq \frac{C}{|L(1)|} \|f\|.$$

Next we assume that $L(1) = 0$. Clearly, we may assume that X contains a nonconstant function. By Lemma 3.18, X must contain all the polynomials. We show that there exists some $\varphi \in \mathrm{Aut}(\mathbb{B}_n)$ such that the linear functional $L_\varphi : X \to \mathbb{C}$ defined by $L_\varphi(f) = L(f \circ \varphi)$ satisfies $L_\varphi(z_1) \neq 0$. In fact, for any $0 < r < 1$, the involution φ_a, where $a = (r, 0, \cdots, 0)$, is given by

$$\varphi_a(z) = \left(\frac{r - z_1}{1 - rz_1}, -\frac{\sqrt{1 - r^2}z_2}{1 - rz_1}, \cdots, -\frac{\sqrt{1 - r^2}z_n}{1 - rz_1} \right), \tag{3.11}$$

so that

$$z_1 \circ \varphi_a(z) = \frac{r - z_1}{1 - rz_1} = r + (r^2 - 1) \sum_{k=1}^{\infty} r^{k-1} z_1^k.$$

If $L_\varphi(z_1) = 0$ for all $\varphi \in \mathrm{Aut}(\mathbb{B}_n)$, then applying L to the above equation shows that

$$\sum_{k=1}^{\infty} r^{k-1} L(z_1^k) = 0$$

for all $r \in (0, 1)$. It follows that $L(z_1^k) = 0$ for all $k \geq 1$. Replacing L by L_φ shows that $L_\varphi(z_1^k) = 0$ for all $k \geq 1$. This implies that $L(z^m) = 0$ for every multi-index $m = (m_1, \cdots, m_n)$ of nonnegative integers with $|m| > 0$. Combining this with $L(1) = 0$, we conclude that $L = 0$ on $H(\mathbb{B}_n)$, a contradiction.

So we may assume that $L(z_1) \neq 0$. For $f \in X$ we consider

$$F(f) = \frac{1}{2\pi} \int_0^{2\pi} L(f(e^{it}z))e^{-it} \, dt.$$

The continuity of L on X shows that $|F(f)| \leq C\|f\|$ for all $f \in X$. On the other hand, a calculation using the homogeneous expansion of f shows that

$$F(f) = \sum_{k=1}^{n} L(z_k) \frac{\partial f}{\partial z_k}(0).$$

Rewrite this as

$$F(f) = \delta \sum_{k=1}^{n} \overline{w}_k \frac{\partial f}{\partial z_k}(0) = \delta \langle \nabla f(0), w \rangle,$$

where δ is a positive constant independent of f and $w = (w_1, \cdots, w_n)$ is a unit vector in $\mathbb{C}^n$. For every $1 \leq k \leq n$ there exists a unitary matrix U_k such that $U_k(w) = e_k$, where $\{e_1, \cdots, e_n\}$ is the standard orthonormal basis of $\mathbb{C}^n$. We have

$$F(f \circ U_k) = \delta \langle \nabla(f \circ U_k)(0), w \rangle = \delta \langle U_k \nabla f(0), w \rangle$$

$$= \delta \langle \nabla f(0), U_k w \rangle = \delta \frac{\partial f}{\partial z_k}(0)$$

for every $f \in X$. It follows that

$$\left|\frac{\partial(f\circ\varphi)}{\partial z_k}(0)\right| \le \frac{|F(f\circ\varphi\circ U_k)|}{\delta} \le \frac{C}{\delta}\|f\circ\varphi\circ U_k\| = \frac{C}{\delta}\|f\|$$

for all $f \in X$, $\varphi \in \mathrm{Aut}(\mathbb{B}_n)$, and $1 \le k \le n$. Therefore,

$$|\widetilde{\nabla} f(z)| = |\nabla(f\circ\varphi_z)(0)| \le \frac{Cn}{\delta}\|f\|$$

for all $f \in X$ and $z \in \mathbb{B}_n$. This shows that f belongs to the Bloch space and $\|f\|_{\mathcal{B}} \le C'\|f\|$ for some positive constant C'. $\square$

3.5 Pointwise Multipliers

A function f is called a pointwise multiplier of a space X if for every $g \in X$ the pointwise product fg also belongs to X. Thus we often denote a pointwise multiplier f of a space X by $fX \subset X$.

In this section we characterize the pointwise multipliers of the Bloch space and the little Bloch space. Throughout this section we use the following norm on $\mathcal{B}$:

$$\|g\| = |g(0)| + \sup\{(1 - |z|^2)|\nabla g(z)| : z \in \mathbb{B}_n\}, \qquad g \in \mathcal{B}.$$

Lemma 3.20. *Suppose X is a Banach space of holomorphic functions in $\mathbb{B}_n$. If X contains the constant functions and if each point evaluation is a bounded linear functional on X, then every pointwise multiplier of X is in H^∞.*

Proof. Suppose f is a pointwise multiplier of X. Since X contains the constant function 1, we have $f \in X$. In particular, f is holomorphic. An application of the closed graph theorem then shows that $T = M_f$, the operator of multiplication by f on X, is bounded on X.

The adjoint of T, T^*, is a bounded linear operator on X^*. We consider the action of T^* on each e_z, where $z \in \mathbb{B}_n$ and e_z is the point evaluation at z. By assumption, each $e_z \in X^*$.

For each $z \in \mathbb{B}_n$ and $g \in X$, the definition of X^* and T^* gives

$$T^*(e_z)(g) = e_z(Tg) = g(z)f(z) = f(z)e_z(g).$$

This shows that $T^*e_z = f(z)e_z$, and hence $|f(z)| \le \|T^*\|$ for all $z \in \mathbb{B}_n$. $\square$

In particular, if f is a pointwise multiplier of $\mathcal{B}$ or $\mathcal{B}_0$, then f must be bounded in $\mathbb{B}_n$.

Theorem 3.21. *For a holomorphic function f in $\mathbb{B}_n$ the following conditions are equivalent:*

(a) $f\mathcal{B} \subset \mathcal{B}$.
(b) $f\mathcal{B}_0 \subset \mathcal{B}_0$.

(c) $f \in H^\infty(\mathbb{B}_n)$ *and the function*

$$(1 - |z|^2)|\nabla f(z)| \log \frac{1}{1 - |z|^2}$$

is bounded in $\mathbb{B}_n$.

Proof. Suppose $f\mathcal{B} \subset \mathcal{B}$. Then $f \in H^\infty(\mathbb{B}_n)$ and there exists a positive constant $C > 0$ such that $\|fg\| \le C\|g\|$ for all $g \in \mathcal{B}$. Since

$$\nabla(fg)(z) = f(z)\nabla g(z) + g(z)\nabla f(z), \qquad z \in \mathbb{B}_n,$$

we have

$$|g(z)||\nabla f(z)|(1 - |z|^2) \le \|f\|_\infty \|g\| + C\|g\|$$

for all $g \in \mathcal{B}$ and $z \in \mathbb{B}_n$. Taking the supremum over all $g \in \mathcal{B}$ with $\|g\| \le 1$ and $g(0) = 0$, and applying Theorem 3.9, we conclude that the function

$$(1 - |z|^2)|\nabla f(z)| \log \frac{1}{1 - |z|^2}$$

must be bounded in $\mathbb{B}_n$. This shows that (a) implies (c), and, with the help of Theorem 3.14 instead of Theorem 3.9, it also proves that (b) implies (c).

Next suppose that f satisfies the conditions in (c). For every $g \in \mathcal{B}$ we have

$$(1 - |z|^2)|\nabla(fg)(z)| \le |f(z)|(1 - |z|^2)|\nabla g(z)| + |g(z)|(1 - |z|^2)|\nabla f(z)|.$$

The first term of the right hand side above is bounded in $\mathbb{B}_n$, because $f \in H^\infty(\mathbb{B}_n)$ and $g \in \mathcal{B}$; the second term is bounded in $\mathbb{B}_n$ because of the inequality in (c) and the fact that g grows at most as fast as $-\log(1 - |z|^2)$. This shows that (c) implies (a). Note that if g is in the little Bloch space, then

$$\lim_{|z| \to 1^-} \frac{g(z)}{\log(1 - |z|^2)} = 0.$$

So the above argument also shows that (c) implies (b). $\qquad\square$

3.6 Atomic Decomposition

In this section we show that the Bloch space also admits an atomic decomposition similar to that of the Bergman spaces.

Fix a parameter $b > n$ and fix a sequence $\{a_k\}$ satisfying the conditions in Theorem 2.23. The sequence $\{a_k\}$ induces a partition $\{D_k\}$ of $\mathbb{B}_n$ according to Lemma 2.28. We shall also need to further partition each D_k into a finite number of disjoint pieces $D_{k1}, \cdots, D_{kJ}$. See the description of all these following Lemma 2.28.

We are going to use two operators from Section 2.5. The first operator acts on $L^1(\mathbb{B}_n, dv_\alpha)$, where $\alpha = b - (n+1)$, and is denoted by T,

$$Tf(z) = \int_{\mathbb{B}_n} \frac{(1-|w|^2)^{b-n-1}}{|1-\langle z,w\rangle|^b} f(w)\,dv(w).$$

The second operator acts on $H(\mathbb{B}_n)$ and is defined by

$$Sf(z) = \sum_{k=1}^{\infty} \sum_{j=1}^{J} \frac{v_\alpha(D_{kj})f(a_{kj})}{(1-\langle z,a_{kj}\rangle)^b}.$$

Here $\{a_{kj}\}$ is a finer "lattice" in the Bergman metric than $\{a_k\}$.

Lemma 3.22. *There exists a constant $C > 0$, independent of the separation constant r for $\{a_k\}$ and the separation constant η for $\{a_{kj}\}$, such that*

$$|f(z) - Sf(z)| \le C\sigma T(|f|)(z)$$

for all $f \in H(\mathbb{B}_n)$ and $z \in \mathbb{B}_n$, where σ is the constant from Lemma 2.29.

Proof. Let $p = 1$ and $\alpha = 0$ in Lemma 2.29. We obtain

$$|f(z) - Sf(z)| \le C\sigma \sum_{k=1}^{\infty} \frac{(1-|a_k|^2)^{b-(n+1)}}{|1-\langle z,a_k\rangle|^b} \int_{D(a_k,2r)} |f(w)|\,dv(w).$$

According to (2.20), we can find a constant $C_1 > 0$ such that

$$|f(z) - Sf(z)| \le C_1\sigma \sum_{k=1}^{\infty} \int_{D(a_k,2r)} \frac{(1-|w|^2)^{b-(n+1)}}{|1-\langle z,w\rangle|^b} |f(w)|\,dv(w).$$

Since each point of $\mathbb{B}_n$ belongs to at most N of $D(a_k, 2r)$, we must have

$$|f(z) - Sf(z)| \le C_1 N\sigma \int_{\mathbb{B}_n} \frac{(1-|w|^2)^{b-(n+1)}}{|1-\langle z,w\rangle|^b} |f(w)|\,dv(w),$$

which is the desired estimate. $\qquad\square$

We can now prove the main result this section.

Theorem 3.23. *For any $b > n$ there exists a sequence $\{a_k\}$ in $\mathbb{B}_n$ such that the Bloch space $\mathcal{B}$ consists exactly of functions of the form*

$$f(z) = \sum_k c_k \frac{(1-|a_k|^2)^b}{(1-\langle z,a_k\rangle)^b}, \tag{3.12}$$

where $\{c_k\} \in l^\infty$ and the series converges in the weak-star topology of $\mathcal{B}$ when $\mathcal{B}$ is identified as the dual space of A_α^1 for $\alpha = b - n - 1$.

Proof. Let $\{a_k\}$ be a sequence satisfying the conditions of Theorem 2.23 and let f be a function defined by (3.12).

First observe that the series in (3.12) converges uniformly on every compact subset of $\mathbb{B}_n$ whenver $\{c_k\}$ is bounded. In fact, by Lemmas 1.24, 2.20, and 2.21, there exists a constant $C_1 > 0$ such that

$$\sum_k |c_k|(1 - |a_k|^2)^b \leq C_1 \sum_k \int_{D(a_k, r/4)} (1 - |z|^2)^\alpha \, dv(z)$$

$$< C \int_{\mathbb{B}_n} (1 - |z|^2)^\alpha \, dv(z) < \infty.$$

Next observe that if $\{c_k\}$ is bounded, then the series in (3.12) actually converges in the norm topology of A_α^1. In fact, by part (2) of Theorem 1.12, there exists a constant $C_2 > 0$ such that

$$\sum_k |c_k|(1 - |a_k|^2)^b \int_{\mathbb{B}_n} \frac{dv_\alpha(z)}{|1 - \langle z, a_k \rangle|^b} \leq C_2 \sum_k |c_k|(1 - |a_k|^2)^b \log \frac{2}{1 - |a_k|^2}.$$

It follows that for any $b' \in (n, b)$ there exists a constant $C_3 > 0$ such that

$$\sum_k |c_k|(1 - |a_k|^2)^b \int_{\mathbb{B}_n} \frac{dv_\alpha(z)}{|1 - \langle z, a_k \rangle|^b} \leq C_3 \sum_k |c_k|(1 - |a_k|^2)^{b'} < \infty,$$

where the last inequality follows from the estimate in the previous paragraph.

Let g be a function in H^∞, which is dense in A_α^1. Then

$$\langle g, f \rangle_\alpha = \int_{\mathbb{B}_n} g(z)\overline{f(z)} \, dv_\alpha(z) = \sum_k \bar{c}_k (1 - |a_k|^2)^b g(a_k).$$

By Lemmas 2.24 and 1.24, there exists a constant $C_4 > 0$ such that

$$|\langle g, f \rangle_\alpha| \leq C_4 \sum_k \int_{D(a_k, r/4)} |g(z)| \, dv_\alpha(z) \leq C_4 \|g\|_{1,\alpha}.$$

This shows that f induces a bounded linear functional on A_α^1 under the integral pairing $\langle \, , \, \rangle_\alpha$. According to Theorem 3.17, we must have $f \in \mathcal{B}$.

It is easy to see that, with some obvious minor adjustments, the above argument also works when the sequence $\{a_k\}$ is replaced by the more dense sequence $\{a_{kj}\}$. See Exercise 3.23.

To prove the other half of the theorem, we consider the space X consisting of holomorphic functions f such that

$$\|f\|_X = \sup\{(1 - |z|^2)|f(z)| : z \in \mathbb{B}_n\} < \infty.$$

It is easy to see that X is a Banach space with the norm defined above.

Let S and T be the operators corresponding to the parameter $b + 1$. If $f \in X$, then by Lemma 3.22,

$$|f(z) - Sf(z)| \leq C_5 \sigma \int_{\mathbb{B}_n} \frac{(1 - |w|^2)^{b-n} |f(w)| \, dv(w)}{|1 - \langle z, w \rangle|^{b+1}}$$

for all $z \in \mathbb{B}_n$, where C_5 is a positive constant independent of the separation constant r for $\{a_k\}$ and the separation constant η for $\{a_{kj}\}$. By Theorem 1.12, there exists another constant $C_6 > 0$ such that

$$\|f - Sf\|_X \leq C_6 \sigma \|f\|_X, \qquad f \in X.$$

If we choose the separation constants η and r so that $C_6 \sigma < 1$, then the operator $I - S$ has norm less than 1 on X, where I is the identity operator, and so the operator S is invertible on X.

Given any $f \in \mathcal{B}$, we consider the function $g = R^{\alpha,1} f$, where

$$\alpha = b - (n + 1).$$

Since $R^{\alpha,1}$ is a differential operator of order 1 with polynomial coefficients (see Proposition 1.15), we have $g \in X$. With $h = S^{-1} g \in X$ we then have the representation

$$g(z) = \sum_{k=1}^{\infty} \sum_{j=1}^{J} \frac{v_\beta(D_{kj}) h(a_{kj})}{(1 - \langle z, a_{kj} \rangle)^{b+1}}, \tag{3.13}$$

where

$$\beta = (b + 1) - (n + 1) = b - n.$$

Apply the inverse of $R^{\alpha,1}$, $R_{\alpha,1}$, to both sides of (3.13) and use Proposition 1.14. We arrive at the representation

$$f(z) = \sum_{k=1}^{\infty} \sum_{j=1}^{J} c_{kj} \frac{(1 - |a_{kj}|^2)^b}{(1 - \langle z, a_{kj} \rangle)^b},$$

where

$$c_{kj} = \frac{v_\beta(D_{kj}) h(a_{kj})}{(1 - |a_{kj}|^2)^b}.$$

Since $h \in X$ and

$$v_\beta(D_{kj}) \leq v_\beta(D_k) \sim (1 - |a_k|^2)^{n+1+\beta}$$
$$= (1 - |a_k|^2)^{b+1} \sim (1 - |a_{kj}|^2)^{b+1},$$

we have $\{c_{kj}\} \in l^\infty$ and the proof of the theorem is complete. □

It is clear from the proof of the above theorem that the Bloch norm of a function f is comparable to

$$\inf\{\|\{c_k\}\|_\infty : f \text{ is represented by (3.12)}\}.$$

We can also adopt the proof of the preceding theorem to obtain an atomic decomposition theorem for the little Bloch space.

Theorem 3.24. *For any $b > n$ there exists a sequence $\{a_k\}$ in $\mathbb{B}_n$ such that $\mathcal{B}_0$ consists exactly of functions of the form*

$$f(z) = \sum_{k=1}^{\infty} c_k \frac{(1 - |a_k|^2)^b}{(1 - \langle z, a_k \rangle)^b},$$

where $c_k \to 0$ as $k \to \infty$.

Proof. By the proof of Theorem 3.23, there exists a constant $C > 0$ such that

$$\left\| \sum_{k=1}^{\infty} c_k \left(\frac{1 - |a_k|^2}{1 - \langle z, a_k \rangle} \right)^b \right\| \leq C \sup_{k \geq 1} |c_k|,$$

where $\| \ \|$ is the norm in $\mathcal{B}$. It follows that if $c_k \to 0$, then

$$\lim_{N \to \infty} \|f - f_N\| = 0,$$

where f is given by the series (3.12) and $\{f_N\}$ is its partial sum sequence. Since each f_N is obviously in $\mathcal{B}_0$, we see that $f \in \mathcal{B}_0$ whenever $c_k \to 0$.

To prove the other direction, we consider the action of the operator S (corresponding to the parameter $b + 1$) on the space

$$X_0 = \{f \in H(\mathbb{B}_n) : (1 - |z|^2)f(z) \in \mathbb{C}_0(\mathbb{B}_n)\}.$$

The differential operator $R^{\alpha,1}$ is a bounded invertible operator from $\mathcal{B}_0$ onto X_0. When the seperation constant r for $\{a_k\}$ is small enough, S is invertible on X_0, and the proof of Theorem 3.23 shows every function $f \in \mathcal{B}_0$ admits a representation

$$f(z) = \sum_{k=1}^{\infty} \sum_{j=1}^{J} c_{kj} \frac{(1 - |a_{kj}|^2)^b}{(1 - \langle z, a_{kj} \rangle)^b}$$

with

$$c_{kj} = \frac{v_\beta(D_{kj})h(a_{kj})}{(1 - |a_{kj}|^2)^b},$$

where $h \in X_0$. Since

$$v_\beta(D_{kj}) \leq v_\beta(D_k) \sim (1 - |a_k|^2)^{b+1}$$

and $1 - |a_{kj}|^2$ is comparable to $1 - |a_k|^2$, the condition

$$\lim_{k \to \infty} (1 - |a_{kj}|^2)h(a_{kj}) = 0$$

implies that $c_{kj} \to 0$ as $k \to \infty$, and the proof is complete. $\square$

3.7 Complex Interpolation

In this section we give further evidence that the Bloch space behaves like the limit of A_α^p when $p \to \infty$. We illustrate this behavior with the complex interpolation between $\mathcal{B}$ and the weighted Bergman spaces A_α^p.

Theorem 3.25. *Suppose $\alpha > -1$ and*

$$\frac{1}{p} = \frac{1 - \theta}{p'},$$

where $\theta \in (0, 1)$ and $1 \le p' < \infty$. Then

$$\left[A_\alpha^{p'}, \mathcal{B}\right]_\theta = A_\alpha^p$$

with equivalent norms.

Proof. Fix some real number β such that $\beta > \alpha$. According to Theorem 2.11, P_β is a bounded projection from $L^q(\mathbb{B}_n, dv_\alpha)$ onto A_α^q for any $1 \le q < \infty$. Also, by Theorem 3.4, P_β maps $L^\infty(\mathbb{B}_n)$ boundedly onto $\mathcal{B}$.

If $f \in A_\alpha^p \subset L^p(\mathbb{B}_n, dv_\alpha)$, then by the well-known complex interpolation of L^p spaces, there exists a family of functions h_ζ in

$$L^p(\mathbb{B}_n, dv_\alpha) + L^\infty(\mathbb{B}_n) = L^p(\mathbb{B}_n, dv_\alpha)$$

such that

(a) h_ζ depends on the parameter ζ continuously in $0 \le \operatorname{Re} \zeta \le 1$ and analytically in $0 < \operatorname{Re} \zeta < 1$.

(b) $h_\zeta \in L^{p'}(\mathbb{B}_n, dv_\alpha)$ for $\operatorname{Re} \zeta = 0$ and $h_\zeta \in L^\infty(\mathbb{B}_n)$ for $\operatorname{Re} \zeta = 1$, with

$$\sup\{\|h_\zeta\|_{p',\alpha}^{p'} : \operatorname{Re} \zeta = 0\} \le \|h\|_{p,\alpha}^p,$$

and

$$\sup\{\|h\|_\infty : \operatorname{Re} \zeta = 1\} \le \|h\|_{p,\alpha}.$$

(c) $f = h_\theta$.

Let $f_\zeta = P_\beta h_\zeta$. Then $f_\zeta \in A_\alpha^{p'}$ for $\operatorname{Re} \zeta = 0$, $f_\zeta \in \mathcal{B}$ for $\operatorname{Re} \zeta = 1$, and $f_\theta = f$. Appropriate norm estimates also hold for $\operatorname{Re} \zeta = 0$ and $\operatorname{Re} \zeta = 1$. This shows that $f \in [A_\alpha^{p'}, \mathcal{B}]_\theta$.

Conversely, if $f \in [A_\alpha^{p'}, \mathcal{B}]_\theta$, then there exists a family of functions f_ζ in

$$A_\alpha^{p'} + \mathcal{B} = A_\alpha^{p'},$$

where the parameter ζ satisfies $0 \le \operatorname{Re} \zeta \le 1$, such that

(i) f_ζ depends on the parameter ζ continuously in $0 \le \operatorname{Re} \zeta \le 1$ and analytically in $0 < \operatorname{Re} \zeta < 1$.

(ii) $\|f_\zeta\|_{p',\alpha} \leq \|f\|_\theta$ for all $\operatorname{Re}\zeta = 0$ and $\|f_\zeta\|_\mathcal{B} \leq \|f\|_\theta$ for all $\operatorname{Re}\zeta = 1$.

(iii) $f = f_\theta$.

Define

$$h_\zeta(z) = \frac{c_{\beta+1}}{c_\beta}(1 - |z|^2)\left(f_\zeta(z) + \frac{Rf_\zeta(z)}{n + 1 + \beta}\right),$$

where $0 \leq \operatorname{Re}\zeta \leq 1$. By Theorem 2.16,

$$\|h_\zeta\|_{p',\alpha} \leq C\|f_\zeta\|_{p',\alpha}, \qquad \operatorname{Re}\zeta = 0,$$

and by Theorem 3.4,

$$\|h_\zeta\|_\infty \leq C\|f_\zeta\|_\mathcal{B}, \qquad \operatorname{Re}\zeta = 1.$$

By the complex interpolation for L^p spaces, we have $h_\theta \in L^p(\mathbb{B}_n, dv_\alpha)$. Since $f_\theta = P_\beta h_\theta$ (see the proof of Theorem 3.4), we conclude that f belongs to A^p_α. This completes the proof of the theorem. $\qquad\square$

Notes

The Bloch space of the unit disk plays an important role in classical geometric function theory. Several excellent surveys exist, including [7], [8], [14], [15], and [124].

Serious research on the Bloch space of the unit ball began with Timoney's papers [111] and [112]. In particular, Theorem 3.4 was proved in [111], although the proof here is less technical and more constructive. Theorems 3.6 and 3.9 are from [134].

The integral representation for the Bloch space and the duality between A^1 and $\mathcal{B}$ can be found in many different papers, including [99] and [100], and any attempt to find their first appearance would prove difficult. The explicit identification of the Bloch space as the dual of A^p_α, $0 < p \leq 1$, using an appropriate weighted integral pairing, was done in [133].

Pointwise multipliers of the Bloch space was characterized in [9] and later, independently, in [123]. The maximality of the Bloch space among Möbius invariant function spaces was proved in [93] and [113].

Atomic decomposition for the Bloch space can be found in [30] and [91]. The complex interpolation spaces between the Bloch space and a Bergman space A^p_α, $1 \leq p < \infty$, have been well known to be just the Bergman spaces A^q_α, $p < q < \infty$, although a precise reference for the result is lacking.

Exercises

3.1. Suppose $m = (m_1, \cdots, m_n)$ is any given multi-index of nonnegative integers with $|m| > 0$. Show that a holomorphic function f in $\mathbb{B}_n$ belongs to the Bloch space if and only if

$$\sup\left\{\left|\frac{\partial^m(f \circ \varphi)}{\partial z^m}(0)\right| : \varphi \in \operatorname{Aut}(\mathbb{B}_n)\right\} < \infty.$$

3.2. Suppose L is any nonzero linear functional on $H(\mathbb{B}_n)$. If a holomorphic function f in $\mathbb{B}_n$ satisfies

$$\sup\{|L(f \circ \varphi)| : \varphi \in \mathrm{Aut}(\mathbb{B}_n)\} < \infty,$$

then f belongs to the Bloch space. If, in addition, $L(1) \neq 0$, then f belongs to $H^\infty(\mathbb{B}_n)$.

3.3. Construct an unbounded function in $\mathcal{B}_0$ and a bounded function not in $\mathcal{B}_0$.

3.4. Show that the Bloch space is not separable.

3.5. Show that $H^\infty \subset \mathcal{B}$ with $\|f\|_\mathcal{B} \leq \|f\|_\infty$.

3.6. Show that for every point $\zeta \in \mathbb{S}_n$, the function $f(z) = \log(1 - \langle z, \zeta \rangle)$ belongs to the Bloch space, but not to the little Bloch space.

3.7. Suppose $s > 0$ and f is holomorphic in $\mathbb{B}_n$. If the function

$$g(z) = (1 - |z|^2)^s f(z)$$

belongs to $C(\overline{\mathbb{B}}_n)$, show that g actually belongs to $C_0(\mathbb{B}_n)$.

3.8. Show that for any real a and b there exists a positive constant C such that

$$\left| 1 - \frac{(1 - |z|^2)^a}{(1 - |w|^2)^a} \frac{(1 - \langle w, u \rangle)^b}{(1 - \langle z, u \rangle)^b} \right| \leq C\beta(z, w)$$

for all z, w, and u in $\mathbb{B}_n$ with $\beta(z, w) \leq 1$.

3.9. Suppose f is holomorphic in $\mathbb{B}_n$ and $t > 0$. Show that $f \in \mathcal{B}$ if and only if the function $(1 - |z|^2)^t R^t f(z)$ is bounded in $\mathbb{B}_n$.

3.10. Suppose f is holomorphic in $\mathbb{B}_n$, $t > 0$, and

$$g(z) = (1 - |z|^2)^t R^t f(z).$$

Show that $f \in \mathcal{B}_0$ if and only if $g \in C_0(\mathbb{B}_n)$ if and only if $g \in C(\overline{\mathbb{B}}_n)$.

3.11. If $\Phi : \mathcal{B}_0 \to \mathcal{B}_0$ is a linear operator satisfying $\|\Phi(f)\|_\mathcal{B} = \|f\|_\mathcal{B}$ for all $f \in \mathcal{B}_0$, show that there exists some $\varphi \in \mathrm{Aut}(\mathbb{B}_n)$ such that

$$\Phi(f) = f \circ \varphi - f(\varphi(0)), \qquad f \in \mathcal{B}_0.$$

See [62].

3.12. Show that the Taylor coefficients $\{a_m\}$ of a function $f \in \mathcal{B}$ are bounded. Similarly, if $f \in \mathcal{B}_0$, then its Taylor coefficients $\{a_m\}$ tend to 0 as $|m| \to \infty$.

3.13. If $f \in \mathcal{B}$ and $a = (a_1, \cdots, a_n) \in \mathbb{B}_n$. Show that there exist functions $f_k \in \mathcal{B}$, $1 \le k \le n$, such that

$$f(z) - f(a) = \sum_{k=1}^{n} (z_k - a_k) f_k(z), \qquad z \in \mathbb{B}_n.$$

Do the same with $\mathcal{B}$ replaced by $\mathcal{B}_0$.

3.14. Show that

$$\sup\{|\nabla f(z)|^2 : \|f\|_\mathcal{B} \le 1\} = \frac{n+1}{2(1-|z|^2)^2}$$

for every $z \in \mathbb{B}_n$.

3.15. Show that there exists a function f in $\mathcal{B}_0$ such that f cannot be approximated by its Taylor polynomials in the norm topology of $\mathcal{B}$.

3.16. For any $b \ge 0$ there exists a positive constant C with the following property. If h is any bounded function of the form

$$h(z) = (1 - |z|^2)^b f(z), \qquad z \in \mathbb{B}_n,$$

where f is holomorphic in $\mathbb{B}_n$, then

$$|h(z) - h(w)| \le C\|h\|_\infty \beta(z, w)$$

for all z and w in $\mathbb{B}_n$.

3.17. Let $\alpha = b - (n+1)$ and define an operator A on $\mathcal{B}$ as follows.

$$Af(z) = \sum_k \frac{v_\alpha(D_k) h(a_k)}{(1 - \langle z, a_k \rangle)^b},$$

where

$$h(z) = \frac{c_{\alpha+1}}{c_\alpha}(1 - |z|^2) \left(f(z) + \frac{Rf(z)}{n+1+\alpha} \right).$$

Show that A is invertible on $\mathcal{B}$ when the separation constant r for $\{a_k\}$ is small enough. Use this to give an alternative proof of the atomic decomposition of $\mathcal{B}$.

3.18. Show that

$$|\varphi_a(z) - a|^2 = \frac{(1 - |a|^2)(|z|^2 - |\langle z, a \rangle|^2)}{|1 - \langle z, a \rangle|^2}$$

for all a and z in $\mathbb{B}_n$.

3.19. For any $p > 0$ there exists a constant $C > 0$ (depending on p) such that

$$\int_{\mathbb{S}_n} |f(r\zeta)|^p \, d\sigma(\zeta) \leq C\|f\|_{\mathcal{B}}^p \left[\log \frac{1}{1-r}\right]^{p/2}$$

for all $f \in \mathcal{B}$ with $f(0) = 0$.

3.20. Show that a holomorphic function f in $\mathbb{B}_n$ belongs to $\mathcal{B}$ if and only if

$$\sup \frac{|f(z) - f(w)|(1 - |z|^2)^{1/2}(1 - |w|^2)^{1/2}}{|w - P_w(z) - (1 - |w|^2)^{1/2} Q_w(z)|} < \infty,$$

where the supremum is taken over all $z \in \mathbb{B}_n$ and all $w \in \mathbb{B}_n - \{0\}$. Recall that P_w is the orthogonal projection from $\mathbb{C}^n$ onto the one-dimensional subspace spanned by w, and Q_w is the orthogonal projection from $\mathbb{C}^n$ onto the orthogonal complement of w. See [72].

3.21. Show that

$$\limsup_{w \to z} \frac{\beta(z, w)}{|z - w|} = \sup\{|\nabla f(z)| : \|f\|_{\mathcal{B}} \leq 1\}$$

for all $z \in \mathbb{B}_n$.

3.22. Let M be the Banach space of all finite Borel measures μ on $\overline{\mathbb{B}_n}$ equipped with the norm $\|\mu\| = |\mu|(\overline{\mathbb{B}_n})$. Then M is the Banach dual of $C(\overline{\mathbb{B}_n})$. Show that A_α^1 is weak-star closed in M.

3.23. Reprove the first part of Theorem 3.23 when the sequence $\{a_k\}$ is replaced by the more dense sequence $\{a_{kj}\}$.

3.24. For any $\alpha > 0$ formulate and prove an atomic decomposition for the space $A^{-\alpha}$ consisting of functions $f \in H(\mathbb{B}_n)$ such that

$$\|f\|_{-\alpha} = \sup\{(1 - |z|^2)^\alpha |f(z)| : z \in \mathbb{B}_n\} < \infty.$$

Show that this cannot be done for $H^\infty(\mathbb{B}_n)$.

3.25. Formulate and prove a duality theorem between A_α^1 and $A^{-\alpha}$.

3.26. Is the Bloch space isomorphic to $H^\infty(\mathbb{B}_n)$?

3.27. Suppose φ is holomorphic in $\mathbb{B}_n$. Show that the following two conditions are equivalent:

(a) $\varphi \in \mathcal{B}$.
(b) There exists a constant $C > 0$ such that

$$\int_{\mathbb{B}_n} |P_\alpha(\varphi \bar{f})(z)|^2 \, dv_\alpha(z) \leq C \int_{\mathbb{B}_n} |f(z)|^2 \, dv_\alpha(z)$$

for all $f \in A_\alpha^2$.

3.28. Suppose φ is holomorphic in $\mathbb{B}_n$. Show that the following two conditions are equivalent:

(a) $\varphi \in \mathcal{B}$.
(b) There exists a constant $C > 0$ such that

$$\int_{\mathbb{B}_n} |\overline{\varphi(z)} f(z) - P_\alpha(\overline{\varphi} f)(z)|^2 \, dv_\alpha(z) \leq C \int_{\mathbb{B}_n} |f(z)|^2 \, dv_\alpha(z)$$

for all $f \in A_\alpha^2$.

3.29. Formulate and prove the little oh versions of Problems 3.27 and 3.28.

4

Hardy Spaces

In this chapter we study holomorphic Hardy spaces and the associated Cauchy-Szegö projection. Main topics covered include the existence of boundary values, the boundedness of the Cauchy-Szegö projection on $L^p(\mathbb{S}_n, d\sigma)$ for $1 < p < \infty$, Littlewood-Paley identities based on the the radial derivative and the invariant gradient, embedding theorems, and complex interpolation. Since the boundary behavior of functions in Hardy spaces are usually studied via the Poisson integral, the first section of the chapter contains the basic properties of the Poisson transform.

4.1 The Poisson Transform

The Bergman kernel plays an essential role in the study of Bergman spaces. Two integral kernels are fundamental in the theory of Hardy spaces; they are the Cauchy-Szegö kernel,

$$C(z, \zeta) = \frac{1}{(1 - \langle z, \zeta \rangle)^n},$$
(4.1)

and the (invariant) Poisson kernel,

$$P(z, \zeta) = \frac{(1 - |z|^2)^n}{|1 - \langle z, \zeta \rangle|^{2n}}.$$
(4.2)

Note that the Poisson kernel here is different from the associated Poisson kernel when $\mathbb{B}_n$ is thought of as the unit ball in $\mathbb{R}^{2n}$, unless $n = 1$.

The following result will be referred to as Cauchy's formula.

Proposition 4.1. *If f belongs to the ball algebra, then*

$$f(z) = \int_{\mathbb{S}_n} C(z, \zeta) f(\zeta) \, d\sigma(\zeta)$$

for all $z \in \mathbb{B}_n$.

Proof. For any given $z \in \mathbb{B}_n$ the Cauchy kernel $C(z, \zeta)$ is bounded in $\zeta \in \mathbb{S}_n$. Approximating f by f_r uniformly on $\overline{\mathbb{B}}_n$, and then approximating each f_r uniformly by its Taylor polynomials, we reduce the proof of Cauchy's formula to the case where f is a monomial.

When $f(z) = z^m$, where $m = (m_1, \cdots, m_n)$ is a multi-index of nonnegative integers, we have

$$\int_{\mathbb{S}_n} \frac{\zeta^m \, d\sigma(\zeta)}{(1 - \langle z, \zeta \rangle)^n} = \sum_{k=0}^{\infty} \frac{\Gamma(k+n)}{k! \, \Gamma(n)} \int_{\mathbb{S}_n} \zeta^m \langle z, \zeta \rangle^k \, d\sigma(\zeta)$$

$$= \frac{\Gamma(n + |m|)}{|m|! \, \Gamma(n)} \int_{\mathbb{S}_n} \zeta^m \langle z, \zeta \rangle^{|m|} \, d\sigma(\zeta).$$

Since

$$\langle z, \zeta \rangle^{|m|} = \sum_{|\alpha| = |m|} \frac{|m|!}{\alpha!} z^\alpha \overline{\zeta}^\alpha$$

by the multi-nomial formula, where $\alpha = (\alpha_1, \cdots, \alpha_n)$ is a multi-index of nonnegative integers, we have

$$\int_{\mathbb{S}_n} \frac{\zeta^m \, d\sigma(\zeta)}{(1 - \langle z, \zeta \rangle)^n} = \frac{\Gamma(n + |m|)}{|m|! \, \Gamma(n)} \frac{|m|!}{m!} z^m \int_{\mathbb{S}_n} |\zeta^m|^2 \, d\sigma(\zeta).$$

An application of Lemma 1.11 then gives us

$$\int_{\mathbb{S}_n} \frac{\zeta^m \, d\sigma(\zeta)}{(1 - \langle z, \zeta \rangle)^n} = z^m.$$

$\square$

As a consequence of Cauchy's formula we obtain the following Poisson integral representation for functions in the ball algebra.

Proposition 4.2. *If f is in the ball algebra, then*

$$f(z) = \int_{\mathbb{S}_n} P(z, \zeta) f(\zeta) \, d\sigma(\zeta)$$

for all $z \in \mathbb{B}_n$.

Proof. Fix f in the ball algebra and z in $\mathbb{B}_n$. The function

$$g(w) = f(w) C(w, z), \qquad w \in \overline{\mathbb{B}}_n,$$

also belongs to the ball algebra. Applying Cauchy's formula to g at the point z, we obtain

$$\frac{f(z)}{(1 - |z|^2)^n} = \int_{\mathbb{S}_n} \frac{f(\zeta) \, d\sigma(\zeta)}{|1 - \langle z, \zeta \rangle|^{2n}},$$

or

$$f(z) = \int_{\mathbb{S}_n} P(z, \zeta) f(\zeta) \, d\sigma(\zeta),$$

which is the desired integral representation.

$\square$

For every function $f \in L^1(\mathbb{S}_n, d\sigma)$ we can define a function $P[f]$ on $\mathbb{B}_n$ as follows.

$$P[f](z) = \int_{\mathbb{S}_n} P(z, \zeta) f(\zeta) \, d\sigma(\zeta). \tag{4.3}$$

The function $P[f]$ will be called the Poisson transform, or the Poisson integral, of f.

More generally, if μ is any finite complex Borel measure on $\mathbb{S}_n$, we define

$$P[\mu](z) = \int_{\mathbb{S}_n} P(z, \zeta) \, d\mu(\zeta), \qquad z \in \mathbb{B}_n. \tag{4.4}$$

The function $P[\mu]$ is called the Poisson transform, or the Poisson integral of μ.

The Cauchy transform of a function in $L^1(\mathbb{S}_n, d\sigma)$ is defined as

$$C[f](z) = \int_{\mathbb{S}_n} C(z, \zeta) f(\zeta) \, d\sigma(\zeta), \qquad z \in \mathbb{B}_n, \tag{4.5}$$

and more generally, the Cauchy transform of a finite complex Borel measure μ on $\mathbb{S}_n$ is defined as

$$C[\mu](z) = \int_{\mathbb{S}_n} C(z, \zeta) \, d\mu(\zeta), \qquad z \in \mathbb{B}_n. \tag{4.6}$$

The Cauchy transform will be studied in Section 4.3.

In this section, we prove several fundamental properties of the Poisson transform, including the Möbius invariance and the existence of boundary values almost everywhere on $\mathbb{S}_n$.

First recall that every automorphism φ is the composition of a unitary U and a symmetry φ_a. It is clear from the definition that each φ_a extends holomorphically to the closed unit ball. Furthermore, the restriction of φ_a to the unit sphere $\mathbb{S}_n$ is a homeomorphism. The same can be said for each unitary and hence for each automorphism of $\mathbb{B}_n$.

The following result states that the Poisson transform is invariant under the action of the automorphism group.

Theorem 4.3. *If* $f \in L^1(\mathbb{S}_n, d\sigma)$ *and* $\varphi \in \mathrm{Aut}(\mathbb{B}_n)$, *then*

$$P[f \circ \varphi](z) = P[f](\varphi(z))$$

for all $z \in \mathbb{B}_n$.

Proof. By an approximation argument, we may assume that f is continuous on the unit sphere $\mathbb{S}_n$.

Write $\varphi = U\varphi_a$, where U is a unitary and $a = \varphi^{-1}(0)$. By Lemmas 1.2 and 1.3,

$$P(\varphi(z), \varphi(\zeta)) = \frac{P(z, \zeta)}{P(a, \zeta)}.$$

It follows that

$$\int_{\mathbb{S}_n} P(\varphi(r\eta), \varphi(\zeta)) \, d\sigma(\eta) = P(\varphi(0), \varphi(\zeta))$$

for $\varphi \in \mathrm{Aut}(\mathbb{B}_n)$, $\zeta \in \mathbb{S}_n$, and $0 \le r < 1$. Replacing $\varphi(\zeta)$ by ζ in the above equation, and then using Fubini's theorem, we obtain

$$P[f](\varphi(0)) = \int_{\mathbb{S}_n} P[f](\varphi(r\zeta)) \, d\sigma(\zeta)$$

for all $0 \le r < 1$. For f continuous on $\mathbb{S}_n$, it is easy to see that

$$\lim_{r \to 1^-} P[f](\varphi(r\zeta)) = f(\varphi(\zeta))$$

uniformly for $\zeta \in \mathbb{S}_n$. Thus

$$P[f](\varphi(0)) = \int_{\mathbb{S}_n} f(\varphi(\zeta)) \, d\sigma(\zeta),$$

or

$$P[f](\varphi(0)] = P[f \circ \varphi](0).$$

In general, we apply the above identity twice to obtain

$$P[f](\varphi(z)) = P[f](\varphi \circ \varphi_z(0)) = P[f \circ \varphi \circ \varphi_z](0)$$
$$= P[f \circ \varphi](\varphi_z(0)) = P[f \circ \varphi](z).$$

This proves the Möbius invariance of the Poisson transform. $\square$

Corollary 4.4. *If $f \in L^1(\mathbb{S}_n, d\sigma)$, then we have the change of variables formula*

$$\int_{\mathbb{S}_n} f \circ \varphi(\zeta) \, d\sigma(\zeta) = \int_{\mathbb{S}_n} P(a, \zeta) f(\zeta) \, d\sigma(\zeta) \tag{4.7}$$

for every $\varphi \in \mathrm{Aut}(\mathbb{B}_n)$, where $a = \varphi(0)$.

Proof. Simply set $z = 0$ in Theorem 4.3. $\square$

It follows from Corollary 4.4 that for any fixed $z \in \mathbb{B}_n$, the Poisson kernel $P(z, \zeta)$ is the real Jacobian determinant of the mapping

$$\varphi_z : \mathbb{S}_n \to \mathbb{S}_n$$

with respect to the surface measure σ. In particular, we have

$$\int_{\mathbb{S}_n} f \circ \varphi_z(\zeta) \, d\sigma(\zeta) = \int_{\mathbb{S}_n} P(z, \zeta) f(\zeta) \, d\sigma(\zeta) \tag{4.8}$$

for $f \in L^1(\mathbb{S}_n, d\sigma)$ and $z \in \mathbb{B}_n$.

Corollary 4.5. *If f is in the ball algebra, then*

$$|f(z)|^p \leq \int_{\mathbb{S}_n} P(z,\zeta)|f(\zeta)|^p \, d\sigma(\zeta)$$

for all $z \in \mathbb{B}_n$ and $0 < p < \infty$.

Proof. The special case $z = 0$ follows from the subharmonicity of $|f|^p$. For a general $z \in \mathbb{B}_n$ we apply the special case to the function

$$g(w) = f \circ \varphi_z(w), \qquad w \in \overline{\mathbb{B}_n},$$

and make a change of variables according to (4.8). The desired result then follows. $\square$

We proceed to show that the Poisson transform of any finite Borel measure on $\mathbb{S}_n$ has a very strong boundary value at almost every point of $\mathbb{S}_n$. This will be used in the next section to show that every function in any Hardy space must have a boundary value at almost every point of $\mathbb{S}_n$.

We begin with some geometric considerations and several notions of maximal functions.

The first maximum function is defined for Borel measures on $\mathbb{S}_n$ in terms of a so-called nonisotropic metric on $\mathbb{S}_n$. Thus for z and w in $\overline{\mathbb{B}_n}$ we define

$$d(z,w) = |1 - \langle z, w \rangle|^{1/2}. \tag{4.9}$$

Elementary calculations (see Exercise 4.1) show that

$$d(z,w) \leq d(z,u) + d(u,w) \tag{4.10}$$

for all z, w, and u in $\overline{\mathbb{B}_n}$. Furthermore, if z and w are points on $\mathbb{S}_n$, then $d(z,w) = 0$ if and only if $z = w$. It follows that the restriction of d on $\mathbb{S}_n$ is a metric.

For $\zeta \in \mathbb{S}_n$ and $\delta > 0$ we let

$$Q(\zeta,\delta) = \{\eta \in \mathbb{S}_n : d(\zeta,\eta) < \delta\} \tag{4.11}$$

be the nonisotropic metric ball at ζ with radius δ. It is obvious that $Q(\zeta,\delta) = \mathbb{S}_n$ when $\delta \geq \sqrt{2}$.

Lemma 4.6. *There exist positive constants A_1 and A_2 (depending on n only) such that*

$$A_1 \leq \frac{\sigma(Q(\zeta,\delta))}{\delta^{2n}} \leq A_2$$

for all $\zeta \in \mathbb{S}_n$ and all $\delta \in (0, \sqrt{2})$.

Proof. The result is obvious when $n = 1$. Also, it follows from symmetry that $\sigma(Q(\zeta,\delta))$ is independent of ζ.

So we may assume that $n > 1$ and $\zeta = e_1$. Applying (1.13), we obtain

$$\sigma(Q(\zeta, \delta)) = (n-1) \int_{E(\delta)} (1 - |z|^2)^{n-2} \, dA(z),$$

where

$$E(\delta) = \{z \in \mathbb{C} : |z| < 1, |1 - z| < \delta^2\}$$

and dA is the normalized area measure on $\mathbb{C}$. From

$$1 - |z|^2 = (1 - |z|)(1 + |z|) \leq 2(1 - |z|)$$

and the definition of $E(\delta)$ we deduce

$$\sigma(Q(\zeta, \delta)) \leq (n-1)2^{n-2} \int_{E(\delta)} (1 - |z|)^{n-2} \, dA(z)$$

$$\leq (n-1)2^{n-2}\delta^{2(n-2)} A(E(\delta))$$

$$< (n-1)2^{n-2}\delta^{2n-4}\delta^4 = (n-1)2^{n-2}\delta^{2n}.$$

On the other hand, for $\delta \leq \sqrt{2}$, the set $E(\delta)$ clearly contains

$$\Omega(\delta) = \{1 + re^{i\theta} : 0 < r < \delta^2, |\theta - \pi| \leq \pi/4\}.$$

For $z = 1 + re^{i\theta} \in \Omega(\delta)$, we have

$$1 - |z|^2 = r(-2\cos\theta - r) \geq r(\sqrt{2} - 1),$$

so

$$\sigma(Q(\zeta, \delta)) \geq (n-1) \int_{\Omega(\delta)} (1 - |z|^2)^{n-2} \, dA(z)$$

$$\geq \frac{n-1}{2} \int_0^{\delta^2} r^{n-2}(\sqrt{2} - 1)^{n-2} \, r \, dr$$

$$= \frac{n-1}{2n}(\sqrt{2} - 1)^{n-2}\delta^{2n}.$$

This completes the proof of the lemma.   □

We will need the following elementary covering lemma on several occasions.

Lemma 4.7. *Suppose N is a natural number and*

$$E = \bigcup_{k=1}^{N} Q(\zeta_k, \delta_k).$$

There exists a subsequence $\{k_i\}$, $1 \leq i \leq M$, such that

(a) The balls $Q(\zeta_{k_i}, \delta_{k_i})$ are disjoint.

(b) The balls $Q(\zeta_{k_i}, 3\delta_{k_i})$ cover E.

(c) The inequality

$$\sigma(E) \leq C \sum_{i=1}^{M} \sigma(Q(\zeta_{k_i}, \delta_{k_i}))$$

holds for

$$C = \sup\left\{\frac{\sigma(Q(\zeta, 3\delta))}{\sigma(Q(\zeta, \delta))} : \zeta \in \mathbb{S}_n, \delta > 0\right\}.$$

Proof. First notice that C is a finite positive constant in view of Lemma 4.6.

Without loss of generality we may assume that the finite sequence $\{\delta_k\}$ is non-increasing. We let $k_1 = 1$ and construct a subsequence $\{k_i\}$ inductively as follows.

Suppose $i \geq 1$ and k_i has been chosen. If $Q(\zeta_{k_i}, \delta_{k_i})$ intersects $Q(\zeta_k, \delta_k)$ for every $k > k_i$, we stop. Otherwise, let k_{i+1} be the first index k after k_i such that $Q(\zeta_{k_{i+1}}, \delta_{k_{i+1}})$ is disjoint from $Q(\zeta_{k_i}, \delta_{k_i})$. Since the original collection is finite, this process stops after a finite number of steps, and we obtain a subsequence $\{k_i\}$ satisfying (a).

If $k_i \leq k < k_{i+1}$, then $\delta_k \leq \delta_{k_i}$, and $Q(\zeta_k, \delta_k)$ intersects $Q(\zeta_{k_i}, \delta_{k_i})$. It follows from the triangle inequality that

$$Q(\zeta_k, \delta_k) \subset Q(\zeta_{k_i}, 3\delta_{k_i}).$$

Similarly, if $k_M < k$, then

$$Q(\zeta_k, \delta_k) \subset Q(\zeta_{k_M}, 3\delta_{k_M}).$$

This proves (b). Part (c) follows from (b) and the definition of C. $\square$

We now introduce the first maximal operator.

For a complex Borel measure μ on $\mathbb{S}_n$ we let $|\mu|$ denote the total variation of μ, so $|\mu|$ becomes a positive Borel measure on $\mathbb{S}_n$, and we write $\|\mu\| = |\mu|(\mathbb{S}_n)$. The function

$$(M\mu)(\zeta) = \sup_{\delta>0} \frac{|\mu|(Q(\zeta, \delta))}{\sigma(Q(\zeta, \delta))}, \qquad \zeta \in \mathbb{S}_n, \tag{4.12}$$

is called the maximal function of μ on $\mathbb{S}_n$. It is clear that μ and $|\mu|$ have the same maximal function.

When $d\mu = f\, d\sigma$, where $f \in L^1(\mathbb{S}_n, d\sigma)$, we use Mf to denote the resulting maximal function. Thus

$$(Mf)(\zeta) = \sup_{\delta>0} \frac{1}{\sigma(Q(\zeta, \delta))} \int_{Q(\zeta, \delta)} |f|\, d\sigma, \qquad \zeta \in \mathbb{S}_n. \tag{4.13}$$

For each fixed $\delta > 0$ the function

$$\zeta \mapsto \frac{|\mu|(Q(\zeta, \delta))}{\sigma(Q(\zeta, \delta))}$$

is lower semi-continuous on $\mathbb{S}_n$. It follows that $M\mu$ is always lower semi-continuous, that is, the set

$$\{M\mu > t\} = \{\zeta \in \mathbb{S}_n : (M\mu)(\zeta) > t\}$$

is open in $\mathbb{S}_n$ for every t.

Lemma 4.8. *There exists a constant $C > 0$ such that*

$$\sigma(M\mu > t) \leq \frac{C\|\mu\|}{t}$$

for every complex Borel measure μ on $\mathbb{S}_n$ and every $t > 0$.

Proof. Fix μ and $t > 0$. If K is a compact subset of the open set $\{M\mu > t\}$, then K is covered by a finite collection Φ of open balls $Q = Q(\zeta, \delta)$ such that $|\mu|(Q) > t\sigma(Q)$. Let Φ_0 be a subcollection of Φ chosen according to Lemma 4.7. Then

$$\sigma\left(\bigcup_\Phi Q\right) \leq C \sum_{\Phi_0} \sigma(Q),$$

and so

$$\sigma(K) \leq C \sum_{Q \in \Phi_0} \sigma(Q) < \frac{C}{t} \sum_{Q \in \Phi_0} |\mu|(Q) \leq \frac{C\|\mu\|}{t},$$

where the disjointness of Φ_0 was used in the last inequality above. Taking the supremum over all compact K in $\{M\mu > t\}$ gives the desired result. $\square$

If $d\mu = f d\sigma$ is absolutely continuous with respect to σ, then the preceding lemma states that

$$t\sigma(Mf > t) \leq C \int_{\mathbb{S}_n} |f| \, d\sigma \tag{4.14}$$

for all $t > 0$. In general, a measurable function g on $\mathbb{S}_n$ satisfying

$$\sup_{t>0}(t\sigma(|g| > t)) < \infty$$

is said to belong to *weak* $L^1(\mathbb{S}_n, d\sigma)$.

Theorem 4.9. *For each $p \in (1, \infty)$ there exists a constant $C_p > 0$ such that*

$$\int_{\mathbb{S}_n} |Mf|^p \, d\sigma \leq C_p \int_{\mathbb{S}_n} |f|^p \, d\sigma$$

for all $f \in L^p(\mathbb{S}_n, d\sigma)$.

Proof. It is obvious that the maximal operator M is sub-additive, that is,

$$M(f + g) \leq Mf + Mg.$$

It is also obvious that M maps $L^\infty(\mathbb{S}_n)$ into $L^\infty(\mathbb{S}_n)$. The desired result then follows from (4.14) and the Marcinkiewicz interpolation theorem. $\square$

The second maximal operator we will introduce makes use of certain approach regions in $\mathbb{B}_n$ near $\mathbb{S}_n$. Thus for $\zeta \in \mathbb{S}_n$ and $\alpha > 1$ we let $D_\alpha(\zeta)$ denote the set of points z in $\mathbb{B}_n$ such that

$$|1 - \langle z, \zeta \rangle| < \frac{\alpha}{2}(1 - |z|^2).$$

If $\alpha \leq 1$, the above condition defines the empty set. For any fixed $\zeta \in \mathbb{S}_n$, the regions $D_\alpha(\zeta)$ fill $\mathbb{B}_n$ as $\alpha \to \infty$.

For $\alpha > 1$ and f continuous in $\mathbb{B}_n$, we define a function $M_\alpha f$ on $\mathbb{S}_n$ by

$$(M_\alpha f)(\zeta) = \sup\{|f(z)| : z \in D_\alpha(\zeta)\}. \tag{4.15}$$

Since the continuity of f implies that $\{M_\alpha f \leq t\}$ is closed in $\mathbb{S}_n$ for every t, the maximal function $M_\alpha f$ is lower semi-continuous.

Theorem 4.10. *For every $\alpha > 1$ there exists a constant $C = C_\alpha > 0$ such that*

$$M_\alpha P[\mu] \leq CM\mu$$

for every finite complex Borel measure μ on $\mathbb{S}_n$.

Proof. Since $M|\mu| = M\mu$ and $|P[\mu]| \leq P[|\mu|]$, we may as well assume that μ is positive. By Lemma 4.6, there exists a positive constant C such that $\sigma(Q(\zeta, \delta)) \leq C\delta^{2n}$ for all $\zeta \in \mathbb{S}_n$ and $\delta > 0$.

Fix a point $\zeta \in \mathbb{S}_n$ such that $M\mu(\zeta) < \infty$, and fix a point $z \in D_\alpha(\zeta)$. Let $r = |z|, t = 8\alpha(1 - r)$, and

$$V_0 = \{\eta \in \mathbb{S}_n : |1 - \langle \eta, \zeta \rangle| < t\}.$$

For $1 \leq k \leq N$, where N is the first natural number with $2^N t > 2$, define

$$V_k = \{\eta \in \mathbb{S}_n : 2^{k-1}t \leq |1 - \langle \eta, \zeta \rangle| < 2^k t\}.$$

Since $V_k \subset Q(\zeta, \sqrt{2^k t})$, we have

$$\mu(V_k) \leq \mu\big(Q(\zeta, \sqrt{2^k t})\big) \leq M\mu(\zeta)\sigma\big(Q(\zeta, \sqrt{2^k t})\big) \leq C(2^k t)^n M\mu(\zeta) \tag{4.16}$$

for $0 \leq k \leq N$.

Since $P(z, \eta) < 2^n(1 - r)^{-n}$, we have

$$\int_{V_0} P(z, \eta) \, d\mu(\eta) \leq \frac{2^n \mu(V_0)}{(1-r)^n} \leq C\left(\frac{2t}{1-r}\right)^n M\mu(\zeta) = C(16\alpha)^n M\mu(\zeta).$$

For any $\eta \in \mathbb{S}_n$, we have

$$|1 - \langle z, \zeta \rangle| < \frac{\alpha}{2}(1 - |z|^2) \leq \alpha(1 - |z|) \leq \alpha|1 - \langle z, \eta \rangle|,$$

or

$$d(z, \zeta) < \sqrt{\alpha}\, d(z, \eta).$$

It follows from the triangle inequality for the nonisotropic "metric" d on $\overline{\mathbb{B}}_n$ that

$$d(\zeta, \eta) \le d(\zeta, z) + d(z, \eta) \le (1 + \sqrt{\alpha})\, d(z, \eta) \le 2\sqrt{\alpha}\, d(z, \eta).$$

If $1 \le k \le N$ and $\eta \in V_k$, then

$$|1 - \langle \zeta, \eta \rangle| \le 4\alpha |1 - \langle z, \eta \rangle|,$$

and so

$$P(z, \eta) \le \frac{(4\alpha t)^n}{|1 - \langle \zeta, \eta \rangle|^{2n}} \le \left(\frac{16\alpha}{4^k t} \right)^n.$$

This along with (4.16) shows that

$$\int_{V_k} P(z, \eta)\, d\mu(\eta) \le \left(\frac{16\alpha}{4^k t} \right)^n \mu(V_k) \le C \left(\frac{16\alpha}{2^k} \right)^n M\mu(\zeta)$$

for $1 \le k \le N$.

From the decomposition

$$\int_{\mathbb{S}_n} P(z, \eta)\, d\mu(\eta) = \int_{V_0} P(z, \eta)\, d\mu(\eta) + \sum_{k=1}^{N} \int_{V_k} P(z, \eta)\, d\mu(\eta)$$

we now deduce that

$$\int_{\mathbb{S}_n} P(z, \eta)\, d\mu(\eta) \le 2C(16\alpha)^n M\mu(\zeta).$$

Since $z \in D_\alpha(\zeta)$ is arbitrary, we conclude that

$$M_\alpha P[\mu](\zeta) \le 2C(16\alpha)^n M\mu(\zeta),$$

completing the proof of the theorem.   □

Corollary 4.11. *For every $1 < p < \infty$ and $\alpha > 1$ there exists a constant $C > 0$ such that*

$$\int_{\mathbb{S}_n} |M_\alpha P[f]|^p\, d\sigma \le C \int_{\mathbb{S}_n} |f|^p\, d\sigma$$

for all $f \in L^p(\mathbb{S}_n, d\sigma)$.

Proof. This is a direct consequence of Theorems 4.9 and 4.10.   □

We need two Lebesgue differentiation type results before we can prove the existence of boundary values of the Poisson transform of a measure on $\mathbb{S}_n$. The phrase "for almost every $\zeta \in \mathbb{S}_n$" always refers to Lebesgue measure σ, unless otherwise specified.

Lemma 4.12. *If $f \in L^1(\mathbb{S}_n, d\sigma)$, then*

$$\lim_{\delta \to 0} \frac{1}{\sigma(Q(\zeta, \delta))} \int_{Q(\zeta, \delta)} |f - f(\zeta)| \, d\sigma = 0 \tag{4.17}$$

for almost every $\zeta \in \mathbb{S}_n$. Consequently,

$$f(\zeta) = \lim_{\delta \to 0} \frac{1}{\sigma(Q(\zeta, \delta))} \int_{Q(\zeta, \delta)} f \, d\sigma \tag{4.18}$$

for almost every $\zeta \in \mathbb{S}_n$.

Proof. Define

$$T_f(\zeta) = \limsup_{\delta \to 0} \frac{1}{\sigma(Q(\zeta, \delta))} \int_{Q(\zeta, \delta)} |f - f(\zeta)| \, d\sigma, \qquad \zeta \in \mathbb{S}_n.$$

For any $\epsilon > 0$ we can find a function g, continuous on $\mathbb{S}_n$, such that

$$\|f - g\|_1 = \int_{\mathbb{S}_n} |f - g| \, d\sigma < \epsilon.$$

Let $h = f - g$. Then $T_f \leq T_g + T_h$, $T_g = 0$, and $T_h \leq |h| + Mh$. It follows that $T_f \leq |h| + Mh$, and so for any $t > 0$, $\{T_f > t\}$ is a subset of

$$E_t = \left\{ |h| > \frac{t}{2} \right\} \bigcup \left\{ Mh > \frac{t}{2} \right\}.$$

Note that

$$\int_{\mathbb{S}_n} |h| \, d\sigma \geq \frac{t}{2} \sigma \left(|h| > \frac{t}{2} \right).$$

Combining this with (4.14), we find a constant $C > 0$, independent of t and ϵ, such that

$$\sigma(E_t) \leq \frac{C\|h\|_1}{t} \leq \frac{C\epsilon}{t}.$$

This implies that

$$\sigma(T_f > t) \leq \frac{C\epsilon}{t}.$$

Since ϵ is arbitrary, we must have $\sigma\{T_f > t\} = 0$ for any $t > 0$. Since t is arbitrary, we conclude that $T_f(\zeta) = 0$ for almost every $\zeta \in \mathbb{S}_n$. $\qquad \square$

Corollary 4.13. *If $f \in L^1(\mathbb{S}_n, d\sigma)$, then $|f(\zeta)| \leq Mf(\zeta)$ for almost every $\zeta \in \mathbb{S}_n$.*

Proof. Since

$$\left| \frac{1}{\sigma(Q(\zeta, \delta))} \int_{Q(\zeta, \delta)} f \, d\sigma \right| \leq Mf(\zeta),$$

the desired inequality follows from (4.18) in Lemma 4.12. $\qquad \square$

Lemma 4.14. *Suppose μ is a finite Borel measure on $\mathbb{S}_n$. If μ is singular with respect to σ, then*

$$\limsup_{\delta \to 0} \frac{\mu(Q(\zeta, \delta))}{\sigma(Q(\zeta, \delta))} = 0$$

for almost every $\zeta \in \mathbb{S}_n$.

Proof. Without loss of generality we may assume that μ is positive. Define

$$D\mu(\zeta) = \limsup_{\delta \to 0} \frac{\mu(Q(\zeta, \delta))}{\sigma(Q(\zeta, \delta))}, \qquad \zeta \in \mathbb{S}_n.$$

For any positive number ϵ we can decompose $\mu = \mu_1 + \mu_2$, where μ_1 is the restriction of μ to some compact set K with $\sigma(K) = 0$ and $\|\mu_2\| < \epsilon$. Off the compact set K, we have $D\mu_1 = 0$, or $D\mu = D\mu_2$. It follows that for any positive number t,

$$K \cup \{D\mu > t\} = K \cup \{D\mu_2 > t\} \subset K \cup \{M\mu_2 > t\},$$

so the σ-measure of the last set is at most $C\|\mu_2\|/t$, where C is the positive constant from Lemma 4.8. Since $\|\mu_2\| < \epsilon$ and ϵ is arbitrary, we must have $\sigma\{D\mu > t\} = 0$ for any $t > 0$. Letting $t \to 0$, we conclude that $D\mu(\zeta) = 0$ for almost every $\zeta \in \mathbb{S}_n$. $\square$

For any finite complex Borel measure μ on $\mathbb{S}_n$, we consider the Lebesgue decomposition $d\mu = f\, d\sigma + d\mu_s$, where $f \in L^1(\mathbb{S}_n, d\sigma)$ and μ_s is singular with respect to σ. Combining (4.18) and Lemma 4.14, we conclude that

$$\lim_{\delta \to 0} \frac{\mu(Q(\zeta, \delta))}{\sigma(Q(\zeta, \delta))} = f(\zeta) \tag{4.19}$$

for almost every $\zeta \in \mathbb{S}_n$.

In what follows we are going to write

$$\frac{d\mu}{d\sigma}(\zeta) = D\mu(\zeta) = \lim_{\delta \to 0} \frac{\mu(Q(\zeta, \delta))}{\sigma(Q(\zeta, \delta))} \tag{4.20}$$

for $\zeta \in \mathbb{S}_n$, whenever the limit exists. If $d\mu = f\, d\sigma + d\mu_s$ is the Lebesgue decomposition of μ, then

$$\frac{d\mu}{d\sigma} = f \tag{4.21}$$

almost everywhere on $\mathbb{S}_n$.

Suppose f is a function on $\mathbb{B}_n$. We say that f has K-limit L at some $\zeta \in \mathbb{S}_n$, and write

$$\mathrm{Klim} f(\zeta) = L,$$

if for every $\alpha > 1$ we have

$$\lim_{z \to \zeta, z \in D_\alpha(\zeta)} f(z) = L.$$

We can now prove the existence of boundary values of the Poisson transform of any finite complex Borel measure on $\mathbb{S}_n$.

Theorem 4.15. *If μ is a finite complex Borel measure on $\mathbb{S}_n$, then*

$$\mathrm{Klim}P[\mu](\zeta) = \frac{d\mu}{d\sigma}(\zeta)$$

for almost every point $\zeta \in \mathbb{S}_n$.

Proof. First assume that μ is positive and $\zeta \in \mathbb{S}_n$ is a point such that

$$\frac{d\mu}{d\sigma}(\zeta) = 0.$$

For any $\epsilon > 0$ we can find a positive number δ_ϵ such that

$$\mu(Q(\zeta, \delta)) < \epsilon\sigma(Q(\zeta, \delta))$$

for all $\delta \in (0, \delta_\epsilon)$. Let μ_ϵ be the restriction of μ to $Q(\zeta, \delta_\epsilon)$ and let $\nu = \mu - \mu_\epsilon$. It is clear that $P[\nu]$ has K-limit 0 at ζ and $M\mu_\epsilon(\zeta) \le \epsilon$. If a sequence $\{z_k\} \subset D_\alpha(\zeta)$ converges to ζ, then by Theorem 4.10,

$$\limsup_{k\to\infty} P[\mu](z_k) = \limsup_{k\to\infty} \left(P[\nu](z_k) + P[\mu_\epsilon](z_k)\right) \le C\epsilon.$$

Since ϵ is arbitrary, we must have

$$\mathrm{Klim}_{k\to\infty} P[\mu](\zeta) = 0. \tag{4.22}$$

Since $|P[\mu]| \le P[|\mu|]$, a combination of this and Lemma 4.14 proves the theorem in the case when μ is singular.

Next assume that μ is absolutely continuous with respect to σ, say, $d\mu = f\,d\sigma$. By Lemma 4.12,

$$\lim_{\delta\to 0} \frac{1}{\sigma(Q(\zeta, \delta))} \int_{Q(\zeta, \delta)} |f - f(\zeta)|\,d\sigma = 0$$

for almost every $\zeta \in \mathbb{S}_n$. Fix a point $\zeta \in \mathbb{S}_n$ such that the above limit is zero and define a finite positive Borel measure μ' on $\mathbb{S}_n$ by

$$\mu'(E) = \int_E |f - f(\zeta)|\,d\sigma.$$

By Lemma 4.12, we have

$$\frac{d\mu'}{d\sigma}(\eta) = |f(\eta) - f(\zeta)|$$

for almost all $\eta \in \mathbb{S}_n$. It follows that

$$\frac{d\mu'}{d\sigma}(\zeta) = 0$$

and

$$P[\mu'](z) = \int_{\mathbb{S}_n} P(z, \eta) \, d\mu'(\eta)$$

$$= \int_{\mathbb{S}_n} P(z, \eta) |f(\eta) - f(\zeta)| \, d\sigma(\eta)$$

$$\geq |P[f](z) - f(\zeta)|.$$

Combining this with (4.22), we see that the K-limit of $P[f] - f(\zeta)$ at ζ is 0, that is,

$$\mathrm{Klim}\, P[f](\zeta) = f(\zeta).$$

The general case then follows from the Lebesgue decomposition of μ. $\square$

Corollary 4.16. *If $g \in L^1(\mathbb{S}_n, d\sigma)$ and $f = P[g]$, then*

$$\mathrm{Klim}\, f(\zeta) = g(\zeta)$$

for almost every $\zeta \in \mathbb{S}_n$. If μ is a singular measure on $\mathbb{S}_n$ and $f = P[\mu]$, then

$$\mathrm{Klim}\, f(\zeta) = 0$$

for almost every $\zeta \in \mathbb{S}_n$.

4.2 Hardy Spaces

For $0 < p < \infty$ the Hardy space H^p consists of holomorphic functions f in $\mathbb{B}_n$ such that

$$\|f\|_p^p = \sup_{0 < r < 1} \int_{\mathbb{S}_n} |f(r\zeta)|^p \, d\sigma(\zeta) < \infty. \tag{4.23}$$

We begin with the maximum rate of growth for a function in H^p near the boundary of $\mathbb{B}_n$.

Theorem 4.17. *Suppose $0 < p < \infty$ and $f \in H^p$. Then*

$$|f(z)| \leq \frac{\|f\|_p}{(1 - |z|^2)^{n/p}}$$

for all $z \in \mathbb{B}_n$. Furthermore, the exponent n/p is best possible.

Proof. Fix $f \in H^p$ and $z \in \mathbb{B}_n$. For any $0 < r < 1$ consider

$$F_r(w) = f_r(\varphi_z(w)) \frac{(1 - |z|^2)^{n/p}}{(1 - \langle w, z \rangle)^{2n/p}}, \qquad w \in \mathbb{B}_n,$$

where $f_r(w) = f(rw)$ for $w \in \mathbb{B}_n$ and φ_z is the involutive automorphism of $\mathbb{B}_n$ that interchanges 0 and z. By the subharmonicity of $|F_r|^p$ and the change of variables formula (4.8), we have

$$|F_r(0)|^p \leq \int_{\mathbb{S}_n} |F_r(\zeta)|^p \, d\sigma(\zeta) = \int_{\mathbb{S}_n} |f_r(\zeta)|^p \, d\sigma(\zeta) \leq \|f\|_p^p.$$

The desired result then follows from letting $r \to 1^-$.   □

It is easy to see that the exponent n/p in the theorem above is best possible. However, by approximating functions in H^p by polynomials (see Corollary 4.26) we can show that

$$f(z) = o\left(\frac{1}{(1 - |z|^2)^{n/p}}\right), \qquad |z| \to 1^-,$$

for every $f \in H^p$.

As a consequence of Theorem 4.17 we see that point evaluations are bounded linear functionals on each of the Hardy spaces. Actually, point evaluations are uniformly bounded on compact subsets of $\mathbb{B}_n$.

Corollary 4.18. *Suppose $0 < p < \infty$ and K is a compact subset of $\mathbb{B}_n$. Then there exists a constant $C > 0$ such that*

$$|f(z)| \leq C\|f\|_p$$

for all $f \in H^p$ and $z \in K$.

Next we show that the Hardy spaces are all complete.

Corollary 4.19. *If $1 \leq p < \infty$, the Hardy space H^p is a Banach space with the norm $\| \ \|_p$. If $0 < p < 1$, H^p is a complete metric space with the distance function*

$$d(f, g) = \|f - g\|_p^p.$$

Proof. It suffices to show that each H^p is complete in $\| \ \|_p$. So assume that $\{f_k\}$ is a Cauchy sequence in H^p. By Corollary 4.18, the sequence $\{f_k(z)\}$ is uniformly Cauchy on every compact subset of $\mathbb{B}_n$. It follows that $\{f_k\}$ converges to a holomorphic function f in $\mathbb{B}_n$, and the convergence is uniform on every compact subset of $\mathbb{B}_n$.

Given any $\epsilon > 0$ there is a natural number N such that

$$\|f_k - f_l\|_p < \epsilon, \qquad k > N, l > N.$$

For any $0 < r < 1$ we have

$$\int_{\mathbb{S}_n} |f_k(r\zeta) - f(r\zeta)|^p \, d\sigma(\zeta) = \lim_{l \to \infty} \int_{\mathbb{S}_n} |f_k(r\zeta) - f_l(r\zeta)|^p \, d\sigma(\zeta) \leq \epsilon^p$$

whenever $k > N$. It follows that $f_k - f \in H^p$ for all $k > N$, so $f \in H^p$, and

$$\|f_k - f\|_p \leq \epsilon$$

for all $k > N$. This shows that H^p is complete .   □

We are going to prove two Littlewood-Paley type idenities for H^p in terms of the radial derivative and the invariant gradient. Thus for $0 < p < \infty$ and f holomorphic in $\mathbb{B}_n$ we introduce the integral means

$$M_p(r, f) = \left(\int_{\mathbb{S}_n} |f(r\zeta)|^p \, d\sigma(\zeta) \right)^{1/p},$$

where $0 \le r < 1$. We shall see that $M_p(r, f)^p$, as a function of r, is differentiable on $[0, 1)$.

We begin with the case $n = 1$.

Let f be a nonconstant analytic function in the unit disk $\mathbb{D}$ with $f(0) = 0$ and $p > 0$. Fix any $r \in (0, 1)$ such that f does not have any zero on $|z| = r$. Let $\{z_1, \cdots, z_m\}$ be the zeros of f in $0 < |z| < r$. Define

$$\Omega = \Omega_\epsilon = \{z \in \mathbb{D} : |z| < r\} - \{z \in \mathbb{D} : |z| \le \epsilon\} - \bigcup_{k=1}^m \{z \in \mathbb{D} : |z - z_k| \le \epsilon\},$$

where ϵ is a small enough positive number such that the disks being removed from $|z| < r$ to form Ω_ϵ are disjoint.

The classical Green's formula states that

$$\int_\Omega (v\Delta u - u\Delta v) \, dx \, dy = \int_{\partial\Omega} \left(v\frac{\partial u}{\partial n} - u\frac{\partial v}{\partial n} \right) ds,$$

where $u(z) = |f(z)|^p$, $v(z) = \log(r/|z|)$, and $z = x + iy$. Since $\Delta v = 0$ in Ω, we have

$$\int_\Omega (\Delta|f|^p) \log\frac{r}{|z|} \, dx \, dy = \int_{|z|=r} \left(\frac{\partial|f|^p}{\partial n} \log\frac{r}{|z|} - |f|^p\frac{\partial}{\partial n} \log\frac{r}{|z|} \right) ds$$

$$- \int_{|z|=\epsilon} \left(\frac{\partial|f|^p}{\partial n} \log\frac{r}{|z|} - |f|^p\frac{\partial}{\partial n} \log\frac{r}{|z|} \right) ds$$

$$- \sum_{k=1}^m \int_{|z-z_k|=\epsilon} \left(\frac{\partial|f|^p}{\partial n} \log\frac{r}{|z|} - |f|^p\frac{\partial}{\partial n} \log\frac{r}{|z|} \right) ds.$$

On the circle $|z| = r$, with $z = re^{i\theta}$, we have

$$\log\frac{r}{|z|} = 0, \qquad \frac{\partial}{\partial n} \log\frac{r}{|z|} = -\frac{1}{r}, \qquad ds = r \, d\theta.$$

It is easy to see that

$$\lim_{\epsilon \to 0} \int_{|z|=\epsilon} \left(\frac{\partial|f|^p}{\partial n} \log\frac{r}{|z|} - |f|^p\frac{\partial}{\partial n} \log\frac{r}{|z|} \right) ds = 0,$$

and for $1 \le k \le m$,

$$\lim_{\epsilon \to 0} \int_{|z-z_k|=\epsilon} \left(\frac{\partial |f|^p}{\partial n} \log \frac{r}{|z|} - |f|^p \frac{\partial}{\partial n} \log \frac{r}{|z|} \right) ds = 0.$$

Since

$$\Delta |f(z)|^p = 4 \frac{\partial^2}{\partial z \partial \bar{z}} |f(z)|^p = p^2 |f'(z)|^2 |f(z)|^{p-2}$$

whenever $f(z) \neq 0$, and since the singularity of $|f'(z)|^2 |f(z)|^{p-2}$ at each zero in $\mathbb{D}$ is integrable with respect to area measure, we let ϵ approach 0 and conclude that

$$p^2 \int_{|z|<r} |f'(z)|^2 |f(z)|^{p-2} \log \frac{r}{|z|} \, dx \, dy = \int_0^{2\pi} |f(re^{i\theta})|^p \, d\theta.$$

We established this under the assumption that f is nonvanishing on $|z| = r$. A continuity argument then shows that the above identity holds for every $r \in [0, 1)$. Using polar coordinates on $|z| < r$, we can write

$$\frac{1}{2\pi} \int_0^{2\pi} |f(re^{i\theta})|^p \, d\theta = \frac{p^2}{2\pi} \int_0^r \left(\log \frac{r}{t} \right) t \, dt \int_0^{2\pi} |f'(te^{i\theta})|^2 |f(te^{i\theta})|^{p-2} \, d\theta$$

$$= \frac{p^2}{2\pi} \log r \int_0^r t \, dt \int_0^{2\pi} |f'(te^{i\theta})|^2 |f(te^{i\theta})|^{p-2} \, d\theta$$

$$+ \frac{p^2}{2\pi} \int_0^r t \log \frac{1}{t} \, dt \int_0^{2\pi} |f'(te^{i\theta})|^2 |f(te^{i\theta})|^{p-2} \, d\theta.$$

Differentiating with respect to r, we arrive at

$$r \frac{d}{dr} \left(\frac{1}{2\pi} \int_0^{2\pi} |f(re^{i\theta})|^p \, d\theta \right) = \frac{p^2}{2} \int_{|z|<r} |f'(z)|^2 |f(z)|^{p-2} \, dA(z).$$

We now generalize this formula to higher dimensions.

Theorem 4.20. *Suppose f is holomorphic in $\mathbb{B}_n$ and $f(0) = 0$. Then*

$$r \frac{d}{dr} M_p(r, f)^p = \frac{p^2}{2n} \int_{|z|<r} |Rf(z)|^2 |f(z)|^{p-2} |z|^{-2n} \, dv(z) \qquad (4.24)$$

for all $0 < p < \infty$ and $0 \leq r < 1$.

Proof. We have just proved the case $n = 1$. When $n > 1$ and f is holomorphic in $\mathbb{B}_n$, we consider the slice functions

$$f_\zeta(w) = f(w\zeta), \qquad w \in \mathbb{D}, \zeta \in \mathbb{S}_n.$$

Each f_ζ is then an analytic function in the open unit disk $\mathbb{D}$. We apply the one-dimensional result to each slice function to obtain

$$r \frac{d}{dr} M_p(r, f_\zeta)^p = \frac{p^2}{2\pi} \int_0^r t \, dt \int_0^{2\pi} |f'_\zeta(te^{i\theta})|^2 |f_\zeta(te^{i\theta})|^{p-2} \, d\theta.$$

It is easy to see that

$$w f_\zeta'(w) = (Rf)(w\zeta), \qquad w \in \mathbb{D}, \zeta \in \mathbb{S}_n.$$

It follows that

$$r \frac{d}{dr} M_p(r, f_\zeta)^p = \frac{p^2}{2\pi} \int_0^r \frac{dt}{t} \int_0^{2\pi} |Rf(t\zeta e^{i\theta})|^2 |f(t\zeta e^{i\theta})|^{p-2} \, d\theta.$$

Integrate both sides of the above equation over $\zeta \in \mathbb{S}_n$ with respect to the measure σ and apply Lemma 1.10. We obtain

$$r \frac{d}{dr} M_p(r, f)^p = p^2 \int_0^r \frac{dt}{t} \int_{\mathbb{S}_n} |Rf(t\zeta)|^2 |f(t\zeta)|^{p-2} \, d\sigma(\zeta).$$

By Lemma 1.8, the above identity can be rewritten as

$$r \frac{d}{dr} M_p(r, f)^p = \frac{p^2}{2n} \int_{|z|<r} |Rf(z)|^2 |f(z)|^{p-2} |z|^{-2n} \, dv(z).$$

This completes the proof of the theorem. □

Corollary 4.21. *If $0 < p < \infty$ and $f \in H^p$, then the integral means $M_p(r, f)$ are increasing in r, and*

$$\|f\|_p = \lim_{r \to 1^-} M_p(r, f).$$

Proof. Obvious. □

Theorem 4.22. *Suppose $0 < p < \infty$. Then*

$$\|f - f(0)\|_p^p = \frac{p^2}{2n} \int_{\mathbb{B}_n} |Rf(z)|^2 |f(z) - f(0)|^{p-2} |z|^{-2n} \log \frac{1}{|z|} \, dv(z)$$

for all $f \in H^p$.

Proof. Without loss of generality we may assume that $f(0) = 0$. In this case, we divide both sides of the equation in (4.24) by r and then integrate with respect to r from 0 to t, where $0 < t < 1$. The result is

$$M_p(t, f)^p = \frac{p^2}{2n} \int_0^t \frac{dr}{r} \int_{|z|<r} |Rf(z)|^2 |f(z)|^{p-2} |z|^{-2n} \, dv(z).$$

By Fubini's theorem,

$$M_p(t, f)^p = \frac{p^2}{2n} \int_{|z|<t} |f(z)|^{p-2} |Rf(z)|^2 |z|^{-2n} \, dv(z) \int_{|z|}^t \frac{dr}{r}$$

$$= \frac{p^2}{2n} \int_{|z|<t} |f(z)|^{p-2} |Rf(z)|^2 |z|^{-2n} \log \frac{t}{|z|} \, dv(z).$$

Let $t \to 1$ and the desired result follows from Corollary 4.21 and the monotone convergence theorem. □

Note that a consequence of the theorem above is that a holomorphic function f in $\mathbb{B}_n$ belongs to H^p if and only if

$$\int_{\mathbb{B}_n} |Rf(z)|^2 |f(z)|^{p-2} |z|^{-2n} \log \frac{1}{|z|} \, dv(z) < \infty,$$

which is clearly equivalent to

$$\int_{\mathbb{B}_n} |Rf(z)|^2 |f(z)|^{p-2} (1 - |z|^2) \, dv(z) < \infty.$$

We now derive a similar representation of the H^p norm in terms of the invariant gradient of f.

Theorem 4.23. *Suppose $0 < p < \infty$ and f is holomorphic in $\mathbb{B}_n$. Then*

$$\|f\|_p^p = |f(0)|^p + \left(\frac{p}{2}\right)^2 \int_{\mathbb{B}_n} |\widetilde{\nabla} f(z)|^2 |f(z)|^{p-2} G(z) \, d\tau(z), \tag{4.25}$$

where G is the invariant Green function.

Proof. It suffices to prove the result for f holomorphic up to the boundary; the general case then follows from an approximation argument. Note also that Theorems 4.23 and 4.22 are the same when $n = 1$. So we assume $n > 1$.

For any $\epsilon > 0$ we consider the function

$$f_\epsilon(z) = (|f(z)|^2 + \epsilon)^{p/2}, \qquad z \in \mathbb{B}_n.$$

It is clear that f_ϵ is real-analytic on $\overline{\mathbb{B}}_n$. If r is a small enough positive number, we apply the invariant Green's formula (see Theorem 1.25) in the shell $r \leq |z| \leq 1 - r$ to write the integral

$$\int_{r < |z| < 1-r} G \widetilde{\Delta} f_\epsilon \, d\tau$$

as

$$\int_{|z|=1-r} \left(G \frac{\partial f_\epsilon}{\partial \widetilde{n}} - f_\epsilon \frac{\partial G}{\partial \widetilde{n}} \right) d\widetilde{\sigma} - \int_{|z|=r} \left(G \frac{\partial f_\epsilon}{\partial \widetilde{n}} - f_\epsilon \frac{\partial G}{\partial \widetilde{n}} \right) d\widetilde{\sigma},$$

where $\widetilde{n}$ is the unit outward normal vector in the Bergman metric and $\widetilde{\sigma}$ is the surface area element in the Bergman metric; see Section 1.6. Let $r \to 0^+$ and use the same arguments from the proof of Theorem 1.27. We obtain

$$\int_{\mathbb{B}_n} G \widetilde{\Delta} f_\epsilon \, d\tau = \int_{\mathbb{S}_n} f_\epsilon \, d\sigma - f_\epsilon(0),$$

or

$$\int_{\mathbb{S}_n} (|f(\zeta)|^2 + \epsilon)^{p/2} \, d\sigma(\zeta) = (|f(0)|^2 + \epsilon)^{p/2} + \int_{\mathbb{B}_n} G(z) \widetilde{\Delta} f_\epsilon(z) \, d\tau(z). \tag{4.26}$$

A straightforward computation shows that

$$\tilde{\Delta} f_\epsilon(z) = \left(\frac{p}{2}\right)^2 (|f(z)|^2 + \epsilon)^{p/2-1} |\tilde{\nabla} f(z)|^2 \frac{|f(z)|^2 + 2\epsilon/p}{|f(z)|^2 + \epsilon},$$

so

$$\lim_{\epsilon \to 0^+} \tilde{\Delta} f_\epsilon(z) = \left(\frac{p}{2}\right)^2 |f(z)|^{p-2} |\tilde{\nabla} f(z)|^2$$

for almost every $z \in \mathbb{B}_n$. Let $\epsilon \to 0$ in (4.26) and use Fatou's lemma. We conclude that

$$\int_{\mathbb{B}_n} |\tilde{\nabla} f(z)|^2 |f(z)|^{p-2} G(z) \, d\tau(z) < \infty.$$

It is easy to see that

$$\frac{|f(z)|^2 + 2\epsilon/p}{|f(z)|^2 + \epsilon} \le \max\left(\frac{2}{p}, 1\right)$$

for all $z \in \mathbb{B}_n$. Since f is assumed to be holomorphic up to the boundary, for every $p \ge 2$ there exists a constant $C > 0$ such that

$$0 \le \tilde{\Delta} f_\epsilon(z) \le C |\tilde{\nabla} f(z)|^2, \qquad z \in \mathbb{B}_n, 0 < \epsilon < 1.$$

For every $0 < p < 2$ there exists a constant $C > 0$ such that

$$0 \le \tilde{\Delta} f_\epsilon(z) \le C |\tilde{\nabla} f(z)|^2 |f(z)|^{p-2}, \qquad z \in \mathbb{B}_n, 0 < \epsilon < 1.$$

Since

$$\int_{\mathbb{B}_n} G(z) |\tilde{\nabla} f(z)|^2 \, d\tau(z) < \infty, \qquad \int_{\mathbb{B}_n} |\tilde{\nabla} f(z)|^2 |f(z)|^{p-2} G(z) \, d\tau(z) < \infty,$$

we can let $\epsilon \to 0$ in (4.26) and apply the dominated convergence theorem to obtain

$$\int_{\mathbb{S}_n} |f(\zeta)|^p \, d\sigma(\zeta) = |f(0)|^p + \left(\frac{p}{2}\right)^2 \int_{\mathbb{B}_n} |\tilde{\nabla} f(z)|^2 |f(z)|^{p-2} G(z) \, d\tau(z).$$

This completes the proof of the theorem.   □

A consequence of the preceding theorem is that a holomorphic function f in $\mathbb{B}_n$ belongs to H^p if and only if

$$\int_{\mathbb{B}_n} |\tilde{\nabla} f(z)|^2 |f(z)|^{p-2} G(z) \, d\tau(z) < \infty, \qquad (4.27)$$

which is clearly equivalent to

$$\int_{\mathbb{B}_n} |\tilde{\nabla} f(z)|^2 |f(z)|^{p-2} \frac{dv(z)}{1 - |z|^2} < \infty. \qquad (4.28)$$

Also notice that if we replace f by $f - f(0)$ in Theorem 4.23, then the identity (4.25) becomes

$$\|f - f(0)\|_p^p = \left(\frac{p}{2}\right)^2 \int_{\mathbb{B}_n} |\widetilde{\nabla} f(z)|^2 |f(z) - f(0)|^{p-2} G(z)\, d\tau(z) \qquad (4.29)$$

for all holomorphic functions f in $\mathbb{B}_n$.

The following result is the generalization to $\mathbb{B}_n$ of the classical Hardy-Littlewood maximal theorem.

Theorem 4.24. *For every $\alpha > 1$ there exists a constant $C = C(\alpha) > 0$ such that*

$$\int_{\mathbb{S}_n} |M_\alpha f|^p \, d\sigma \le C\|f\|_p^p$$

for all $p > 0$ and $f \in H^p$.

Proof. Fix some $p \in (0, \infty)$ and $f \in H^p$.

For $g = |f|^{p/2}$ we can apply Corollary 4.5 to obtain

$$g_r(z) \le \int_{\mathbb{S}_n} P(z, \zeta) g_r(\zeta)\, d\sigma(\zeta), \qquad z \in \mathbb{B}_n, \qquad (4.30)$$

where $g_r(z) = g(rz)$ for $r \in (0, 1)$ and $z \in \overline{\mathbb{B}}_n$. Since $\{g_r\}$ is a bounded set in $L^2(\mathbb{S}_n, d\sigma)$, Alaoglu's theorem tells us that there exists a sequence $\{r_k\} \subset (0, 1)$, increasing to 1, such that $\{g_{r_k}\}$ weakly converges to some h in $L^2(\mathbb{S}_n, d\sigma)$. Furthermore, Fatou's lemma gives

$$\int_{\mathbb{S}_n} |h|^2 \, d\sigma \le \liminf_{k \to \infty} \int_{\mathbb{S}_n} |g_{r_k}|^2 \, d\sigma = \|f\|_p^p.$$

For any fixed $z \in \mathbb{B}_n$, the function $\zeta \mapsto P(z, \zeta)$ is in $L^2(\mathbb{S}_n, d\sigma)$, so we can replace r by r_k in (4.30) and let $k \to \infty$. The result is

$$g(z) \le \int_{\mathbb{S}_n} P(z, \zeta) h(\zeta)\, d\sigma(\zeta), \qquad z \in \mathbb{B}_n.$$

Since $|M_\alpha f|^p = |M_\alpha g|^2$, we have

$$|M_\alpha f|^p \le |M_\alpha P[h]|^2$$

on $\mathbb{S}_n$. By Corollary 4.11, there exists a constant $C > 0$, independent of p and f, such that

$$\int_{\mathbb{S}_n} |M_\alpha f|^p \, d\sigma \le \int_{\mathbb{S}_n} |M_\alpha P[h]|^2 \, d\sigma \le C \int_{\mathbb{S}_n} |h|^2 \, d\sigma = C\|f\|_p^p.$$

This completes the proof of the theorem. $\qquad\qquad\square$

We now show that functions in Hardy spaces have K-limits almost everywhere on $\mathbb{S}_n$, and each H^p can naturally be identified with a closed subspace of $L^p(\mathbb{S}_n, d\sigma)$.

Theorem 4.25. *Suppose $f \in H^p$ with $0 < p < \infty$. Then the limit*

$$f^*(\zeta) = \mathrm{Klim} f(\zeta)$$

exists for almost all $\zeta \in \mathbb{S}_n$. Moreover,

$$\|f\|_p^p = \int_{\mathbb{S}_n} |f^*(\zeta)|^p \, d\sigma(\zeta),$$

and

$$\lim_{r \to 1^-} \int_{\mathbb{S}_n} |f(r\zeta) - f^*(\zeta)|^p \, d\sigma(\zeta) = 0.$$

Proof. We prove the case $p \geq 1$ here. The case $p < 1$ is more involved and less complex analytic; we refer the reader to [94] for a full proof.

If $1 < p < \infty$ and $f \in H^p$, then the set $\{f_r : 0 < r < 1\}$ is bounded in $L^p(\mathbb{S}_n, d\sigma)$, where $f_r(\zeta) = f(r\zeta)$ for $\zeta \in \mathbb{B}_n$ and $0 < r < 1$. By Alaoglu's theorem, there exists a sequence r_k that increases to 1 such that the sequence

$$f_k(\zeta) = f_{r_k}(\zeta), \qquad \zeta \in \mathbb{S}_n,$$

weakly converges in $L^p(\mathbb{S}_n, d\sigma)$ to some $g \in L^p(\mathbb{S}_n, d\sigma)$. For each fixed $z \in \mathbb{B}_n$, the function

$$\zeta \mapsto P(z, \zeta)$$

is in $L^q(\mathbb{S}_n, d\sigma)$, where $1/p + 1/q = 1$. Therefore,

$$\lim_{k \to \infty} P[f_k](z) = P[g](z)$$

for every $z \in \mathbb{B}_n$. Since $P[f_k](z) = f(r_k z)$, we obtain $f(z) = P[g](z)$ for every $z \in \mathbb{B}_n$. Furthermore, Theorem 4.15 tells us that

$$\mathrm{Klim} f(\zeta) = g(\zeta)$$

for almost all $\zeta \in \mathbb{S}_n$.

If $p = 1$, the above argument can be modified to produce a finite complex Borel measure μ such that $f = P[\mu]$. According to Theorem 4.15, we still have

$$\mathrm{Klim} f(\zeta) = g(\zeta)$$

for almost every $\zeta \in \mathbb{S}_n$, where $g \, d\sigma$ is the absolutely continuous part of μ.

Since each region $D_\alpha(\zeta)$ contains $r\zeta$ for r sufficiently close to 1^-, we have

$$\lim_{r \to 1^-} f_r(\zeta) = g(\zeta)$$

for almost every $\zeta \in \mathbb{S}_n$. Fatou's lemma then gives $\|g\|_p \leq \|f\|_p$.

It is easy to see that $|f_r| \leq M_\alpha f$ for r sufficiently close to 1. According to Theorem 4.24, $M_\alpha f \in L^p(\mathbb{S}_n, d\sigma)$. So we can apply the dominated convergence theorem to obtain

$$\lim_{r \to 1^-} \int_{\mathbb{S}_n} |f(r\zeta) - g(\zeta)|^p \, d\sigma(\zeta) = 0.$$

This completes the proof of the theorem for $1 \leq p \leq \infty$. $\qquad \square$

Let $H^p(\mathbb{S}_n)$ denote the space of all functions f^*, where $f \in H^p$. Since H^p is complete, and the mapping $f \mapsto f^*$ is an isometry from H^p into $L^p(\mathbb{S}_n, d\sigma)$, we conclude that $H^p(\mathbb{S}_n)$ is a closed subspace of $L^p(\mathbb{S}_n, d\sigma)$.

Corollary 4.26. *Suppose $0 < p < \infty$ and $f \in H^p$. Then*

$$\lim_{r \to 1^-} \|f_r - f\|_p = 0.$$

Also, the set of polynomials is dense in each H^p.

Proof. The first assertion follows from Theorem 4.25. The second assertion follows from approximating each f_r uniformly by its Taylor polynomials.   □

Recall that the ball algebra $A(\mathbb{B}_n)$ consists of holomorphic functions f in $\mathbb{B}_n$ that are continuous up to the boundary. Let $A(\mathbb{S}_n)$ denote the space of functions that are restrictions of functions in $A(\mathbb{B}_n)$ to the sphere $\mathbb{S}_n$. By the maximum principle, the space $A(\mathbb{S}_n)$ is a closed subspace of $C(\mathbb{S}_n)$ with the sup-norm.

Since each function in H^p can be approximated by functions in the ball algebra, we see that the space $H^p(\mathbb{S}_n)$ is the closure of $A(\mathbb{S}_n)$ in $L^p(\mathbb{S}_n, d\sigma)$.

Corollary 4.27. *Suppose $p \geq 1$ and $f \in H^p$. If f^* is the boundary function of f, then $f = P[f^*] = C[f^*]$.*

Proof. For $r \in (0, 1)$ consider the dilation $f_r(z) = f(rz)$, $z \in \overline{\mathbb{B}}_n$. Then by Propositions 4.1 and 4.2, we have $f_r = P[f_r] = C[f_r]$ for all $r \in (0, 1)$. The kernels $C(z, \zeta)$ and $P(z, \zeta)$ are bounded in ζ for any fixed $z \in \mathbb{B}_n$. Since

$$\lim_{r \to 1^-} \int_{\mathbb{S}_n} |f_r(\zeta) - f^*(\zeta)|^p \, d\sigma(\zeta) = 0,$$

we can let $r \to 1$ and take the limit inside the Poisson and Cauchy integrals to obtain $f = P[f^*] = C[f^*]$.   □

From now on, we are going to use the same symbol f to denote a function in H^p and its boundary function in $L^p(\mathbb{S}_n)$.

4.3 The Cauchy-Szegö Projection

Since H^2 can be identified with a closed subspace of $L^2(\mathbb{S}_n, d\sigma)$, there exists an orthogonal projection from $L^2(\mathbb{S}_n, d\sigma)$ onto H^2. We denote this projection by C and call it the Cauchy-Szegö projection.

As a Hilbert space of holomorphic functions in $\mathbb{B}_n$, H^2 has every point evaluation in $\mathbb{B}_n$ as a bounded linear functional; see Theorem 4.17. Therefore, H^2 possesses a reproducing kernel, that is, for every $z \in \mathbb{B}_n$, there exists a function $K_z \in H^2$ such that $f(z) = \langle f, K_z \rangle$ for all $f \in H^2$, where $\langle \, , \, \rangle$ is the inner product in H^2.

Proposition 4.28. *The Cauchy-Szegö kernel is the reproducing kernel of H^2 and the Cauchy-Szegö projection $C : L^2(\mathbb{S}_n, d\sigma) \to H^2$ is simply the Cauchy transform, that is,*

$$(Cf)(z) = \int_{\mathbb{S}_n} C(z, \zeta) f(\zeta)\, d\sigma(\zeta) = \int_{\mathbb{S}_n} \frac{f(\zeta)\, d\sigma(\zeta)}{(1 - \langle z, \zeta \rangle)^n}, \qquad (4.31)$$

where $f \in L^2(\mathbb{S}_n, d\sigma)$ and $z \in \mathbb{B}_n$.

Proof. Recall from Proposition 4.1 that the Cauchy-Szegö kernel reproduces functions in the ball algebra. Since the ball algebra is dense in each H^p, the Cauchy-Szegö kernel also reproduces functions in H^1. In particular, the Cauchy-Szegö kernel reproduces functions in H^2. By the uniqueness of reproducing kernels, the Cauchy-Szegö kernel is the reproducing kernel of H^2.

For each $z \in \mathbb{B}_n$ let

$$h_z(\zeta) = \overline{C(z, \zeta)}, \qquad \zeta \in \mathbb{S}_n.$$

Then $h_z \in H^2$, and for $f \in L^2(\mathbb{S}_n, d\sigma)$, we have

$$(Cf)(z) = \langle Cf, h_z \rangle = \langle f, Ch_z \rangle = \langle f, h_z \rangle,$$

where $\langle\, ,\, \rangle$ is the inner product in $L^2(\mathbb{S}_n, d\sigma)$. This shows that the Cauchy-Szegö projection C is given by

$$(Cf)(z) = \int_{\mathbb{S}_n} C(z, \zeta) f(\zeta)\, d\sigma(\zeta).$$

$\square$

Our goal in this section is to show that the Cauchy transform maps $L^p(\mathbb{S}_n, d\sigma)$ boundedly onto the Hardy space H^p when $1 < p < \infty$. This will be done via the maximal operator M_α introduced in Section 4.1.

Recall from Section 4.1 that d denotes a certain nonisotropic metric on $\mathbb{S}_n$ and $D_\alpha(\zeta)$ is a certain approach region in $\mathbb{B}_n$ for each $\zeta \in \mathbb{S}_n$ and $\alpha > 1$. We begin with a few technical estimates of the Cauchy-Szegö kernel.

Lemma 4.29. *Suppose ζ, η, and ω are points on $\mathbb{S}_n$ satisfying*

$$d(\omega, \eta) < \delta, \qquad d(\omega, \zeta) > 2\delta,$$

where δ is some positive number. Then

$$|C(z, \eta) - C(z, \omega)| < (16\alpha)^{n+1} \delta |1 - \langle \zeta, \omega \rangle|^{-(n+1/2)}$$

for all $z \in D_\alpha(\zeta)$.

Proof. Write $z = z_1 + z_2$ and $\eta = \eta_1 + \eta_2$, where z_1 and η_1 are parallel to ω, while z_2 and η_2 are perpendicular to ω. Then

$$\langle z, \eta \rangle - \langle z, \omega \rangle = \langle z_2, \eta_2 \rangle + \langle z_1, \eta_1 - \omega \rangle,$$

and so

$$|\langle z, \eta \rangle - \langle z, \omega \rangle| \le |z_2||\eta_2| + |\eta_1 - \omega|.$$

Since

$$|z_2|^2 = |z|^2 - |z_1|^2 < (1 + |z_1|)(1 - |z_1|) \le 2|1 - \langle z_1, \omega \rangle| = 2|1 - \langle z, \omega \rangle|,$$

and similarly,

$$|\eta_2|^2 \le 2|1 - \langle \eta, \omega \rangle|,$$

we have

$$|\langle z, \eta \rangle - \langle z, \omega \rangle| \le 2|1 - \langle z, \omega \rangle|^{\frac{1}{2}}|1 - \langle \eta, \omega \rangle|^{\frac{1}{2}} + |1 - \langle \eta, \omega \rangle|.$$

From the assumption $d(\eta, \omega) < \delta$ we then deduce that

$$|\langle z, \eta \rangle - \langle z, \omega \rangle| \le \left(2|1 - \langle z, \omega \rangle|^{\frac{1}{2}} + |1 - \langle \eta, \omega \rangle|^{\frac{1}{2}} \right) \delta. \tag{4.32}$$

By the proof of Theorem 4.10, we have

$$|1 - \langle \zeta, \omega \rangle| \le 4\alpha|1 - \langle z, \omega \rangle|, \quad |1 - \langle \zeta, \eta \rangle| \le 4\alpha|1 - \langle z, \eta \rangle|. \tag{4.33}$$

On the other hand, the triangle inequality gives

$$|1 - \langle \zeta, \eta \rangle|^{1/2} \ge |1 - \langle \zeta, \omega \rangle|^{1/2} - |1 - \langle \omega, \eta \rangle|^{1/2} \ge 2\delta - \delta = \delta,$$

so by the triangle inequality again,

$$|1 - \langle \zeta, \omega \rangle|^{1/2} \le |1 - \langle \zeta, \eta \rangle|^{1/2} + |1 - \langle \eta, \omega \rangle|^{1/2}$$
$$\le |1 - \langle \zeta, \eta \rangle|^{1/2} + \delta < 2|1 - \langle \zeta, \eta \rangle|^{1/2}.$$

It follows that

$$|1 - \langle \zeta, \omega \rangle| \le 4|1 - \langle \zeta, \eta \rangle| \le 16\alpha|1 - \langle z, \eta \rangle|. \tag{4.34}$$

Also,

$$|1 - \langle \eta, \omega \rangle| < \delta^2 < \frac{1}{4}|1 - \langle \zeta, \omega \rangle| < \alpha|1 - \langle z, \omega \rangle|.$$

Combining this with (4.32), we obtain

$$|\langle z, \eta \rangle - \langle z, \omega \rangle| \le \delta(2 + \sqrt{\alpha})|1 - \langle z, \omega \rangle|^{\frac{1}{2}} \le 3\delta\sqrt{\alpha}|1 - \langle z, \omega \rangle|^{\frac{1}{2}}. \tag{4.35}$$

It is clear that

$$C(z, \eta) - C(z, \omega) = \sum_{k=0}^{n-1} \frac{\langle z, \eta \rangle - \langle z, \omega \rangle}{(1 - \langle z, \eta \rangle)^{k+1}(1 - \langle z, \omega \rangle)^{n-k}}. \tag{4.36}$$

Applying (4.35) first, then applying (4.33) and (4.34), we obtain

$$|C(z,\eta) - C(z,\omega)| \le \frac{3\delta\sqrt{\alpha} \sum_{k=0}^{n-1} (16\alpha)^{k+1}(4\alpha)^{n-k-\frac{1}{2}}}{|1 - \langle \zeta, \omega \rangle|^{n+\frac{1}{2}}}.$$

An elementary estimate of the above series then completes the proof of the lemma. $\square$

Lemma 4.30. *There exists a constant $C > 0$ such that*

$$\int_{d(\eta,\zeta)>2r} \frac{d\sigma(\eta)}{|1 - \langle \zeta, \eta \rangle|^{n+1/2}} \le \frac{C}{r}$$

for all $\zeta \in \mathbb{S}_n$ and $r > 0$.

Proof. Let $I(r)$ denote the concerned integral.

If $n = 1$, it is easy to see that

$$I(r) = 2 \int_{4r^2}^{\pi} |1 - e^{i\theta}|^{-3/2}\, d\theta < \left(\frac{\pi}{2}\right)^{3/2} \frac{2}{r}.$$

If $n > 1$, we can apply (1.13) to obtain

$$I(r) = (n-1) \int_{E(r)} \frac{(1 - |z|^2)^{n-2}}{|1 - z|^{n+\frac{1}{2}}}\, dA(z),$$

where

$$E(r) = \{z \in \mathbb{C} : |z| < 1, |1 - z| > 4r^2\}$$

is a subset of the unit disk in the complex plane. Estimate the numerator of the integrand of $I(r)$ by

$$1 - |z|^2 = (1 + |z|)(1 - |z|) \le 2|1 - z|,$$

and then evaluate the resulting integral via $1 - z = te^{i\theta}$. We get

$$I(r) \le 2^{n-2}(n-1) \int_{4r^2}^{\infty} t^{-3/2}\, dt = \frac{2^{n-2}(n-1)}{r}.$$

This completes the proof of the lemma. $\square$

Corollary 4.31. *For $\zeta \in \mathbb{S}_n$, $\omega \in \mathbb{S}_n$, $\alpha > 1$, and $\delta > 0$ let*

$$\Delta(\zeta, \omega, \alpha, \delta) = \sup\{|C(z,\eta) - C(z,\omega)| : z \in D_\alpha(\zeta), d(\eta, \omega) < \delta\}.$$

Then there exists a positive constant $C = C(\alpha)$ (independent of ω and δ) such that

$$\int_{d(\zeta,\omega)>2\delta} \Delta(\zeta, \omega, \alpha, \delta)\, d\sigma(\zeta) \le C$$

for all $\omega \in \mathbb{S}_n$ and $\delta > 0$.

Proof. This follows directly from Lemmas 4.29 and 4.30.   □

The following covering lemma will be important for our next maximal theorem, which in turn gives the boundedness of the Cauchy-Szegö projection on $L^p(\mathbb{S}_n, d\sigma)$ for $p > 1$.

Lemma 4.32. *Suppose μ is a finite complex Borel measure on $\mathbb{S}_n$ and $t > \|\mu\|$. Then there exists a constant $C > 0$, d-balls Q_k, and disjoint Borel sets $V_k \subset Q_k$, such that*

(a) $\{M\mu > t\} \subset \bigcup Q_k = \bigcup V_k$.
(b) $\sigma(Q_k) \leq Ct^{-1}|\mu|(Q_k)$.
(c) $\sum \sigma(Q_k) \leq Ct^{-1}\|\mu\|$.
(d) $|\mu|(V_k) < Ct\sigma(V_k)$.

Proof. Without loss of generality, we may assume that μ is positive. Let

$$C = \sup\left\{\frac{\sigma(Q(\zeta, 4r))}{\sigma(Q(\zeta, r))} : r > 0, \zeta \in \mathbb{S}_n\right\}.$$

By Lemma 4.6, C is a finite positive constant.

For any $t > \|\mu\|$ we write $E_t = \{M\mu > t\}$. Each point $\zeta \in E_t$ gives rise to a largest r such that

$$\mu(Q(\zeta, r)) \geq t\sigma(Q(\zeta, r)). \tag{4.37}$$

Since $t > \|\mu\|$, we have $Q(\zeta, r) \neq \mathbb{S}_n$ and so

$$\mu(Q(\zeta, 4r)) < t\sigma(Q(\zeta, 4r)). \tag{4.38}$$

Thus E_t is covered by a collection Γ_1 of balls $Q(\zeta, r)$ that satisfy (4.37) and (4.38).

Let R_1 be the supremum of the radii of the members of Γ_1, and choose a ball $Q(\zeta_1, r_1)$ from Γ_1 such that $r_1 > 3R_1/4$. Discard all members of Γ_1 that intersect $Q(\zeta_1, r_1)$ and call the remaining collection Γ_2.

Let R_2 be the supremum of the radii of the members of Γ_2 and choose a ball $Q(\zeta_2, r_2)$ from Γ_2 such that $r_2 > 3R_2/4$. Discard all members of Γ_2 that intersect $Q(\zeta_2, r_2)$ and call the remaining collection Γ_3.

We continue this process infinitely or until some Γ_k becomes empty. The result is a sequence $\{Q(\zeta_k, r_k)\}$ of disjoint balls satisfying (4.37) and (4.38). For each k we let $Q_k = Q(\zeta_k, 4r_k)$, and choose a set V_k between $Q(\zeta_k, r_k)$ and Q_k such that $\cup V_k = \cup Q_k$ and $V_l \cap V_k = \emptyset$ for $k \neq l$.

If some $Q \in \Gamma_1$ was discarded at the k-th stage, then Q intersects $Q(\zeta_k, r_k)$ and the radius of Q, $r(Q)$, is less than $4r_k/3$. Since $1+4/3+4/3 < 4$, we have $Q \subset Q_k$. Thus $E_t \subset \cup Q_k$ and (a) is proved.

Next, by (4.37), we have

$$\sigma(Q_k) \leq C\sigma(Q(\zeta_k, r_k)) \leq Ct^{-1}\mu(Q(\zeta_k, r_k)).$$

This proves (b), and (c) follows from adding the inequalities in (b) and the fact that $\{Q_k\}$ are disjoint.

Finally, it follows from (4.38) that

$$\mu(V_k) \leq \mu(Q_k) < t\sigma(Q_k) \leq Ct\sigma(Q(\zeta_k, r_k)),$$

which proves (d).   □

Before we prove the general maximal theorem for Cauchy transforms of functions in $L^p(\mathbb{S}_n, d\sigma)$, we establish the following special case first.

Lemma 4.33. *For any* $\alpha > 1$ *there exists a constant* $C > 0$ *such that*

$$\int_{\mathbb{S}_n} |M_\alpha C[f]|^2 \, d\sigma \leq C \int_{\mathbb{S}_n} |f|^2 \, d\sigma$$

for all $f \in L^2(\mathbb{S}_n, d\sigma)$.

Proof. Given $f \in L^2(\mathbb{S}_n, d\sigma)$, we write $f = g + h$, where $g \in H^2(\mathbb{S}_n)$ and $h \perp H^2(\mathbb{S}_n)$. For any fixed $z \in \mathbb{B}_n$, the function $\zeta \mapsto \overline{C(z, \zeta)}$ is in $H^2(\mathbb{S}_n)$. It follows that $C[h] = 0$, so $C[f] = C[g] = g$ (see Corollary 4.27). By Theorem 4.24,

$$\int_{\mathbb{S}_n} |M_\alpha C[f]|^p \, d\sigma = \int_{\mathbb{S}_n} |M_\alpha g|^2 \, d\sigma \leq C \int_{\mathbb{S}_n} |g|^2 \, d\sigma \leq C \int_{\mathbb{S}_n} |f|^2 \, d\sigma.$$

    □

We are now ready to prove the crucial weak L^1 estimate for the Cauchy transform.

Theorem 4.34. *For every* $\alpha > 1$ *there exists a constant* $C > 0$ *such that*

$$\sigma\{M_\alpha C[\mu] > t\} \leq \frac{C\|\mu\|}{t} \tag{4.39}$$

for all finite complex Borel measures μ *on* $\mathbb{S}_n$ *and all* $t > 0$.

Proof. We fix a finite complex measure μ on $\mathbb{S}_n$ and fix a positive number t. We may assume that $t > \|\mu\|$, because (4.39) clearly holds with $C = 1$ if $t \leq \|\mu\|$. Let $E_t = \{M\mu > t\}$.

Choose $\{Q_k\}$ and $\{V_k\}$ in accordance with Lemma 4.32. For each k let $c_k = \mu(V_k)/\sigma(V_k)$ and define a measure β_k by

$$\beta_k(E) = (\mu - c_k\sigma)(V_k \cap E).$$

Let $\beta = \sum_k \beta_k$.

Let $d\mu = f \, d\sigma + d\mu_s$ be the Lebesgue decomposition of μ into its absolutely continuous and singular parts with respect to σ. Let g be the function on $\mathbb{S}_n$ that takes value c_k on each V_k and equals f on $\mathbb{S}_n - \cup V_k$. By considering the cases

$E \subset V_k$ and $E \cap (\cup V_k) = \emptyset$ separately, with $\mu_s(E) = 0$ in the latter case as μ_s is concentrated on $E_t \subset \cup V_k$, we easily check that

$$\mu(E) = \int_E g \, d\sigma + \beta(E)$$

for every Borel set $E \subset \mathbb{S}_n$. Therefore, we arrive at a new decomposition

$$d\mu = g \, d\sigma + d\beta,$$

where g is good in the sense that it is not too large and β captures the bad part of μ.

We now estimate the maximal function of the Cauchy transform of g and β, respectively.

Since $Mf \leq M\mu$, Corollary 4.13 implies that $|f(\zeta)| \leq t$ almost everywhere outside E_t, which implies that $|g|^2 \leq t|f|$ outside E_t, so

$$\int_{\mathbb{S}_n - E_t} |g|^2 \, d\sigma \leq t \int_{\mathbb{S}_n} |f| \, d\sigma \leq t\|\mu\|.$$

By part (d) of Lemma 4.32, $|c_k| \leq At$ for each k, where

$$A = \sup \left\{ \frac{\sigma(Q(\zeta, 4r))}{\sigma(Q(\zeta, r))} : \zeta \in \mathbb{S}_n, r > 0 \right\}.$$

Part (c) of Lemma 4.32 then implies that

$$\sum_k \int_{V_k} |g|^2 \, d\sigma = \sum_k |c_k|^2 \sigma(V_k) \leq A^2 t^2 \sum_k \sigma(Q_k) \leq A^3 \|\mu\| t.$$

Therefore,

$$\int_{\mathbb{S}_n} |g|^2 \, d\sigma \leq (1 + A^3)\|\mu\| t.$$

By Lemma 4.33, there exists a constant $C_1 > 0$, independent of μ and t, such that

$$\int_{\mathbb{S}_n} (M_\alpha C[g])^2 \, d\sigma \leq C_1 \|\mu\| t.$$

Since

$$\sigma\{M_\alpha C[g] > t\} \leq \frac{1}{t^2} \int_{\mathbb{S}_n} (M_\alpha C[g])^2 \, d\sigma,$$

we conclude that

$$\sigma \left\{ M_\alpha C[g] > \frac{t}{2} \right\} \leq \frac{C_1 \|\mu\|}{t}, \tag{4.40}$$

which is the desired estimate for the Cauchy transform of g.

To deal with the bad part β, we let

$$\Omega = \left\{ M_\alpha C[\beta] > \frac{t}{2} \right\}, \qquad W = \bigcup_k Q'_k,$$

where $Q'_k = Q(\omega_k, 2\delta_k)$ if we write $Q_k = Q(\omega_k, \delta_k)$. Clearly,

$$\Omega \subset W \cup (\Omega - W) \subset W \cup \left(\cup_k (\Omega - Q'_k)\right).$$

By part (c) of Lemma 4.32,

$$\sigma(W) \le \sum_k \sigma(Q'_k) \le A \sum_k \sigma(Q_k) \le \frac{A^2 \|\mu\|}{t}.$$

Since each β_k is concentrated on $V_k \subset Q_k$, we have

$$C[\beta_k](z) = \int_{Q_k} (C(z, \eta) - C(z, \omega_k)) \, d\beta_k(\eta).$$

Using the notation introduced in Lemma 4.31, we have

$$|C[\beta_k](z)| \le \Delta(\zeta, \omega_k, \alpha, \delta_k) \|\beta_k\|$$

for all $k \ge 1$, $\zeta \in \mathbb{S}_n$, and $z \in D_\alpha(\zeta)$. Taking the supremum over $z \in D_\alpha(\zeta)$, we obtain

$$M_\alpha C[\beta_k](\zeta) \le \Delta(\zeta, \omega_k, \alpha, \delta_k) \|\beta_k\|$$

for $k \ge 1$ and $\zeta \in \mathbb{S}_n$. By Lemma 4.31, there exists a constant $C_2 > 0$ such that

$$\int_{\mathbb{S}_n - Q'_k} M_\alpha C[\beta_k] \, d\sigma \le C_2 \|\beta_k\|$$

for all $k \ge 1$. Since $\Omega - W \subset \mathbb{S}_n - Q'_k$, we can add the above inequalities and obtain

$$\int_{\Omega - W} M_\alpha C[\beta] \, d\sigma \le C_2 \sum_k \|\beta_k\|.$$

Recall that $c_k \sigma(V_k) = \mu(V_k)$, so $\|\beta_k\| \le 2|\mu|(V_k)$ for each k. Thus

$$\sum_k \|\beta_k\| \le 2\|\mu\|.$$

It follows that

$$\int_{\Omega - W} M_\alpha C[\beta] \, d\sigma \le 2C_2 \|\mu\|.$$

The above integrand exceeds $t/2$ at every point of Ω, so

$$\sigma(\Omega - W) \le \frac{4C_2 \|\mu\|}{t}.$$

Combining this with an earlier estimate for $\sigma(W)$, we find a positive constant C_3 such that

$$\sigma \left\{ M_\alpha C[\beta] > \frac{t}{2} \right\} \le \frac{C_3 \|\mu\|}{t},$$

which is the desired estimate for the Cauchy transform of β.

It is easy to see that

$$M_\alpha C[\mu] \leq M_\alpha C[g] + M_\alpha C[\beta].$$

The desired estimate (4.39) for μ now follows from our individual estimates for g and β. $\qquad\square$

As a consequence of the above weak L^1 estimate, we now derive the Koranyi-Vagi maximal theorem for the Cauchy transform.

Theorem 4.35. *For any $\alpha > 1$ and $1 < p < \infty$ there exists a constant $C > 0$ such that*

$$\int_{\mathbb{S}_n} |M_\alpha C[f]|^p \, d\sigma \leq C \int_{\mathbb{S}_n} |f|^p \, d\sigma$$

for all $f \in L^p(\mathbb{S}_n, d\sigma)$.

Proof. Since the mapping $f \mapsto M_\alpha C[f]$ is subadditive, the case $1 < p \leq 2$ follows from Lemma 4.33, Theorem 4.34, and the Marcinkiewicz interpolation theorem. Also, since

$$|C[f](r\zeta)| \leq M_\alpha C[f](\zeta)$$

for all $\zeta \in \mathbb{S}_n$, where $r \in (0,1)$ is sufficiently close to 1 (depending on α), the Cauchy transform maps $L^p(\mathbb{S}_n)$ boundedly onto H^p when $1 < p \leq 2$.

Suppose $2 < p < \infty$ with $1/p + 1/q = 1$. Let X be the vector space consisting of all finite linear combinations of functions of the form $\zeta^m \overline{\zeta}^{m'}$, where m and m' are multi-indexes of nonnegative integers. If f and g are functions in X, then they are bounded, and their Cauchy transforms are polynomials. By the already proved L^2 case,

$$\int_{\mathbb{S}_n} g(\zeta)\overline{C[f](\zeta)} \, d\sigma(\zeta) = \int_{\mathbb{S}_n} C[g](\zeta)\overline{f(\zeta)} \, d\sigma(\zeta).$$

Since $1 < q < 2$, we have

$$\int_{\mathbb{S}_n} |C[g](\zeta)|^q \, d\sigma(\zeta) \leq C \int_{\mathbb{S}_n} |g(\zeta)|^q \, d\sigma(\zeta),$$

and so an application of Hölder's inequality gives

$$\left| \int_{\mathbb{S}_n} g(\zeta)\overline{C[f](\zeta)} \, d\sigma(\zeta) \right| \leq C' \|g\|_q \|f\|_p.$$

Since X is dense in $L^q(\mathbb{S}_n, d\sigma)$, we deduce from the standard L^p duality theory that

$$\|C[f]\|_p \leq C' \|f\|_p.$$

Since X is also dense in $L^p(\mathbb{S}_n, d\sigma)$, the Cauchy transform maps $L^p(\mathbb{S}_n, d\sigma)$ boundedly into H^p. Now an application of the Hardy-Littlewood maximal theorem (Theorem 4.24) completes the proof of the theorem. $\qquad\square$

The proof of Theorem 4.35 also gives the following.

Theorem 4.36. *If* $1 < p < \infty$, *the Cauchy transform* C *maps* $L^p(\mathbb{S}_n, d\sigma)$ *boundedly onto* H^p.

It is well known that the Cauchy transform C is unbounded on $L^1(\mathbb{S}_n, d\sigma)$. In fact, the boundedness of C on $L^1(\mathbb{S}_n, d\sigma)$ would imply the boundedness of C on $L^\infty(\mathbb{S}_n)$ by duality. To show that this is not possible, consider the bounded function

$$f(\zeta) = 2i \arg(1 - \zeta_1) = \log(1 - \zeta_1) - \overline{\log(1 - \zeta_1)}$$

on $\mathbb{S}_n$, a calculation shows that

$$C[f](z) = \log(1 - z_1),$$

which is clearly unbounded.

The following corollary is now almost obvious and will be used many times later in the book, sometimes without even being mentioned, because it is so natural.

Corollary 4.37. *If* $1 < p < \infty$ *and* $1/p + 1/q = 1$, *then*

$$\int_{\mathbb{S}_n} C(f)\, \overline{g}\, d\sigma = \int_{\mathbb{S}_n} f\, \overline{C(g)}\, d\sigma$$

holds for all $f \in L^p(\mathbb{S}_n, d\sigma)$ *and* $g \in L^q(\mathbb{S}_n, d\sigma)$.

Proof. The desired result clearly holds for f and g in $\mathbb{C}(\mathbb{S}_n)$, because C is an orthogonal projection on $L^2(\mathbb{S}_n, d\sigma)$. The general case then follows from approximating f and g by functions in $\mathbb{C}(\mathbb{S}_n)$ and using the boundedness of C on $L^p(\mathbb{S}_n, d\sigma)$ and $L^q(\mathbb{S}_n, d\sigma)$, respectively. $\square$

Another consequence of the boundedness of the Cauchy-Szegö projection on L^p, $1 < p < \infty$, is the following result concerning complex interpolation of Hardy spaces.

Theorem 4.38. *Suppose* $1 \le p_0 < p_1 < \infty$ *and*

$$\frac{1}{p} = \frac{1 - \theta}{p_0} + \frac{\theta}{p_1},$$

where $\theta \in (0, 1)$. *Then* $[H^{p_0}, H^{p_1}]_\theta = H^p$ *with equivalent norms*.

Proof. We prove the case $1 < p_0$. When $p_0 = 1$, the proof is much more involved and makes use of several real variable methods; we refer the reader to [36].

If we identify each H^p with a closed subspace of $L^p(\mathbb{S}_n)$, then the inclusion $[H^{p_0}, H^{p_1}]_\theta \subset H^p$ follows from the complex interpolation of L^p spaces; see Theorem 1.33.

To prove the other inclusion, fix a function $f \in H^p \subset L^p(\mathbb{S}_n)$ and use Theorem 1.33 to choose g_ζ, $0 \le \mathrm{Re}\,\zeta \le 1$, such that g_ζ depends on ζ continuously in $0 \le \mathrm{Re}\,\zeta \le 1$ and analytically on $0 < \mathrm{Re}\,\zeta < 1$, $g_\theta = f$,

$$\int_{\mathbb{S}_n} |g_\zeta|^{p_0} \, d\sigma \le \|f\|_p^p, \qquad \operatorname{Re}\zeta = 0,$$

and

$$\int_{\mathbb{S}_n} |g_\zeta|^{p_1} \, d\sigma \le \|f\|_p^p, \qquad \operatorname{Re}\zeta = 1.$$

Let $f_\zeta = C[g_\zeta]$. Then $f_\theta = f$, $f_\zeta \in H^{p_0}$ for $\operatorname{Re}\zeta = 0$, and $f_\theta \in H^{p_1}$ for $\operatorname{Re}\zeta = 1$, so $f \in [H^{p_0}, H^{p_1}]_\theta$. $\qquad\square$

4.4 Several Embedding Theorems

In this section we prove several embedding theorems for the Hardy spaces. In particular, one of these results will be used in the next section when we characterize bounded linear functionals on H^p when $0 < p < 1$.

We first discuss restriction and extension operators defined on holomorphic functions. Thus for any integer k satisfying $1 \le k < n$ we define an operator

$$R_k : H(\mathbb{B}_n) \to H(\mathbb{B}_k)$$

by

$$R_k(f)(z_1, \cdots, z_k) = f(z_1, \cdots, z_k, 0, \cdots, 0). \tag{4.41}$$

Similarly, we define an operator

$$E_k : H(\mathbb{B}_k) \to H(\mathbb{B}_n)$$

by

$$E_k(f)(z_1, \cdots, z_n) = f(z_1, \cdots, z_k). \tag{4.42}$$

Theorem 4.39. *Suppose $1 \le k < n$ and $0 < p < \infty$. Then the operator R_k maps H^p boundedly onto the weighted Bergman space $A_{n-k-1}^p(\mathbb{B}_k)$ of the unit ball in $\mathbb{C}^k$.*

Proof. Let $f \in H^p = H^p(\mathbb{B}_n)$. For $z \in \mathbb{C}^n$ we write $z = (w, u)$, where $w \in \mathbb{C}^k$ and $u \in \mathbb{C}^{n-k}$. By (1.15),

$$\int_{\mathbb{S}_n} |f|^p \, d\sigma = c \int_{\mathbb{B}_k} (1 - |w|^2)^{n-k-1} \, dv_k(w) \int_{\mathbb{S}_{n-k}} |f(w, \sqrt{1 - |w|^2}\,\zeta)|^p \, d\sigma_{n-k}(\zeta),$$

where

$$c = \frac{(n-1)!}{(k-1)!(n-k)!}.$$

Since

$$\int_{\mathbb{S}_{n-k}} |f(w, \sqrt{1 - |w|^2}\,\zeta)|^p \, d\sigma_{n-k}(\zeta) \ge |f(w, 0)|^p = |R_k f(w)|^p,$$

we see that R_k maps $H^p(\mathbb{B}_n)$ boundedly into $A^p_{n-k-1}(\mathbb{B}_k)$.

To see that R_k maps $H^p(\mathbb{B}_n)$ onto $A^p_{n-k-1}(\mathbb{B}_k)$, let $f \in A^p_{n-k-1}(\mathbb{B}_k)$. Since $f = R_k E_k(f)$, it suffices for us to show that $E_k(f)$ belongs to $H^p(\mathbb{B}_n)$ of $\mathbb{B}_n$. This follows from applying the identity at the beginning of this proof to the function $E_k(f)$; see Lemma 1.9 as well.   □

Corollary 4.40. *If $1 \leq k < n$ and $0 < p < \infty$, then the operator E_k maps the weighted Bergman space $A^p_{n-k-1}(\mathbb{B}_k)$ into $H^p(\mathbb{B}_n)$.*

Proof. This follows from the proof of Theorem 4.39.   □

Theorem 4.41. *Suppose $t > 0$ and α is real. If neither $n + \alpha$ nor $n + \alpha + t$ is a negative integer, then*

$$R^{\alpha,t} H^p \subset A^p_{pt-1}$$

for $2 \leq p < \infty$, and

$$R_{\alpha,t} A^p_{pt-1} \subset H^p$$

for $0 < p \leq 2$. When $p = 2$, the operator $R^{\alpha,t}$ is a bounded, invertible operator from H^2 onto A^2_{2t-1}.

Proof. First assume that $0 < p \leq 1$ and $f \in A^p_{pt-1}$. Let b be a positive number large enough so that Theorem 2.30 (atomic decompositoin) holds for A^p_{pt-1} and so that $pb > pt + n$. If $f \in A^p_{pt-1}$, then by Theorem 2.30, there exists a sequence $\{c_k\}$ in l^p such that

$$f(z) = \sum_k c_k \frac{(1 - |a_k|^2)^{(pb-n-pt)/p}}{(1 - \langle z, a_k \rangle)^b}.$$

For each k we use Lemma 2.18 to find a constant $C_1 > 0$ such that

$$\int_{\mathbb{S}_n} \left| R_{\alpha,t} \frac{1}{(1 - \langle \zeta, a_k \rangle)^b} \right|^p d\sigma(\zeta) \leq C_1 \int_{\mathbb{S}_n} \left| \frac{1}{(1 - \langle \zeta, a_k \rangle)^{b-t}} \right|^p d\sigma(\zeta).$$

Since $0 < p \leq 1$, we obtain

$$\int_{\mathbb{S}_n} |R_{\alpha,t} f(\zeta)|^p d\sigma(\zeta) \leq C_1 \sum_{k=1}^{\infty} |c_k|^p \int_{\mathbb{S}_n} \frac{(1 - |a_k|^2)^{pb-(n+pt)} d\sigma(\zeta)}{|1 - \langle \zeta, a_k \rangle|^{pb-pt}}.$$

According to Theorem 1.12, each of the above integrals is comparable to the constant 1. This shows that $R_{\alpha,t}$ maps A^p_{pt-1} (boundedly) into H^p when $0 < p \leq 1$.

A computation using Taylor expansion shows that $R^{\alpha,t}$ is a bounded and invertible operator from H^2 onto A^2_{2t-1}. This shows that the linear operator T, defined by

$$Tf(z) = (1 - |z|^2)^t R^{\alpha,t} Cf(z), \qquad z \in \mathbb{B}_n,$$

is bounded from $L^2(\mathbb{S}_n, d\sigma)$ to $L^2(\mathbb{B}_n, d\mu)$, where

$$d\mu(z) = \frac{dv(z)}{1 - |z|^2},$$

and C is the Cauchy-Szegö projection. On the other hand, if f is bounded on $\mathbb{S}_n$, then we can differentiate under the integral sign in

$$Cf(z) = \int_{\mathbb{S}_n} \frac{f(\zeta)\, d\sigma(\zeta)}{(1 - \langle z, \zeta \rangle)^n}$$

and use Theorem 1.12 to show that Cf belongs to the Bloch space. Combining this with Theorem 3.5, we see that the operator T maps $L^\infty(\mathbb{S}_n)$ into $L^\infty(\mathbb{B}_n)$. So, by complex interpolation, the operator T maps $L^p(\mathbb{S}_n, d\sigma)$ boundedly into $L^p(\mathbb{B}_n, d\mu)$ for all $p \in [2, \infty)$. In particular, by restricting T to holomorphic functions, we have shown that T maps $H^p(\mathbb{S}_n)$ boundedly into $L^p(\mathbb{B}_n, d\mu)$ for $2 \le p < \infty$. Equivalently, $R^{\alpha, t}$ maps H^p boundedly into A^p_{pt-1} for $2 \le p < \infty$.

Finally, we assume $1 < p < 2$ and $f \in A^p_{pt-1}$. We need to show that $R_{\alpha, t} f \in H^p$. Obviously, we may assume that f vanishes at the origin to a sufficiently large degree so that the function $|f(z)|^p |z|^{2p(\alpha+1)}$ is integrable near the origin.

If g is a unit vector in H^q, where $1/p + 1/q = 1$, then a computation with Taylor series shows that

$$\int_{\mathbb{S}_n} R_{\alpha, t} f\, \overline{g}\, d\sigma = C \int_{\mathbb{B}_n} f(z) \overline{R^{\alpha', t} g(z)} (1 - |z|^2)^{2t-1} |z|^{2(\alpha+1)}\, dv(z),$$

or

$$\int_{\mathbb{S}_n} R_{\alpha, t} f\, \overline{g}\, d\sigma = C \int_{\mathbb{B}_n} (1 - |z|^2)^t f(z) \overline{(1 - |z|^2)^t R^{\alpha', t} g(z)} \frac{|z|^{2(\alpha+1)}\, dv(z)}{1 - |z|^2},$$

where $\alpha' = t + \alpha$ and

$$C = \frac{\Gamma(n + 1 + \alpha + 2t)}{n\Gamma(2t)\Gamma(n + 1 + \alpha)}.$$

Since $1 < p < 2$, we have $2 < q < \infty$, so

$$\|R^{\alpha', t} g\|_{A^q_{qt-1}} \le C' \|g\|_{H^q},$$

where C' is a positive constant independent of g. By the duality of L^p spaces,

$$\|R_{\alpha, t} f\|_{H^p} = \sup \left\{ \left| \int_{\mathbb{S}_n} R_{\alpha, t} f\, \overline{g}\, d\sigma \right| : \int_{\mathbb{S}_n} |g|^q\, d\sigma = 1 \right\}.$$

Since

$$\int_{\mathbb{S}_n} R_{\alpha, t} f\, \overline{g}\, d\sigma = \int_{\mathbb{S}_n} R_{\alpha, t} f\, \overline{C[g]}\, d\sigma,$$

where $C[g]$ is the Cauchy transform of g, and since the Cauchy transform maps $L^q(\mathbb{S}_n)$ boundedly onto H^q, we can find a constant $C > 0$ such that

$$\|R_{\alpha, t} f\|_{H^p} \le C \sup \left\{ \left| \int_{\mathbb{S}_n} R_{\alpha, t} f\, \overline{g}\, d\sigma \right| : \|g\|_{H^q} = 1 \right\}.$$

This shows that $R_{\alpha, t} f$ belongs to H^p. The proof of the theorem is now complete.

$\square$

As a consequence of Theorem 4.41, Theorem 2.30, and Proposition 1.14, we obtain the following partial decomposition theorem for functions in Hardy spaces.

Theorem 4.42. *Suppose $p \in (0, \infty)$ and*

$$b_0 = n \max\left(1, \frac{1}{p}\right).$$

There exists a sequence $\{a_k\}$ in $\mathbb{B}_n$ such that

(a) If $2 \leq p < \infty$ and $b > b_0$, then every function $f \in H^p$ admits an atomic decomposition

$$f(z) = \sum_k c_k \frac{(1 - |a_k|^2)^{b-n/p}}{(1 - \langle z, a_k \rangle)^b},$$

where $\{c_k\} \in l^p$.

(b) If $0 < p \leq 2$ and $b > b_0$, then for every sequence $\{c_k\} \in l^p$ the series

$$f(z) = \sum_k c_k \frac{(1 - |a_k|^2)^{b-n/p}}{(1 - \langle z, a_k \rangle)^b}$$

defines a function in H^p.

Proof. First assume that $2 \leq p < \infty$ and $f \in H^p$. Fix some b with $b > b_0$ and fix some $t > 0$ such that the operators $R_{\alpha,t}$ and $R^{\alpha,t}$ are well defined, where $\alpha = b - (n+1)$. By Theorem 4.41, $R^{\alpha,t} f \in A^p_{pt-1}$, which, according to Theorem 2.30, implies that there exists a sequence $\{c_k\} \in l^p$ such that

$$R^{\alpha,t} f(z) = \sum_k c_k \frac{(1 - |a_k|^2)^{(pb'-n-pt)/p}}{(1 - \langle z, a_k \rangle)^{b'}},$$

where

$$b' = b + t = n + 1 + \alpha + t.$$

It follows from Proposition 1.14 that

$$f(z) = R_{\alpha,t} R^{\alpha,t} f(z) = \sum_k c_k \frac{(1 - |a_k|^2)^{(pb-n)/p}}{(1 - \langle z, a_k \rangle)^b}.$$

Next assume that $0 < p \leq 2$ and

$$f(z) = \sum_k c_k \frac{(1 - |a_k|^2)^{(pb-n)/p}}{(1 - \langle z, a_k \rangle)^b},$$

where $\{c_k\} \in l^p$ and $b > b_0$. Write $b = n + 1 + \alpha$ and choose $t > 0$ such that $R^{\alpha,t}$ and $R_{\alpha,t}$ are well defined. Then by Proposition 1.14,

$$R^{\alpha,t} f(z) = \sum_k c_k \frac{(1 - |a_k|^2)^{[p(b+t)-(n+pt)]/p}}{(1 - \langle z, a_k \rangle)^{b+t}}.$$

By Theorem 2.30, the function $R^{\alpha,t} f$ belongs to A^p_{pt-1}. It follows from Theorem 4.41 that the function $f = R_{\alpha,t} R^{\alpha,t} f$ is in H^p.   □

Our next goal is to obtain an optimal embedding of the Hardy spaces in weighted Bergman spaces without taking any derivatives.

Lemma 4.43. *Suppose p, q, and r are from $(0, \infty]$ and satisfy*

$$\frac{1}{p} + \frac{1}{q} + \frac{1}{r} = 1.$$

If $f \in L^p(X, d\mu)$, $g \in L^q(X, d\mu)$, and $h \in L^r(X, d\mu)$, then the product fgh belongs to $L^1(X, d\mu)$ with

$$\left| \int_X fgh \, d\mu \right| \le \left(\int_X |f|^p \, d\mu \right)^{1/p} \left(\int_X |g|^q \, d\mu \right)^{1/q} \left(\int_X |h|^r \, d\mu \right)^{1/r}.$$

Proof. If any of p, q, and r is infinite, the result is just the classical Hölder's inequality. So we assume that they are all finite.

By Hölder's inequality,

$$\left| \int_X fgh \, d\mu \right| \le \left(\int_X |f|^p \, d\mu \right)^{1/p} \left(\int_X |gh|^{p'} \, d\mu \right)^{1/p'},$$

where $1/p + 1/p' = 1$, or

$$\frac{1}{q/p'} + \frac{1}{r/p'} = 1. \tag{4.43}$$

Using (4.43) and applying Hölder's inequality again to the integral

$$\int_X |g|^{p'} |h|^{p'} \, d\mu,$$

we obtain the desired inequality.   □

Naturally, the above lemma will also be referred to as Hölder's inequality. It should be obvious that it can be generalized to the case of a product of k functions, where k is any positive integer.

Lemma 4.44. *For $f \in L^1(\mathbb{S}_n, d\sigma)$ define*

$$Tf(z) = \int_{\mathbb{S}_n} \frac{f(\zeta) \, d\sigma(\zeta)}{|1 - \langle z, \zeta \rangle|^n}, \qquad z \in \mathbb{B}_n.$$

Then for any $1 \le p < q < \infty$ there exists a positive constant C such that

$$\left(\int_{\mathbb{S}_n} |Tf(r\zeta)|^q \, d\sigma(\zeta) \right)^{1/q} \le C(1 - r^2)^{-n(1/p-1/q)} \|f\|_p$$

for all $f \in L^p(\mathbb{S}_n, d\sigma)$ and $0 < r < 1$.

Proof. Choose $s > 1$ such that

$$1 - \frac{1}{s} = \frac{1}{p} - \frac{1}{q}.$$

Let $a = q$, $b = s/(s-1)$, and $c = p/(p-1)$. Then

$$\frac{1}{a} + \frac{1}{b} = \frac{1}{p}, \quad \frac{1}{a} + \frac{1}{c} = \frac{1}{s}, \quad \frac{1}{a} + \frac{1}{b} + \frac{1}{c} = 1.$$

For the rest of this proof, we write

$$K(z, \zeta) = \frac{1}{(1 - \langle z, \zeta \rangle)^n}.$$

Since

$$|Tf(z)| \le \int_{\mathbb{S}_n} |f(\zeta)||K(z, \zeta)| \, d\sigma(\zeta)$$

$$= \int_{\mathbb{S}_n} \left(|K(z, \zeta)|^{s/a} |f(\zeta)|^{p/a} \right) \left(|f(\zeta)|^{p/b} \right) \left(|K(z, \zeta)|^{s/c} \right) d\sigma(\zeta),$$

an application of Lemma 4.43 shows that $|Tf(z)|$ is less than or equal to the following product:

$$\left[\int_{\mathbb{S}_n} |K(z, \zeta)|^s |f(\zeta)|^p \, d\sigma(\zeta) \right]^{\frac{1}{a}} \left[\int_{\mathbb{S}_n} |f(\zeta)|^p \, d\sigma(\zeta) \right]^{\frac{1}{b}} \left[\int_{\mathbb{S}_n} |K(z, \zeta)|^s \, d\sigma(\zeta) \right]^{\frac{1}{c}}.$$

By Theorem 1.12,

$$\int_{\mathbb{S}_n} |K(z, \zeta)|^s \, d\sigma(\zeta) \le C(1 - |z|^2)^{-n(s-1)}$$

for all $z \in \mathbb{B}_n$, where C is a positive constant, so

$$|Tf(z)|^q \le C(1 - |z|^2)^{-qn(s-1)/c} \|f\|_p^{pq/b} \int_{\mathbb{S}_n} |K(z, \zeta)|^s |f(\zeta)|^p \, d\sigma(\zeta).$$

Write $z = r\eta$, integrate with respect to η, and use Fubini's theorem. We obtain

$$\int_{\mathbb{S}_n} |Tf(r\eta)|^q \, d\sigma(\eta) \le C^2 (1 - r^2)^{-qn(s-1)/c - n(s-1)} \|f\|_p^{p + pq/b},$$

and hence

$$\left(\int_{\mathbb{S}_n} |Tf(r\zeta)|^q \, d\sigma(\zeta) \right)^{1/q} \le C^{2/q} (1 - r^2)^{-n(s-1)(1/c + 1/q)} \|f\|_p^{p(1/b + 1/q)}.$$

This proves the desired estimate. $\square$

Since the Poisson transform $P[f]$ and the Cauchy transform $C[f]$ satisfy

$$|P[f](z)| \le 2^n T|f|(z), \qquad |C[f](z)| \le T|f|(z), \qquad z \in \mathbb{B}_n,$$

we obtain the following corollary as a consequence of Lemma 4.44.

Corollary 4.45. *If $1 \le p < q < \infty$, then there exists a positive constant C such that*

$$\left(\int_{\mathbb{S}_n} |P[f](r\zeta)|^q \, d\sigma(\zeta) \right)^{1/q} \le C(1 - r^2)^{-n(1/p - 1/q)} \|f\|_p$$

and

$$\left(\int_{\mathbb{S}_n} |C[f](r\zeta)|^q \, d\sigma(\zeta) \right)^{1/q} \le C(1 - r^2)^{-n(1/p - 1/q)} \|f\|_p$$

for all $f \in L^p(\mathbb{S}_n, d\sigma)$ and $0 < r < 1$.

The following result significantly strengthens Lemma 4.44.

Theorem 4.46. *Suppose that $1 < p < q < \infty$ and that T is the operator defined in Lemma 4.44. Then there exists a constant $C > 0$ such that*

$$\left(\int_{\mathbb{B}_n} |Tf(z)|^q (1 - |z|^2)^{qn(1/p - 1/q) - 1} \, dv(z) \right)^{1/q} \le C\|f\|_p$$

for all $f \in L^p(\mathbb{S}_n, d\sigma)$.

Proof. For every function $f \in L^1(\mathbb{S}_n, d\sigma)$ we define

$$Sf(r) = (1 - r^2)^{-n/q} \left(\int_{\mathbb{S}_n} |Tf(r\zeta)|^q \, d\sigma(\zeta) \right)^{1/q}, \qquad 0 < r < 1.$$

It is obvious that S is subadditive, that is, $S(f + g) \le Sf + Sg$.

For any $s \in [1, q)$ and any $t > 0$ we apply Lemma 4.44 to obtain

$$\{r \in (0,1) : Sf(r) > t\} \subset \left\{ r \in (0,1) : C_s(1 - r^2)^{-n/s}\|f\|_s > t \right\}$$
$$= \left\{ r \in (0,1) : (1 - r^2)^n < (C_s t^{-1}\|f\|_s)^s \right\},$$

where C_s is a positive constant. If we define

$$d\mu(r) = 2nr(1 - r^2)^{n-1} \, dr = -d(1 - r^2)^n$$

on $(0, 1)$, then

$$\mu(Sf > t) \le \int_{(1 - r^2)^n < (C_s\|f\|_s/t)^s} d[-(1 - r^2)^n] = (C_s t^{-1}\|f\|_s)^s$$

for all $f \in L^p(\mathbb{S}_n, d\sigma)$, where the last equality follows from the change of variables $x = (1 - r^2)^n$ and the fundamental theorem of calculus.

In particular, we can find a constant $C > 0$, dependent on p and q, such that

$$\mu(Sf > t) \le \frac{C}{t}\|f\|_1$$

and

$$\mu(Sf > t) \le \left(\frac{C\|f\|_s}{t}\right)^s$$

for all $t > 0$, where $s = (p+q)/2$. By the Marcinkiewicz interpolation theorem, the operator S is bounded from $L^p(\mathbb{S}_n, d\sigma)$ into $L^p((0,1), d\mu)$, that is,

$$\int_0^1 (1-r)^{n-1-np/q} \, dr \left(\int_{\mathbb{S}_n} |Tf(r\zeta)|^q \, d\sigma(\zeta)\right)^{p/q} \le C' \int_{\mathbb{S}_n} |f|^p \, d\sigma$$

for all $f \in L^p(\mathbb{S}_n, d\sigma)$, where C' is a positive constant depending on p and q. Write

$$\left[\int_{\mathbb{S}_n} |Tf(r\zeta)|^q \, d\sigma(\zeta)\right]^{\frac{p}{q}} = \left[\int_{\mathbb{S}_n} |Tf(r\zeta)|^q \, d\sigma(\zeta)\right]\left[\int_{\mathbb{S}_n} |Tf(r\zeta)|^q \, d\sigma(\zeta)\right]^{\frac{p}{q}-1},$$

and estimate the second factor above by Lemma 4.44. We then obtain

$$\int_0^1 (1-r)^{qn(1/p-1/q)-1} \, dr \int_{\mathbb{S}_n} |Tf(r\zeta)|^q \, d\sigma(\zeta) \le C''\|f\|_p^q.$$

The desired result now follows easily from polar coordinates.   □

Once again, since the operator T dominates the Poisson transform as well as the Cauchy transform, Theorem 4.46 gives rise to the following.

Corollary 4.47. *If* $1 < p < q < \infty$, *there exists a constant* $C > 0$ *such that*

$$\left(\int_{\mathbb{B}_n} |P[f](z)|^q (1-|z|^2)^{qn(1/p-1/q)-1} \, dv(z)\right)^{1/q} \le C\|f\|_p$$

and

$$\left(\int_{\mathbb{B}_n} |C[f](z)|^q (1-|z|^2)^{qn(1/p-1/q)-1} \, dv(z)\right)^{1/q} \le C\|f\|_p$$

for all $f \in L^p(\mathbb{S}_n, d\sigma)$.

We are now ready to prove the following imbedding of Hardy spaces into weighted Bergman spaces.

Theorem 4.48. *Suppose* $0 < p < q < \infty$ *and*

$$\alpha = nq\left(\frac{1}{p} - \frac{1}{q}\right) - 1 = \frac{nq}{p} - (n+1).$$

Then $H^p \subset A_\alpha^q$ *and the inclusion is continuous.*

Proof. For $f \in H^p$ we let $g = |f|^{p/2}$. Then $g \in L^2(\mathbb{S}_n, d\sigma)$ and $g \le P[g]$ on $\mathbb{B}_n$, because g is subharmonic. In Corollary 4.47, if we replace f by g, p by 2, and q by the number $r = 2q/p$, the result is

$$\left(\int_{\mathbb{B}_n} |P[g](z)|^r (1 - |z|^2)^{nr(1/2 - 1/r) - 1} \, dv(z) \right)^{1/r} \le C\|g\|_2,$$

where C is a positive constant independent of g. Since $|g| \le P[g]$, we also have

$$\left(\int_{\mathbb{B}_n} |g(z)|^r (1 - |z|^2)^{nr(1/2 - 1/r) - 1} \, dv(z) \right)^{1/r} \le C\|g\|_2,$$

which is the same as

$$\int_{\mathbb{B}_n} |f(z)|^q (1 - |z|^2)^{nq(1/p - 1/q) - 1} \, dv(z) \le C^{2q/p} \|f\|_p^q.$$

This completes the proof of the theorem. $\square$

The following embedding result will be crucial in the next section when we identify the dual space of H^p for $0 < p < 1$.

Corollary 4.49. *Suppose* $0 < p < 1$ *and*

$$\alpha = \frac{n}{p} - (n + 1).$$

Then $H^p \subset A_\alpha^1$. *Furthermore, there exists a constant* $C > 0$ *such that*

$$\int_{\mathbb{B}_n} |f(z)| \, dv_\alpha(z) \le C\|f\|_p$$

for all $f \in H^p$.

Proof. Let $q = 1$ in Theorem 4.48. $\square$

4.5 Duality

In this section we give a characterization of the bounded linear functionals on the Hardy spaces H^p for $p \ne 1$. The case of $1 < p < \infty$ is a consequence of the boundedness of the Cauchy-Szegö projection on $L^p(\mathbb{S}_n, d\sigma)$, and will be in terms of the natural integral pairing using the surface measure σ on $\mathbb{S}_n$. The case $0 < p < 1$ will be handled by using the weighted volume measures to form a pairing; another more natural pairing using the surface measure will be discussed in the chapter on Lipschitz spaces. The dual space of H^1 can be identified with BMOA using the natural pairing; this will be proved in the next chapter.

For $0 < p < \infty$, the dual space of H^p, denoted $(H^p)^*$, consists of all linear functionals $F : H^p \to \mathbb{C}$ such that

$$|F(f)| \leq C\|f\|_p$$

for all $f \in H^p$, where C is a positive constant dependent on F and p.

For $F \in (H^p)^*$, we use $\|F\|$ to denote the smallest possible constant C above. It follows from general functional analysis that $(H^p)^*$ is a Banach space equipped with this norm, even in the case $0 < p < 1$ when H^p itself is not a Banach space.

Theorem 4.50. *Suppose $1 < p < \infty$ and $1/p + 1/q = 1$. Then the dual space of H^p can be identified with H^q (with equivalent norms) under the integral pairing*

$$\langle f, g \rangle = \int_{\mathbb{S}_n} f(\zeta) \overline{g(\zeta)} \, d\sigma(\zeta),$$

where $f \in H^p$ and $g \in H^q$.

Proof. It is obvious that every function $g \in H^q$ induces a bounded linear functional F on H^p via the formula

$$F(f) = \int_{\mathbb{S}_n} f \, \overline{g} \, d\sigma, \qquad f \in H^p.$$

By Hölder's inequality, $\|F\| \leq \|g\|_q$.

On the other hand, if F is a bounded linear functional on H^p, then F can be extended to a bounded linear functional on $L^p(\mathbb{S}_n, d\sigma)$ (without increasing its norm) by the Hahn-Banach extension theorem. So there exists a function $h \in L^q(\mathbb{S}_n, d\sigma)$ such that

$$F(f) = \int_{\mathbb{S}_n} f \, \overline{h} \, d\sigma, \qquad f \in H^p.$$

Since $f = C(f)$, where C is the Cauchy-Szegö projection, an application of Corollary 4.37 gives

$$F(f) = \int_{\mathbb{S}_n} f \, \overline{C(h)} \, d\sigma, \qquad f \in H^p.$$

Let $g = C(h)$. Then $g \in H^q$ by Theorem 4.36, and

$$F(f) = \int_{\mathbb{S}_n} f \, \overline{g} \, d\sigma, \qquad f \in H^p.$$

Furthermore,

$$\|g\|_q \leq C\|h\|_q = C\|F\|,$$

with C being the norm of the Cauchy transform on $L^q(\mathbb{S}_n, d\sigma)$. This completes the proof. $\square$

The following result shows that the dual spaces of H^p, $0 < p < 1$, are all isomorphic to each other. Also, the dual space of H^p is isomorphic to that of A_t^p for $0 < p < 1$, although the spaces H^p and A_t^p are not isomorphic.

Theorem 4.51. *Suppose* $0 < p < 1$ *and*

$$\alpha = \frac{n}{p} - (n+1).$$

Then the dual space of H^p *can be identified with the Bloch space* $\mathcal{B}$ *(with equivalent norms) under the integral pairing*

$$\langle f, g \rangle = \lim_{r \to 1^-} \int_{\mathbb{B}_n} f(rz) \overline{g(z)} \, dv_\alpha(z),$$

where $f \in H^p$ *and* $g \in \mathcal{B}$.

Proof. If $g \in \mathcal{B}$, then $g = P_\alpha h$ by Theorem 3.4, where h is some function in $L^\infty(\mathbb{B}_n)$, and $\|h\|_\infty \le C\|g\|$, where C is a positive constant independent of g and $\|g\|$ is the norm of g in the Bloch space $\mathcal{B}$. Since P_α is self-adjoint with respect to the integral pairing induced by dv_α, we have

$$\int_{\mathbb{B}_n} f_r \overline{g} \, dv_\alpha = \int_{\mathbb{B}_n} f_r \overline{P_\alpha h} \, dv_\alpha = \int_{\mathbb{B}_n} P_\alpha f_r \overline{h} \, dv_\alpha = \int_{\mathbb{B}_n} f_r \overline{h} \, dv_\alpha$$

for every holomorphic f in $\mathbb{B}_n$ and $0 < r < 1$. If $f \in H^p$, then $f \in L^1(\mathbb{B}_n, dv_\alpha)$ by Corollary 4.49, and so

$$\lim_{r \to 1^-} \int_{\mathbb{B}_n} f_r \, \overline{g} \, dv_\alpha = \int_{\mathbb{B}_n} f \, \overline{h} \, dv_\alpha.$$

This shows that the limit

$$F(f) = \lim_{r \to 1^-} \int_{\mathbb{B}_n} f_r \overline{g} \, dv_\alpha, \qquad f \in H^p,$$

always exists and defines a bounded linear functional on H^p. In fact,

$$|F(f)| \le \|h\|_\infty \|f\|_{1,\alpha} \le C\|g\|\|f\|_p,$$

with C being a positive constant independent of f.

Conversely, if F is a bounded linear functional on H^p, then

$$F(f) = \lim_{r \to 1^-} F(f_r), \qquad f \in H^p.$$

Define a function $g \in H(\mathbb{B}_n)$ by

$$\overline{g(w)} = F_z \left[\frac{1}{(1 - \langle z, w \rangle)^{n+1+\alpha}} \right], \qquad w \in \mathbb{B}_n.$$

Differentiating inside F_z gives

$$\left| \frac{\partial g}{\partial w_k}(w) \right|^p \le (n+1+\alpha)^p \|F\|^p \int_{\mathbb{S}_n} \frac{|\zeta_k|^p \, d\sigma(\zeta)}{|1 - \langle \zeta, w \rangle|^{p(n+2+\alpha)}}$$

for $w \in \mathbb{B}_n$ and $1 \leq k \leq n$. This together with Theorems 1.12 and 3.4 shows that g is in the Bloch space. Since each f_r is in $H^\infty(\mathbb{B}_n)$, we have

$$f_r(z) = \int_{\mathbb{B}_n} \frac{f_r(w)\, dv_\alpha(w)}{(1 - \langle z, w \rangle)^{n+1+\alpha}}, \qquad z \in \mathbb{B}_n.$$

Taking F inside the integral, we obtain

$$F(f_r) = \int_{\mathbb{B}_n} f_r \bar{g}\, dv_\alpha.$$

This shows that

$$F(f) = \lim_{r \to 1^-} \int_{\mathbb{B}_n} f_r \bar{g}\, dv_\alpha, \qquad f \in H^p,$$

completing the proof of the theorem. □

Notes

Standard references for the theory of Hardy spaces in one complex variable are [33] and [42].

Our coverage of the (invariant) Poisson transform, estimates of various maximal functions, the existence of K-limits for functions in Hardy spaces, and the boundedness of the Cauchy-Szegö projection on $L^p(\mathbb{S}_n, d\sigma)$ ($1 < p < \infty$), all follow the presentation in [94].

Theorems 4.20, 4.22, and 4.23 are not as well known as they should be; see [73] and references there. In particular, Theorem 4.22 comes in handy when we identify the dual space of H^1 as BMOA in the next chapter.

Theorems 4.39, 4.40, and 4.41 are from [6]. The proof of Theorem 4.48 is from [18]. Corollary 4.49, which first appeared in [69], is critical for Theorem 4.51, which explicitly identifies the dual space of H^p, $0 < p < 1$, as the Bloch space under a certain weighted volume integral pairing; see [133].

Exercises

4.1. Show that the function

$$d(z, w) = |1 - \langle z, w \rangle|^{1/2}$$

satisfies the triangle inequality on $\overline{\mathbb{B}}_n$.

4.2. Show that the invariant Poisson kernel $P(z, \zeta)$ is M-harmonic in $z \in \mathbb{B}_n$ (for every fixed $\zeta \in \mathbb{S}_n$).

4.3. Suppose $-1 < \alpha < \infty$ and $1 \leq k < n$. Show that the restriction operator R_k maps $A_\alpha^p(\mathbb{B}_n)$ boundedly onto $A_{\alpha+n-k}^p(\mathbb{B}_k)$.

4.4. If μ is a finite complex Borel measure, then the function $C[\mu]$ belongs to H^p for $0 < p < 1$.

4.5. Suppose $0 < p < \infty$ and f is holomorphic in $\mathbb{B}_n$. Show that

$$\int_{\mathbb{B}_n} |f(z)|^p \, dv(z)$$

is equal to

$$|f(0)|^p + \frac{p^2}{2n} \int_{\mathbb{B}_n} |\widetilde{\nabla} f(z)|^2 |f(z)|^{p-2} \, d\tau(z) \int_{|z|}^1 \frac{(1-t^2)^{n-1}(1-t^{2n})}{t^{2n-1}} \, dt.$$

4.6. Suppose $0 < p < \infty$ and f is holomorphic in $\mathbb{B}_n$. Show that

$$\int_{\mathbb{B}_n} |f(z) - f(0)|^p \, dv(z)$$

is equal to

$$p^2 \int_{\mathbb{B}_n} |Rf(z)|^2 |f(z) - f(0)|^{p-2} |z|^{-2n} \, dv(z) \int_{|z|}^1 t^{2n-1} \log \frac{t}{|z|} \, dt.$$

4.7. Derive similar formulas for the norm of f in A_α^p, where $\alpha > -1$.

4.8. Show that $M_p(r, f)^p$ is equal to

$$|f(0)|^p + \left(\frac{p}{2}\right)^2 \int_{|z|<r} |f(z)|^{p-2} |\widetilde{\nabla} f(z)|^2 \, d\tau(z) \int_{|z|}^r \frac{(1-t^2)^{n-1}}{t^{2n-1}} \, dt,$$

where $0 \leq r < 1$ and f is holomorphic in $\mathbb{B}_n$.

4.9. Let $0 < p < \infty$. Show that for any multi-index $m = (m_1, \cdots, m_n)$ of non-negative integers and any compact set K in $\mathbb{B}_n$ there exists a constant $C > 0$ such that

$$\left| \frac{\partial^m f}{\partial z^m}(z) \right| \leq C \|f\|_{H^p}^p$$

for all $f \in H^p$ and all $z \in K$.

4.10. If f is holomorphic on the closed unit ball, show that

$$\int_{\mathbb{B}_n} |\widetilde{\nabla} f(z)|^2 G(z) \, d\tau(z) < \infty.$$

4.11. If $f \in C(\mathbb{S}_n)$ and $F = P[f]$, show that F extends to a function in $C(\overline{\mathbb{B}}_n)$, which we still denote by F, with $F = f$ on $\mathbb{S}_n$.

4.12. If $f \in L^p(\mathbb{B}_n, d\sigma)$, where $1 \leq p \leq \infty$, and $F = P[f]$, show that the functions F_r, where

$$F_r(\zeta) = F(r\zeta), \qquad 0 < r < 1, \zeta \in \mathbb{S}_n,$$

satisfy the inequality $\|F_r\|_p \leq \|f\|_p$, where $\|\ \|_p$ is the norm in $L^p(\mathbb{S}_n, d\sigma)$. If $1 \leq p < \infty$, then we also have

$$\lim_{r \to 1^-} \|F_r - f\|_p = 0.$$

4.13. If μ is a finite complex Borel measure on $\mathbb{S}_n$ and $P[\mu](z) = 0$ for all $z \in \mathbb{B}_n$, show that $\mu = 0$. Therefore, the Poisson transform is one-to-one.

4.14. Suppose μ is a finite complex Borel measure on $\mathbb{S}_n$ and $F = P[\mu]$. Show that $\|F_r\|_1 \leq \|\mu\|$ for all $r \in (0, 1)$, where $\|\mu\|$ is the total variation of μ. It is also true that

$$\lim_{r \to 1^-} F_r \, d\sigma = d\mu$$

in the weak-star topology of the dual space of $C(\mathbb{S}_n)$.

4.15. Suppose $t > 0$ and $f(z) = \sum_m a_m z^m$ is holomorphic in $\mathbb{B}_n$. If $2 \leq p < \infty$, show that $f \in H^p$ implies that the function

$$\sum_m |m|^t a_m z^m$$

belongs to A^p_{pt-1}. If $0 < p \leq 2$, then $f \in A^p_{pt-1}$ implies that the function

$$\sum_m \frac{a_m}{(|m| + 1)^t} z^m$$

belongs to H^p.

4.16. Show that the converse of Theorem 4.42 is false unless $p = 2$.

4.17. Show that

$$|f(z)|^p \leq \int_{\mathbb{S}_n} P(z, \zeta) |f(\zeta)|^p \, d\sigma(\zeta)$$

for all $f \in H^p$ and $z \in \mathbb{B}_n$.

4.18. Suppose $p \neq 2$ and $\Phi : H^p \to H^p$ is surjective linear isometry. Show that there exists some $\varphi \in \mathrm{Aut}(\mathbb{B}_n)$ such that

$$\Phi(f)(z) = \lambda f \circ \varphi(z) \left(\frac{(1 - |a|^2)^n}{(1 - \langle z, a \rangle)^{2n}} \right)^{1/p}$$

for all $f \in H^p$, where $a = \varphi(0)$ and λ is a unimodulus constant. See [94].

4.19. Suppose $a = (a_1, \cdots, a_n) \in \mathbb{B}_n$ and $f \in H^p$, where $0 < p \leq \infty$. Show that there exist functions $f_k \in H^p$, $1 \leq k \leq n$, such that

$$f(z) - f(a) = \sum_{k=1}^{n} (z_k - a_k) f_k(z), \qquad z \in \mathbb{B}_n.$$

4.20. Suppose $p > 0$ and $f \in H^p$. Show that

$$\int_{\mathbb{S}_n} |f(\zeta)|^p \, d\sigma(\zeta) = \lim_{\alpha \to -1^+} \int_{\mathbb{B}_n} |f(z)|^p \, dv_\alpha(z).$$

4.21. Show that for every $p \in (1, \infty)$ there exists a positive constant C such that

$$\int_{\mathbb{S}_n} |f(\zeta)|^p \, d\sigma(\zeta) \leq C \int_{\mathbb{S}_n} |\mathrm{Re}\, f(\zeta)|^p \, d\sigma(\zeta)$$

for all $f \in H^p$ with $f(0) = 0$. Show that this becomes false when $p = 1$.

4.22. Find sharp growth estimates for the Taylor coefficients of functions in H^p.

4.23. Show that every function in H^p can be approximated in norm by its Taylor polynomials if and only if $1 < p < \infty$. See [128].

4.24. Show that there exists no bounded projection from $L^1(\mathbb{S}_n, d\sigma)$ onto H^1.

4.25. Suppose $0 < p < \infty$, $f \in L^p(\mathbb{S}_n)$, and $\{f_k\}$ is a sequence of functions in $L^p(\mathbb{S}_n)$. Show that

$$\lim_{k \to \infty} \int_{\mathbb{S}_n} |f_k(\zeta) - f(\zeta)|^p \, d\sigma(\zeta) = 0$$

if and only if $f_k(\zeta) \to f(\zeta)$ for almost every $\zeta \in \mathbb{S}_n$ and

$$\lim_{k \to \infty} \int_{\mathbb{S}_n} |f_k(\zeta)|^p \, d\sigma(\zeta) = \int_{\mathbb{S}_n} |f(\zeta)|^p \, d\sigma(\zeta).$$

4.26. If $p \geq 2$, show that Theorem 4.48 follows from Theorem 4.42 and Theorem 2.30.

4.27. Suppose $p > 0$ and $\alpha > -1$. Show that the restriction operator R_k maps A_α^p of $\mathbb{B}_n$ boundedly onto $A_{\alpha+n-k}^p$ of $\mathbb{B}_k$.

4.28. Suppose p, q, and t are positive numbers related by

$$\frac{n}{p} - \frac{n}{q} = t.$$

If neither $n + \alpha$ nor $n + \alpha + t$ is a negative integer, show that $R_{\alpha,t} H^p \subset H^q$, and the inclusion is continuous. See [6].

4.29. For each $p \in (0,1]$ there exists a constant $C > 0$ such that every function $f \in H^p$ admits a representation

$$f(z) = \sum_{k=1}^{\infty} F_k(z) G_k(z),$$

where

$$\sum_{k=1}^{\infty} \|F_k\|_{H^{2p}}^p \|G_k\|_{H^{2p}}^p \leq C \|f\|_{H^p}^p.$$

See [6] and [30].

4.30. Show that on the unit disk $\mathbb{D}$ we have

$$H^p = H^q H^r, \qquad \frac{1}{p} = \frac{1}{q} + \frac{1}{r}.$$

4.31. If $n > 1$, show that there exists a function $f \in H^1$ such that f cannot be written as a product of two functions in H^2. See [44].

4.32. If h is an analytic function in the unit disk $\mathbb{D}$ and $0 < r < 1$, we define $n_f(r, w)$ to be the number of roots of the equation $h(z) = w$ in $|z| < r$. For a holomorphic function f in $\mathbb{B}_n$ we define

$$n_f(r, w) = \int_{\mathbb{S}_n} n_{f_\zeta}(r, w) \, d\sigma(\zeta), \qquad 0 < r < 1, w \in \mathbb{D},$$

where each f_ζ is the slice function defined on $\mathbb{D}$ by $f_\zeta(z) = f(z\zeta)$. Show that

$$r \frac{d}{dr} M_p(f, r)^p = \frac{p^2}{2\pi} \int_0^\infty R^{p-1} \, dR \int_0^{2\pi} n_f(r, Re^{i\theta}) \, d\theta,$$

where f is holomorphic in $\mathbb{B}_n$. See [73] and references there.

4.33. Show that

$$r \frac{d}{dr} M_p(f, r)^p = p \int_{\mathbb{S}_n} \frac{Rf(r\zeta)}{f(r\zeta)} |f(r\zeta)|^p \, d\sigma(\zeta)$$

for all holomorphic functions f in $\mathbb{B}_n$. See [73] and references there.

4.34. If $f \in H^p$, show that

$$\lim_{r \to 1^-} (1 - r^2)^n \int_{|z| < r} |\widetilde{\nabla} f(z)|^2 |f(z)|^{p-2} \, d\tau(z) = 0.$$

See [108].

4.35. Show that there exists a positive constant c (depending on n and p) such that

$$\|f\|_p^p = |f(0)|^p + c \int_0^1 \frac{(1 - r^2)^{n-1} \, dr}{r^{2n+1}} \int_{|z| < r} |\widetilde{\nabla} f(z)|^2 |f(z)|^{p-2} \, d\tau(z)$$

for all holomorphic functions f in $\mathbb{B}_n$. See [108].

5

Functions of Bounded Mean Oscillation

In this chapter we study the space of holomorphic functions in $\mathbb{B}_n$ that have bounded mean oscillation on $\mathbb{S}_n$ with respect to Lebesgue measure and the non-isotropic metric introduced in the previous chapter. Highlights of this chapter include Fefferman's duality theorem between BMOA and the Hardy space H^1, Hörmander's characterization of Carleson measures, and an atomic decomposition theorem for BMOA due to Rochberg and Semmes. We also discuss the issue of complex interpolation between BMOA and a Hardy space, and characterize the holomorphic functions in $\mathbb{B}_n$ that have bounded mean oscillation in $\mathbb{B}_n$ with respect to the Bergman metric.

5.1 BMOA

Throughout this chapter we write

$$d(z, w) = |1 - \langle z, w \rangle|^{1/2}$$

for z and w in the closed unit ball $\overline{\mathbb{B}}_n$. Recall that the restriction of d to $\mathbb{S}_n$ is a non-isotropic metric. Therefore, for any $\zeta \in \mathbb{S}_n$ and $r > 0$, the set

$$Q(\zeta, r) = \{\xi \in \mathbb{S}_n : |1 - \langle \zeta, \xi \rangle|^{1/2} < r\}$$

is a non-isotropic metric ball with center ζ and radius r. We will simply call Q a d-ball.

Let BMOA denote the space of functions f in H^2 such that

$$\|f\|_{\text{BMO}}^2 = |f(0)|^2 + \sup \frac{1}{\sigma(Q)} \int_Q |f - f_Q|^2 \, d\sigma < \infty, \qquad (5.1)$$

where

$$f_Q = \frac{1}{\sigma(Q)} \int_Q f \, d\sigma \qquad (5.2)$$

is the average of f over Q and the supremum is taken over $Q = Q(\zeta, r)$ for all $\zeta \in \mathbb{S}_n$ and all $r > 0$.

Lemma 5.1. *A function $f \in H^2$ belongs to* BMOA *if and only if there exists a positive constant C with the following property: for any d-ball Q in $\mathbb{S}_n$ there exists a number c such that*

$$\frac{1}{\sigma(Q)} \int_Q |f - c|^2 \, d\sigma \leq C. \tag{5.3}$$

Proof. If f is in BMOA, then for any Q we can take $c = f_Q$.

Conversely, if for any Q there exists a number c such that (5.3) holds, then we apply the triangle inequality to get

$$\left(\frac{1}{\sigma(Q)} \int_Q |f - f_Q|^2 \, d\sigma \right)^{1/2} \leq \left(\frac{1}{\sigma(Q)} \int_Q |f - c|^2 \, d\sigma \right)^{1/2} + |f_Q - c|.$$

The first term on the right is less than or equal to $\sqrt{C}$, while the second term can be estimated as follows.

$$|f_Q - c|^2 = \left| \frac{1}{\sigma(Q)} \int_Q (f - c) \, d\sigma \right|^2 \leq \frac{1}{\sigma(Q)} \int_Q |f - c|^2 \, d\sigma \leq C.$$

It follows that f is in BMOA. $\square$

Note that in the definition of BMOA, we only need to consider d-balls of radius less than δ, where δ is any fixed positive number.

Proposition 5.2. *The space* BMOA *is a Banach space when equipped with the norm* $\| \ \|_{\text{BMO}}$.

Proof. It is clear that $\|f\|_{\text{BMO}}$ is a well-defined norm on BMOA.

Recall that for $r \geq \sqrt{2}$, we have $Q(\zeta, r) = \mathbb{S}_n$ for any $\zeta \in \mathbb{S}_n$. It follows that $\|f\|_{\text{BMO}} \geq \|f\|_2$. So, if $\{f_k\}$ is a Cauchy sequence in BMOA, it is also a Cauchy sequence in H^2. Thus there exists a function $f \in H^2$ such that $f_k \to f$ in H^2. In particular, $f_k(z) \to f(z)$ uniformly on every compact subset of $\mathbb{B}_n$.

Given $\epsilon > 0$, there exists a positive integer N such that

$$\|f_k - f_l\|_{\text{BMO}} < \epsilon$$

whenever $k > N$ and $l > N$. For any d-ball Q we have

$$|f_k(0) - f_l(0)|^2 + \frac{1}{\sigma(Q)} \int_Q |(f_k - f_l) - (f_{kQ} - f_{lQ})|^2 \, d\sigma < \epsilon^2$$

for all $k > N$ and $l > N$. Let $l \to \infty$. Then

$$|f_k(0) - f(0)|^2 + \frac{1}{\sigma(Q)} \int_Q |(f_k - f) - (f_{kQ} - f_Q)|^2 \, d\sigma \leq \epsilon^2$$

for all $k > N$. Taking the supremum over all d-balls Q, we obtain

$$\|f_k - f\|_{\text{BMO}} \leq \epsilon$$

for all $k > N$. This shows that f is in BMOA and $\|f_k - f\|_{\text{BMO}} \to 0$ as $k \to \infty$. $\square$

We proceed to show that BMOA is invariant under the action of automorphisms. This combined with Theorem 3.19 in Chapter 3 will prove that BMOA is contained in the Bloch space $\mathcal{B}$. However, BMOA is not contained in the little Bloch space. In fact, for any $\zeta \in \mathbb{S}_n$, the function

$$f(z) = \log(1 - \langle z, \zeta \rangle), \qquad z \in \mathbb{B}_n,$$

belongs to BMOA but not to the little Bloch space.

Theorem 5.3. *A function $f \in H^2$ belongs to BMOA if and only if*

$$\|f\|_G^2 = \sup_{a \in \mathbb{B}_n} \int_{\mathbb{S}_n} |f \circ \varphi_a(\zeta) - f(a)|^2 \, d\sigma(\zeta) < \infty. \tag{5.4}$$

Furthermore, $\| \ \|_G$ is a complete Möbius invariant semi-norm on BMOA .

Proof. Fix $f \in H^2$ and $a \in \mathbb{B}_n$ with $a \neq 0$. Consider the integral

$$I_a = \int_{\mathbb{S}_n} |f \circ \varphi_a(\zeta) - f(a)|^2 \, d\sigma(\zeta).$$

Changing the variable of integration, we have

$$I_a = \int_{\mathbb{S}_n} |f(\zeta) - f(a)|^2 P(a, \zeta) \, d\sigma(\zeta).$$

For the d-ball

$$Q = Q(a/|a|, \sqrt{1 - |a|^2}),$$

we have

$$I_a \geq \int_Q |f(\zeta) - f(a)|^2 P(a, \zeta) \, d\sigma(\zeta).$$

Since

$$1 - \langle a, \zeta \rangle = 1 - \langle a/|a|, \zeta \rangle + \langle a/|a|, \zeta \rangle (1 - |a|),$$

we see that $\zeta \in Q$ implies

$$|1 - \langle a, \zeta \rangle| \leq (1 - |a|^2) + (1 - |a|) < 2(1 - |a|^2).$$

Therefore,

$$I_a \geq \frac{1}{4^n (1 - |a|^2)^n} \int_Q |f - f(a)|^2 \, d\sigma.$$

By Lemma 4.6, there exists an absolute constant C such that

$$I_a \geq \frac{C}{\sigma(Q)} \int_Q |f - f(a)|^2 \, d\sigma.$$

As a runs over $\mathbb{B}_n - \{0\}$, the above Q runs over all d-balls of radius less than 1. So, by Lemma 5.1 and the remark following it, the inequality $\|f\|_G < \infty$ implies that f is in BMOA.

To prove the other implication, let f be a function in BMOA and let

$$C_* = \sup \frac{1}{\sigma(Q)} \int_Q |f - f_Q|^2 \, d\sigma < \infty.$$

Consider the integrals I_a in the previous paragraph for $a \in \mathbb{B}_n$. Clearly,

$$\sup\{I_a : |a| \leq 3/4\} < \infty.$$

So we fix some $a \in \mathbb{B}_n$ with $|a| > 3/4$, and let

$$Q_k = Q\left(a/|a|, 2^k \sqrt{1 - |a|}\right)$$

for $k = 0, 1, \cdots, N$, where N is the smallest positive integer such that $Q_N = \mathbb{S}_n$. Since

$$1 - \langle a, \zeta \rangle = 1 - |a| + |a|(1 - \langle a/|a|, \zeta \rangle),$$

for $k \geq 1$ and $\zeta \in Q_k - Q_{k-1}$ we have

$$|1 - \langle a, \zeta \rangle| \sim 4^k(1 - |a|),$$

and so

$$P(a, \zeta) \sim 16^{-nk}(1 - |a|)^{-n}. \tag{5.5}$$

On the other hand, Lemma 4.6 tells us that

$$\sigma(Q_k) \sim 4^{nk}(1 - |a|)^n. \tag{5.6}$$

So there exists an absolute constant $C_1 > 0$ such that

$$\int_{Q_k - Q_{k-1}} |f(\zeta) - f_{Q_k}|^2 P(a, \zeta) \, d\sigma(\zeta) \leq \frac{C_1}{4^{nk}\sigma(Q_k)} \int_{Q_k} |f - f_{Q_k}|^2 \, d\sigma \leq \frac{C_1 C_*}{4^{nk}}$$

for $1 \leq k \leq N$.

Also, for $1 \leq k \leq N$, we have

$$f_{Q_k} - f_{Q_{k-1}} = \frac{1}{\sigma(Q_{k-1})} \int_{Q_{k-1}} (f_{Q_k} - f) \, d\sigma,$$

and so

$$|f_{Q_k} - f_{Q_{k-1}}|^2 \leq \frac{1}{\sigma(Q_{k-1})} \int_{Q_k} |f - f_{Q_k}|^2 \, d\sigma.$$

Since $\sigma(Q_k) \sim \sigma(Q_{k-1})$ by (5.6), there exists a constant $C_2 > 0$ such that

$$|f_{Q_k} - f_{Q_{k-1}}|^2 \leq \frac{C_2}{\sigma(Q_k)} \int_{Q_k} |f - f_{Q_k}|^2 \, d\sigma \leq C_2 C_*$$

for all $1 \leq k \leq N$. Combining this with

$$f_{Q_k} - f_{Q_0} = \sum_{i=1}^{k}(f_{Q_i} - f_{Q_{i-1}}),$$

we obtain

$$|f_{Q_k} - f_{Q_0}|^2 \le C_2 C_* k^2$$

for $1 \le k \le N$. Combining this with (5.5) and (5.6), we find a constant $C_3 > 0$ such that

$$\int_{Q_k - Q_{k-1}} |f_{Q_k} - f_{Q_0}|^2 P(a, \zeta)\, d\sigma(\zeta) \le \frac{C_3 C_* k^2}{4^{nk}}$$

for $1 \le k \le N$. Since

$$|f(\zeta) - f_{Q_0}|^2 \le 2(|f(\zeta) - f_{Q_k}|^2 + |f_{Q_k} - f_{Q_0}|^2),$$

we obtain

$$\int_{Q_k - Q_{k-1}} |f(\zeta) - f_{Q_0}|^2 P(a, \zeta)\, d\sigma(\zeta) \le \frac{C_4 C_* (k^2 + 1)}{4^{nk}}$$

for $1 \le k \le N$.

It is easy to see that

$$P(a, \zeta) \sim \frac{1}{\sigma(Q_0)}, \qquad \zeta \in Q_0,$$

so

$$\int_{Q_0} |f(\zeta) - f_{Q_0}|^2 P(a, \zeta)\, d\sigma(\zeta) \le C_5 C_*.$$

Since

$$\int_{\mathbb{S}_n} |f(\zeta) - f_{Q_0}|^2 P(a, \zeta)\, d\sigma(\zeta) = \int_{Q_0} |f(\zeta) - f_{Q_0}|^2 P(a, \zeta)\, d\sigma(\zeta)$$
$$+ \sum_{k=1}^{N} \int_{Q_k - Q_{k-1}} |f(\zeta) - f_{Q_0}|^2 P(a, \zeta)\, d\sigma(\zeta),$$

we obtain a constant $C_6 > 0$ such that

$$\int_{\mathbb{S}_n} |f(\zeta) - f_{Q_0}|^2 P(a, \zeta)\, d\sigma(\zeta) \le C_6 C_*$$

for any $a \in \mathbb{B}_n$ with $3/4 < |a| < 1$.

From the reproducing formula

$$\int_{\mathbb{S}_n} f(\zeta) P(a, \zeta)\, d\sigma(\zeta) = f(a)$$

we easily check that

$$\int_{\mathbb{S}_n} |f(\zeta) - f_{Q_0}|^2 P(a, \zeta)\, d\sigma(\zeta) = \int_{\mathbb{S}_n} |f(\zeta) - f(a)|^2 P(a, \zeta)\, d\sigma(\zeta) + |f(a) - f_{Q_0}|^2.$$

Therefore,

$$\int_{\mathbb{S}_n} |f(\zeta) - f(a)|^2 P(a, \zeta)\, d\sigma(\zeta) \le C_6 C_*$$

for all $3/4 < |a| < 1$. This completes the proof of the theorem. □

The semi-norm $\| \ \|_G$ is referred to as the Garsia semi-norm, which obviously has the property that

$$\|f \circ \varphi\|_G = \|f\|_G, \qquad f \in \mathrm{BMOA}, \varphi \in \mathrm{Aut}(\mathbb{B}_n).$$

It follows from the open mapping theorem that the norm

$$\|f\| = |f(0)| + \|f\|_G$$

is equivalent to the BMOA norm $\| \ \|_{\mathrm{BMO}}$.

It is clear that H^∞ is contained in BMOA. We will show in the next section that BMOA is the image of $L^\infty(\mathbb{S}_n)$ under the Cauchy-Szegö projection, and so by Theorem 4.36, BMOA is contained in H^p for every $p \in (0, \infty)$.

5.2 Carleson Measures

For $\zeta \in \mathbb{S}_n$ and $r > 0$ we introduce the set

$$Q_r(\zeta) = \{ z \in \mathbb{B}_n : d(z, \zeta) < r \}. \tag{5.7}$$

Clearly, the closure of $Q_r(\zeta)$ intersects $\mathbb{S}_n$ at the non-isotropic d-ball $Q(\zeta, r)$ from the previous section. We shall call $Q_r(\zeta)$ a Carleson tube at ζ.

A positive Borel measure μ in $\mathbb{B}_n$ is called a Carleson measure if there exists a constant $C > 0$ such that

$$\mu(Q_r(\zeta)) \le C r^{2n} \tag{5.8}$$

for all $\zeta \in \mathbb{S}_n$ and $r > 0$. It is obvious that every Carleson measure must be finite. Also, a finite positive Borel measure μ is Carleson if and only if

$$\sup \left\{ \frac{\mu(Q_r(\zeta))}{r^{2n}} : \zeta \in \mathbb{S}_n, 0 < r < \delta \right\} < \infty, \tag{5.9}$$

where δ is any fixed positive constant.

Recall from Lemma 4.6 that the power r^{2n} is comparable to the surface area (or σ-measure) of the non-isotropic ball $Q(\zeta, r)$ in $\mathbb{S}_n$. We shall see from Corollary 5.24 that the volume (or v-measure) of $Q_r(\zeta)$ in $\mathbb{B}_n$ is comparable to $r^{2(n+1)}$.

Theorem 5.4. *A positive Borel measure μ in $\mathbb{B}_n$ is a Carleson measure if and only if the quantity*

$$C_* = \sup_{z \in \mathbb{B}_n} \int_{\mathbb{B}_n} P(z, w) \, d\mu(w)$$

is finite. Here

$$P(z, w) = \frac{(1 - |z|^2)^n}{|1 - \langle z, w \rangle|^{2n}}, \qquad z, w \in \mathbb{B}_n,$$

is the Poisson kernel.

Proof. First assume that $C_* < \infty$. For any $\zeta \in \mathbb{S}_n$ and $0 < r < 1$ we consider the point $z = (1 - r^2)\zeta$. Since

$$1 - \langle z, w \rangle = (1 - r^2)(1 - \langle \zeta, w \rangle) + r^2,$$

we have

$$|1 - \langle z, w \rangle| \leq (1 - r^2)r^2 + r^2 \leq 2r^2$$

for all $w \in Q_r(\zeta)$. It follows that

$$P(z, w) = \frac{(1 - |z|^2)^n}{|1 - \langle z, w \rangle|^{2n}} \geq \frac{(1 - |z|)^n}{|1 - \langle z, w \rangle|^{2n}} \geq \frac{r^{2n}}{(2r^2)^{2n}}$$

for $w \in Q_r(\zeta)$, and so

$$C_* \geq \int_{Q_r(\zeta)} P(z, w) \, d\mu(w) \geq \frac{\mu(Q_r(\zeta))}{4^n r^{2n}}.$$

This shows that μ is a Carleson measure.

Next assume that μ is a Carleson measure and let

$$C = \sup_{\zeta, r} \frac{\mu(Q_r(\zeta))}{r^{2n}}.$$

Since μ is finite, an obvious estimate shows that

$$\sup_{|z| \leq 3/4} \int_{\mathbb{B}_n} P(z, w) \, d\mu(w) < \infty.$$

Fix some $z \in \mathbb{B}_n$ with $|z| > 3/4$ and set $\zeta = z/|z|$. For $k \geq 0$ let

$$r_k = \sqrt{2^{k+1}(1 - |z|)}.$$

Let

$$E_0 = Q_{r_0}(\zeta)$$

and

$$E_k = Q_{r_k}(\zeta) - Q_{r_{k-1}}(\zeta)$$

for $k \geq 1$. For each $k \geq 0$ we have

$$\mu(E_k) \leq 2^{n(k+1)}(1 - |z|)^n C.$$

If $k \geq 1$ and $w \in E_k$, then

$$
\begin{aligned}
|1 - \langle z, w \rangle| &= |(1 - |z|) + |z|(1 - \langle \zeta, w \rangle)| \\
&\geq |z||1 - \langle \zeta, w \rangle| - (1 - |z|) \\
&\geq \left(\frac{3}{4} \cdot 2^k - 1 \right)(1 - |z|) \\
&\geq 2^{k-1}(1 - |z|).
\end{aligned}
$$

This clearly holds for $k = 0$ as well. So

$$\int_{E_k} P(z, w) \, d\mu(w) \leq \frac{16^n C}{2^{nk}}, \qquad k \geq 0.$$

It follows that

$$\sup_{|z| > 3/4} \int_{\mathbb{B}_n} P(z, w) \, d\mu(w) \leq 16^n C \sum_{k=0}^{\infty} \frac{1}{2^{nk}}.$$

The proof is complete. $\qquad\qquad\qquad\qquad\qquad\qquad\qquad\qquad\qquad\qquad\square$

The following theorem characterizes the space BMOA in terms of Carleson measures. This is one of the most fundamental results in the theory of BMO.

Theorem 5.5. *Let $f \in H^2$. Then f is in BMOA if and only if the measure*

$$(1 - |z|^2)^n |\widetilde{\nabla} f(z)|^2 \, d\tau(z) = \frac{|\widetilde{\nabla} f(z)|^2 \, dv(z)}{1 - |z|^2}$$

is a Carleson measure. Here $d\tau$ is the Möbius invariant measure on $\mathbb{B}_n$.

Proof. Recall from Theorem 4.23 that

$$\int_{\mathbb{S}_n} |f - f(0)|^2 \, d\sigma = \int_{\mathbb{B}_n} |\widetilde{\nabla} f(z)|^2 G(z) \, d\tau(z)$$

for every $f \in H^2$, where

$$G(z) = \frac{1}{2n} \int_{|z|}^{1} \frac{(1 - t^2)^{n-1}}{t^{2n-1}} \, dt, \qquad z \in \mathbb{B}_n,$$

is the Green function for the invariant Laplacian $\widetilde{\Delta}$ in $\mathbb{B}_n$. As $|z| \to 1^-$, the Green function $G(z)$ is comparable to $(1 - |z|^2)^n$ (see Proposition 1.26). In particular,

$$\int_{\mathbb{S}_n} |f - f(0)|^2 \, d\sigma \sim \int_{\mathbb{B}_n} |\widetilde{\nabla} f(z)|^2 (1 - |z|^2)^n \, d\tau(z)$$

for f in H^2. So for any $a \in \mathbb{B}_n$ we have

$$\int_{\mathbb{S}_n} |f(\zeta) - f(a)|^2 P(a, \zeta) \, d\sigma(\zeta) = \int_{\mathbb{S}_n} |f \circ \varphi_a - f \circ \varphi_a(0)|^2 \, d\sigma$$

$$\sim \int_{\mathbb{B}_n} |\widetilde{\nabla}(f \circ \varphi_a)(z)|^2 (1 - |z|^2)^n \, d\tau(z)$$

$$= \int_{\mathbb{B}_n} |\widetilde{\nabla} f(z)|^2 (1 - |\varphi_a(z)|^2)^n \, d\tau(z),$$

where in the last step we made a change of variables and used the Möbius invariance of both $\widetilde{\nabla}$ and $d\tau$. Since

$$(1 - |\varphi_a(z)|^2)^n = \frac{(1 - |a|^2)^n (1 - |z|^2)^n}{|1 - \langle z, a \rangle|^{2n}},$$

we have

$$\int_{\mathbb{S}_n} |f(\zeta) - f(a)|^2 P(a, \zeta) \, d\sigma(\zeta) \sim \int_{\mathbb{B}_n} \frac{(1 - |a|^2)^n}{|1 - \langle z, a \rangle|^{2n}} |\widetilde{\nabla} f(z)|^2 (1 - |z|^2)^n \, d\tau(z).$$

The desired result now follows from Theorems 5.3 and 5.4. □

Recall from Theorem 3.1 that

$$\widetilde{\Delta}(|f|^2)(z) = 4|\widetilde{\nabla} f(z)|^2 = 4(1 - |z|^2)(|\nabla f(z)|^2 - |Rf(z)|^2)$$

for f holomorphic in $\mathbb{B}_n$. So f belongs to BMOA if and only if the measure

$$(|\nabla f(z)|^2 - |Rf(z)|^2) \, dv(z)$$

is a Carleson measure. We will also prove in Section 5.4 that a holomorphic function f in $\mathbb{B}_n$ is in BMOA if and only if the measure

$$(1 - |z|^2)|\nabla f(z)|^2 \, dv(z)$$

is a Carleson measure if and only if the measure

$$(1 - |z|^2)|Rf(z)|^2 \, dv(z)$$

is a Carleson measure.

Carleson measures also play a significant role in the theory of Hardy spaces. In the rest of this section we prove Hörmander's generalization of Carleson's classical theorem characterizing Carleson measures in terms of H^p functions.

For any $z \in \mathbb{B}_n$, $z \neq 0$, we write

$$Q_z = Q(z/|z|, \sqrt{1 - |z|}) = \{\zeta \in \mathbb{S}_n : |1 - \langle z/|z|, \zeta \rangle| < 1 - |z|\}. \quad (5.10)$$

The following covering lemma is crucial to our analysis.

Lemma 5.6. *Suppose $E \subset \mathbb{B}_n$ has the property that for no infinite sequence $\{z_k\}$ in E the d-balls Q_{z_k} are all disjoint. Then there exists a finite sequence $\{z_k\}$ in E such that the sets Q_{z_k} are disjoint and*

$$E \subset \bigcup_k \left\{ z \in \mathbb{B}_n : Q_z \subset Q\left(z_k/|z_k|, 5\sqrt{1 - |z_k|}\right) \right\}.$$

Proof. Let

$$T_1 = \sup\{\sqrt{1 - |z|} : z \in E\},$$

and choose $z_1 \in E$ such that $2\sqrt{1 - |z_1|} \geq T_1$. If $z_1, \cdots, z_{k-1}$ have aready been chosen, we let T_k be the supremum of $\sqrt{1 - |z|}$ such that $z \in E$ and Q_z does not intersect $Q_{z_1}, \cdots, Q_{z_{k-1}}$, if such points exist. We then choose $z_k \in E$ such that $2\sqrt{1 - |z_k|} \geq T_k$ and Q_{z_k} is disjoint with Q_{z_j} for $j < k$. By hypothesis, this construction must stop after a finite number of steps.

For any fixed $z \in E$, there must exist some k such that $Q_z \cap Q_{z_k} \neq \emptyset$. If j is the smallest such index, we then have $\sqrt{1 - |z|} \leq T_j$, so

$$2\sqrt{1 - |z_j|} \geq T_j \geq \sqrt{1 - |z|}.$$

This implies that

$$Q_z \subset Q\left(\zeta_j, 5\sqrt{1 - |z_j|}\right), \qquad \zeta_j = z_j/|z_j|.$$

In fact, if $\zeta \in Q_z$ and $\zeta' = z/|z|$, and if η is a point in $Q_z \cap Q_{z_j}$, then

$$d(\zeta, \zeta_j) \leq d(\zeta, \zeta') + d(\zeta', \eta) + d(\eta, \zeta_j)$$

$$\leq \sqrt{1 - |z|} + \sqrt{1 - |z|} + \sqrt{1 - |z_j|}$$

$$\leq 5\sqrt{1 - |z_j|}.$$

This proves the lemma. □

For a function f on $\mathbb{S}_n$ we define another maximal function f_* in $\mathbb{B}_n - \{0\}$ as follows:

$$f_*(z) = \sup_Q \frac{1}{\sigma(Q)} \int_Q |f| \, d\sigma, \tag{5.11}$$

where the supremum is taken over all d-balls Q in $\mathbb{S}_n$ such that $Q_z \subset Q$.

Theorem 5.7. *If μ is a Carleson measure on $\mathbb{B}_n$, then there exists a constant $C > 0$ such that*

$$\mu(f_* > t) \leq \frac{C}{t} \int_{\mathbb{S}_n} |f| \, d\sigma$$

for all $f \in L^1(\mathbb{S}_n, d\sigma)$ and all $t > 0$.

Proof. If $Q_z \subset Q_w$, then $1 - |z| \leq 1 - |w|$ and

$$\left| 1 - \left\langle z, \frac{w}{|w|} \right\rangle \right| \leq (1 - |z|) + |z| \left| 1 - \left\langle \frac{z}{|z|}, \frac{w}{|w|} \right\rangle \right|$$

$$\leq (1 - |z|) + |z|(1 - |w|) \leq 2(1 - |w|).$$

In particular, $Q_z \subset Q_w$ implies that $z \in Q_{\sqrt{2(1-|w|)}}(w/|w|)$.

By Lemma 4.6, if μ is a Carleson measure, we can find a constant $C_1 > 0$ such that

$$\mu\{z \in \mathbb{B}_n : Q_z \subset Q_w\} \leq C_1 \sigma(Q_w)$$

for all $w \in \mathbb{B}_n$, and we can also find a constant $C_2 > 0$ such that

$$\sigma\left(Q(z/|z|, 5\sqrt{1 - |z|})\right) \leq C_2 \sigma\left(Q(z/|z|, \sqrt{1 - |z|})\right)$$

for all $z \in \mathbb{B}_n$.

For any $\epsilon > 0$ and $t > 0$ consider the set

$$E_{t,\epsilon} = \left\{ z \in \mathbb{B}_n : z \neq 0, \int_{Q_z} |f| \, d\sigma > t(\epsilon + \sigma(Q_z)) \right\}.$$

If $\{z_k\}$ is a sequence in $E_{t,\epsilon}$ such that the d-balls Q_{z_k} are disjoint, then

$$\sum_k t(\epsilon + \sigma(Q_{z_k})) \leq \sum_k \int_{Q_{z_k}} |f| \, d\sigma \leq \int_{\mathbb{S}_n} |f| \, d\sigma. \tag{5.12}$$

So $\{z_k\}$ must be finite and we can apply Lemma 5.6 to the set $E_{t,\epsilon}$. Let $E'_{t,\epsilon}$ be the set of all $z \in \mathbb{B}_n$ with the property that $Q_z \subset Q_w$ for some $w \in E_{t,\epsilon}$. Then Lemma 5.6 shows that

$$E'_{t,\epsilon} \subset \bigcup_k \{z \in \mathbb{B}_n : Q_z \subset Q(z_k/|z_k|, 5\sqrt{1 - |z_k|})\}.$$

Therefore,

$$\mu(E'_{t,\epsilon}) \leq \sum_k \mu\left\{ z \in \mathbb{B}_n : Q_z \subset Q(z_k/|z_k|, 5\sqrt{1 - |z_k|}) \right\}$$

$$\leq C_1 \sum_k \sigma\left(Q(z_k/|z_k|, 5\sqrt{1 - |z_k|})\right)$$

$$\leq C_1 C_2 \sum_k \sigma(Q_{z_k})$$

$$\leq \frac{C_1 C_2}{t} \int_{\mathbb{S}_n} |f| \, d\sigma.$$

The last inequality above follows from (5.12).

If $f_*(z) > t$, then there exists a d-ball Q in $\mathbb{S}_n$ such that $Q_z \subset Q$ and

$$\int_Q |f| \, d\sigma > t\sigma(Q).$$

Write

$$Q = Q(\zeta, r) = Q\left(\frac{w}{|w|}, 1 - |w|\right), \qquad w = (1-r)\zeta.$$

Then

$$\int_{Q_w} |f| \, d\sigma > t(\epsilon + \sigma(Q_w))$$

for all sufficiently small ϵ. It follows that $w \in E_{t,\epsilon}$ and $z \in E'_{t,\epsilon}$ for ϵ sufficiently small. Therefore,

$$\mu(f_* > t) \le \limsup_{\epsilon \to 0} \mu(E'_{t,\epsilon}) \le \frac{C_1 C_2}{t} \int_{\mathbb{S}_n} |f| \, d\sigma.$$

This completes the proof of the theorem. $\square$

As a consequence of the above theorem and the Marcinkiewicz interpolation theorem, we obtain the following L^p estimate for the maximal function f_*.

Theorem 5.8. *If μ is a Carleson measure on $\mathbb{B}_n$, then for each $1 < p < \infty$ there exists a constant $C = C_p > 0$ such that*

$$\int_{\mathbb{B}_n} |f_*|^p \, d\mu \le C \int_{\mathbb{S}_n} |f|^p \, d\sigma$$

for all $f \in L^p(\mathbb{S}_n, d\sigma)$.

We now demonstrate the close relationship between Carleson measures and Hardy spaces. In particular, the following theorem will be essential for us later when we establish the duality between H^1 and BMOA.

Theorem 5.9. *Let μ be a positive Borel measure on $\mathbb{B}_n$ and $0 < p < \infty$. Then μ is a Carleson measure if and only if there exists a constant $C > 0$ such that*

$$\int_{\mathbb{B}_n} |f(z)|^p \, d\mu(z) \le C \int_{\mathbb{S}_n} |f(\zeta)|^p \, d\sigma(\zeta)$$

for all $f \in H^p$.

Proof. First assume that μ is a Carleson measure. For $3/4 < |z| < 1$ and $k \ge 0$ let

$$Q_k = Q\left(z/|z|, 2^k \sqrt{1 - |z|}\right).$$

If $1 < p < \infty$ and g is a nonnegative function in $L^p(\mathbb{S}_n, d\sigma)$, then the proof of Theorem 4.10 shows that

$$P[g](z) = \int_{Q_0} P(z,\zeta)g(\zeta)\,d\sigma(\zeta) + \sum_k \int_{Q_k - Q_{k-1}} P(z,\zeta)g(\zeta)\,d\sigma(\zeta)$$

$$\leq \frac{2^n}{(1-|z|)^n} \int_{Q_0} g(\zeta)\,d\sigma(\zeta) + \sum_k \frac{C_1}{4^{nk}\sigma(Q_k)} \int_{Q_k} g(\zeta)\,d\sigma(\zeta)$$

$$\leq C_2 g_*(z)$$

for some constant $C_2 > 0$ and all $z \in \mathbb{B}_n$. This also holds for $|z| \leq 3/4$. By Theorem 5.8, we have

$$\int_{\mathbb{B}_n} |P[g](z)|^p \, d\mu(z) \leq C \int_{\mathbb{S}_n} |g(\zeta)|^p \, d\sigma(\zeta)$$

for some other constant $C > 0$ (independent of g).

If $0 < p < \infty$ and $f \in H^p$, then the function $|g| = |f|^{p/2}$ is in $L^2(\mathbb{S}_n, d\sigma)$, so

$$\int_{\mathbb{B}_n} \left| P[|f|^{p/2}](z) \right|^2 \, d\mu(z) \leq C \int_{\mathbb{S}_n} |g|^2 \, d\sigma = C \int_{\mathbb{S}_n} |f|^p \, d\sigma.$$

By Corollary 4.5,

$$|f(z)|^p \leq \left| P[|f|^{p/2}](z) \right|^2, \qquad z \in \mathbb{B}_n.$$

We obtain

$$\int_{\mathbb{B}_n} |f(z)|^p \, d\mu(z) \leq C \int_{\mathbb{S}_n} |f(\zeta)|^p \, d\sigma(\zeta).$$

Next assume that there exists a constant $C > 0$ such that

$$\int_{\mathbb{B}_n} |f(z)|^p \, d\mu(z) \leq C \int_{\mathbb{S}_n} |f(\zeta)|^p \, d\sigma(\zeta)$$

for all $f \in H^p$. For any $a \in \mathbb{B}_n$ let

$$f(z) = \left[\frac{1 - |a|^2}{(1 - \langle z, a \rangle)^2} \right]^{n/p}, \qquad z \in \mathbb{B}_n.$$

Then we obtain

$$\int_{\mathbb{B}_n} P(a, z) \, d\mu(z) \leq C, \qquad a \in \mathbb{B}_n.$$

By Theorem 5.4, μ is a Carleson measure. $\qquad\qquad\square$

5.3 Vanishing Carleson Measures and VMOA

A positive Borel measure μ on $\mathbb{B}_n$ is called a vanishing Carleson measure if

$$\lim_{r \to 0} \frac{\mu(Q_r(\zeta))}{r^{2n}} = 0 \tag{5.13}$$

uniformly for $\zeta \in \mathbb{S}_n$. We say that a sequence $\{f_k\}$ in H^p converges to 0 ultra-weakly if $\{\|f_k\|_p\}$ is bounded and $\{f_k(z)\}$ converges to 0 for every $z \in \mathbb{B}_n$.

Theorem 5.10. *Let μ be a positive Borel measure on $\mathbb{B}_n$ and $p > 0$. Then the following conditions are equivalent:*

(a) μ is a vanishing Carleson measure.
(b) For every sequence $\{f_k\}$ that converges to 0 ultra-weakly in H^p we have

$$\lim_{k \to \infty} \int_{\mathbb{B}_n} |f_k(z)|^p \, d\mu(z) = 0.$$

(c) The measure μ satisfies

$$\lim_{|z| \to 1^-} \int_{\mathbb{B}_n} P(z, \zeta) \, d\mu(\zeta) = 0.$$

Proof. That (a) implies (b) follows from Theorem 5.9 and approximating the measure μ by the measures μ_r, where $0 < r < 1$ and μ_r is μ times the characteristic function of $r\mathbb{B}_n$.

Choosing

$$f_k(\zeta) = \left[\frac{(1 - |z_k|^2)^n}{(1 - \langle \zeta, z_k \rangle)^{2n}} \right]^{1/p},$$

where $\zeta \in \mathbb{S}_n$ and $|z_k| \to 1^-$ as $k \to \infty$, shows that (b) implies (c).

The proof that (c) implies (a) follows from the same arguments used in the proof of Theorem 5.4. $\square$

The space BMOA is easily seen to be non-separable. We consider a separable subspace of BMOA, denoted by VMOA, which is the closure in BMOA of the set of polynomials. We mention in passing that the letters in VMO stand for vanishing mean oscillation. The letter A in VMOA, just like the letter A in BMOA, refers to analytic functions.

Theorem 5.11. *For $f \in H^2$ the following conditions are equivalent:*

(a) f is in VMOA.
(b) f can be approximated in BMOA by functions holomorphic on the closed unit ball of $\mathbb{C}^n$.
(c) f satisfies

$$\lim_{|a| \to 1^-} \int_{\mathbb{S}_n} |f \circ \varphi_a(\zeta) - f(a)|^2 \, d\sigma(\zeta) = 0,$$

or equivalently,

$$\lim_{|a| \to 1^-} \int_{\mathbb{S}_n} |f(\zeta) - f(a)|^2 P(a, \zeta) \, d\sigma(\zeta) = 0.$$

(d) f satisfies

$$\lim_{|a| \to 1^-} \int_{\mathbb{B}_n} P(a, z) \, d\mu(z) = 0,$$

where

$$d\mu(z) = (1 - |z|^2)^n |\widetilde{\nabla} f(z)|^2 \, d\tau(z) = \frac{|\widetilde{\nabla} f(z)|^2}{1 - |z|^2} \, dv(z).$$

(e) f has the property that

$$\lim_{r \to 0^+} \frac{1}{\sigma(Q(\zeta, r))} \int_{Q(\zeta, r)} \left| f - f_{Q(\zeta, r)} \right|^2 d\sigma = 0$$

uniformly for $\zeta \in \mathbb{S}_n$.

Proof. The equivalence of (c), (d), and (e) follows from the proof of Theorems 5.3, 5.4, and 5.5. The equivalence of (a) and (b) is obvious, because a function holomorphic on the closed unit ball can be approximated uniformly on $\mathbb{B}_n$ by polynomials.

It is obvious that every polynomial satisfies the condition in (e), so an approximation argument shows that (a) implies (e).

It is easy to see that the Garsia semi-norm of f in BMOA is equivalent to the Carleson semi-norm

$$\|f\|_c = \sup_{a \in \mathbb{B}_n} \left[\int_{\mathbb{B}_n} P(a, z) \, d\mu_f(z) \right]^{1/2},$$

where $\mu = \mu_f$ is the measure given in condition (d). Also, it is easy to see that if condition (d) holds, then f can be approximated by f_r in the Carleson semi-norm. This shows that (d) implies (b), and the proof is complete. $\square$

It can be shown that VMOA contains unbounded functions. In fact, if

$$f(z_1, \cdots, z_n) = g(z_1),$$

where g is an unbounded analytic function in the unit disk $\mathbb{D}$ in $\mathbb{C}$ satisfying

$$\int_{\mathbb{D}} |g'(z_1)|^2 \, dA(z_1) < \infty,$$

(for example, g can be the Riemann mapping from the unit disk to an unbounded simply connected domain in $\mathbb{C}$ with finite area), then f belongs to VMOA.

The following result is the little oh version of Theorem 5.5.

Theorem 5.12. *Suppose f is holomorphic in $\mathbb{B}_n$ and*

$$d\mu(z) = (1 - |z|^2)^n |\widetilde{\nabla} f(z)|^2 \, d\tau(z) = \frac{|\widetilde{\nabla} f(z)|^2 \, dv(z)}{1 - |z|^2}.$$

Then f belongs to VMOA if and only if μ is a vanishing Carleson measure.

Proof. This follows from Theorems 5.10 and 5.11. $\square$

5.4 Duality

In this chapter we consider BMOA and VMOA as Banach spaces and show that BMOA can be identified with the second dual of VMOA using the natural integral pairing with the surface measure on $\mathbb{S}_n$. The intermediate space in this duality relation turns out to be the Hardy space H^1.

Theorem 5.13. *The Banach dual of H^1 can be identified with* BMOA *under the integral pairing*

$$\langle f, g \rangle = \lim_{r \to 1^-} \int_{\mathbb{S}_n} f(r\zeta)\,\overline{g(\zeta)}\,d\sigma(\zeta),$$

where f is in H^1 and g is in BMOA .

Proof. First let F be a bounded linear functional on H^1. We extend F to a bounded linear functional on $L^1(\mathbb{S}_n, d\sigma)$ by the Hahn-Banach extension theorem. Since the dual space of $L^1(\mathbb{S}_n, d\sigma)$ is $L^\infty(\mathbb{S}_n)$, there exists a function $h \in L^\infty(\mathbb{S}_n)$ such that

$$F(f) = \int_{\mathbb{S}_n} f\,\overline{h}\,d\sigma, \qquad f \in H^1.$$

Since the Cauchy-Szegö projection is an orthogonal projection on $L^2(\mathbb{S}_n, d\sigma)$, we have

$$F(f) = \lim_{r \to 1^-} \int_{\mathbb{S}_n} f_r\,\overline{h}\,d\sigma$$

$$= \lim_{r \to 1^-} \int_{\mathbb{S}_n} C(f_r)\,\overline{h}\,d\sigma$$

$$= \lim_{r \to 1^-} \int_{\mathbb{S}_n} f_r\,\overline{C(h)}\,d\sigma$$

for all $f \in H^1$. We proceed to show that the function $g = C(h)$ is in BMOA.

Fix $Q = Q(\zeta_0, r)$ in $\mathbb{S}_n$. Define h_1 on $\mathbb{S}_n$ by setting $h_1 = h$ on $Q(\zeta_0, 2r)$ and $h_1 = 0$ elsewhere on $\mathbb{S}_n$. Let $h_2 = h - h_1$. Then $h_2 = 0$ on the closure of $Q(\zeta_0, 2r)$ and $h_2 = h$ elsewhere on $\mathbb{S}_n$. We have $g = g_1 + g_2$, where $g_k = C(h_k)$ for $k = 1, 2$.

We first estimate the boundary integral of g_1 on Q using the fact that C is an orthogonal projection on $L^2(\mathbb{S}_n, d\sigma)$:

$$\int_Q |g_1(\zeta)|^2\,d\sigma \le \int_{\mathbb{S}_n} |g_1(\zeta)|^2\,d\sigma \le \int_{\mathbb{S}_n} |h_1(\zeta)|^2\,d\sigma$$

$$= \int_{Q(\zeta_0, 2r)} |h(\zeta)|^2\,d\sigma \le \|h\|_\infty^2\,\sigma(Q(\zeta_0, 2r)).$$

By Lemma 4.6, there exists a constant $C_1 > 0$ (depending on n only) such that

$$\frac{1}{\sigma(Q)} \int_Q |g_1(\zeta)|^2\,d\sigma(\zeta) \le C_1 \|h\|_\infty^2.$$

To estimate the boundary integral of g_2 on Q, we observe that g_2 is continuous on $Q(\zeta_0, 2r)$. In particular,

$$g_2(\zeta_0) = \int_{\mathbb{S}_n} C(\zeta_0, \zeta) h_2(\zeta)\, d\sigma(\zeta),$$

and

$$g_2(\eta) - g_2(\zeta_0) = \int_{\mathbb{S}_n} (C(\eta, \zeta) - C(\zeta_0, \zeta)) h_2(\zeta)\, d\sigma(\zeta)$$

for all $\eta \in Q$. By Lemma 4.29, there exists a constant $C_2 > 0$ (depending on n only) such that

$$|g_2(\eta) - g_2(\zeta_0)| \le C_2 r \|h\|_\infty \int_{d(\zeta,\zeta_0)>2r} \frac{d\sigma(\zeta)}{|1 - \langle \zeta, \zeta_0 \rangle|^{n+1/2}}$$

for $\eta \in Q$. Combining this with Lemma 4.30 we find a constant $C_3 > 0$ (depending on n only) such that

$$|g_2(\eta) - g_2(\zeta_0)| \le C_3 \|h\|_\infty$$

for all $\eta \in Q$, so

$$\frac{1}{\sigma(Q)} \int_Q |g_2(\zeta) - g_2(\zeta_0)|^2\, d\sigma(\zeta) \le C_3^2 \|h\|_\infty^2.$$

It follows that there exists a constant $C_4 > 0$, depending on n only, such that

$$\frac{1}{\sigma(Q)} \int_Q |g(\zeta) - g_2(\zeta_0)|^2\, d\sigma(\zeta) \le C_4^2 \|h\|_\infty^2.$$

Since Q is arbitrary, this along with Lemma 5.1 shows that g is in BMOA with

$$\|g\|_{\mathrm{BMO}} \le C_4 \|h\|_\infty.$$

Next we assume that g is in BMOA and consider the functional

$$F(f) = \int_{\mathbb{S}_n} f\, \overline{g}\, d\sigma,$$

where f is a polynomial (recall that the polynomials are dense in H^1). We proceed to show that F extends to a bounded linear functional on H^1.

Polarizing the formula in Theorem 4.22, we can write

$$F(f) = \int_{\mathbb{S}_n} f\, \overline{g}\, d\sigma = f(0)\overline{g(0)} + \tilde{F}(f),$$

where

$$\tilde{F}(f) = \frac{2}{n} \int_{\mathbb{B}_n} Rf(z)\overline{Rg(z)} |z|^{-2n} \log \frac{1}{|z|}\, dv(z).$$

An application of Hölder's inequality shows that $|\widetilde{F}(f)|^2$ is less than or equal to $4/n^2$ times

$$\left(\int_{\mathbb{B}_n} \frac{|Rf(z)|^2}{|f(z)|} |z|^{-2n} \log \frac{1}{|z|} \, dv(z) \right) \left(\int_{\mathbb{B}_n} |f(z)||Rg(z)|^2 |z|^{-2n} \log \frac{1}{|z|} \, dv(z) \right).$$

The first integral above is dominated by the H^1 norm of f, according to Theoem 4.22. The second integral above is also dominated by the H^1 norm of f, because of Theorem 5.9 and the fact that g belonging to BMOA implies that the measure

$$\frac{|\widetilde{\nabla} g(z)|^2}{1 - |z|^2} \, dv(z) = (|\nabla g(z)|^2 - |Rg(z)|^2) \, dv(z)$$

is Carleson (see Theorem 5.5), which easily implies that the measure

$$|Rg(z)|^2 |z|^{-2n} \log \frac{1}{|z|} \, dv(z)$$

is Carleson. This shows that F extends to a bounded linear functional on H^1.   □

Theorem 5.14. *If f is holomorphic in $\mathbb{B}_n$, then the following conditions are equivalent:*

(a) f is in BMOA .
(b) $[|\widetilde{\nabla} f(z)|^2/(1 - |z|^2)] \, dv(z)$ is a Carleson measure.
(c) $(1 - |z|^2)|\nabla f(z)|^2 \, dv(z)$ is a Carleson measure.
(d) $(1 - |z|^2)|Rf(z)|^2 \, dv(z)$ is a Carleson measure.

Proof. It follows from Lemma 2.14 that

$$(1 - |z|^2)|Rf(z)|^2 \leq (1 - |z|^2)|\nabla f(z)|^2 \leq \frac{|\widetilde{\nabla} f(z)|^2}{1 - |z|^2}.$$

We see that (b) implies (c), and (c) implies (d). That (a) implies (b) was proved in Theorem 5.5. The proof of Theorem 5.13 shows that (d) implies (a).   □

The following result indicates that the space BMOA plays the same role in the Hardy space theory as the Bloch space does in the Bergman space theory.

Theorem 5.15. *The Cauchy transform maps $L^\infty(\mathbb{S}_n)$ boundedly onto BMOA .*

Proof. By the Hahn-Banach extension theorem and the fact that the dual space of $L^1(\mathbb{S}_n, d\sigma)$ is $L^\infty(\mathbb{S}_n)$ with respect to the integral pairing induced by $d\sigma$, we see that the dual space H^1 can be identified with $CL^\infty(\mathbb{S}_n)$.   □

The next result gives further evidence that the space BMOA behaves like the limit space of H^p when $p \to \infty$.

Theorem 5.16. *Suppose* $1 \leq p_0 < \infty$ *and* $\theta \in (0, 1)$. *If* $p = p_0/(1 - \theta)$, *then*

$$[H^{p_0}, \mathrm{BMOA}]_\theta = H^p$$

with equivalent norms.

Proof. We prove the case $1 < p_0$. When $p_0 = 1$, the argument is much more complicated and involves real-variable methods; we refer the reader to [36] and references there.

If we identify H^p with a closed subspace of $L^p(\mathbb{S}_n)$, then for every $f \in H^p \subset L^p(\mathbb{S}_n, d\sigma)$, Theorem 1.33 tells us that there exists a family of functions g_ζ, $0 \leq \mathrm{Re}\,\zeta \leq 1$, such that $g_\theta = f$, $g_\zeta \in L^{p_0}(\mathbb{S}_n)$ for $\mathrm{Re}\,\zeta = 0$, and $g_\zeta \in L^\infty(\mathbb{S}_n)$ for $\mathrm{Re}\,\zeta = 1$. Let $f_\zeta = C[g_\zeta]$. Then $f_\theta = f$, $f_\zeta \in H^{p_0}$ for $\mathrm{Re}\,\zeta = 0$, and $f_\zeta \in \mathrm{BMOA}$ for $\mathrm{Re}\,\zeta = 1$. This shows that $H^p \subset [H^{p_0}, \mathrm{BMOA}]_\theta$.

The other direction follows from duality and Theorem 4.38. In fact, if $f \in [H^{p_0}, \mathrm{BMOA}]_\theta$, then there exists a family of functions f_ζ, $\mathrm{Re}\,\zeta \in [0, 1]$, such that $f_\theta = f$, $\|f_\zeta\|_{p_0} \leq \|f\|_\theta$ for $\mathrm{Re}\,\zeta = 0$, and $\|f_\zeta\|_{\mathrm{BMO}} \leq \|f\|_\theta$ for $\mathrm{Re}\,\zeta = 1$. Fix $g \in H^q$, where $1/p + 1/q = 1$. By Theorem 4.38, we have

$$H^q = [H^{q_0}, H^1]_\theta,$$

where $1/p_0 + 1/q_0 = 1$, because

$$\frac{1}{q} = \frac{1 - \theta}{q_0} + \frac{\theta}{1}.$$

So there exists a family of functions g_ζ, $\mathrm{Re}\,\zeta \in [0, 1]$, such that $g_\theta = g$, $\|g_\zeta\|_{q_0} \leq C\|g\|_q$ for $\mathrm{Re}\,\zeta = 0$, and $\|g_\zeta\|_1 \leq C\|g\|_q$ for $\mathrm{Re}\,\zeta = 1$, where C is a positive constant. It follows from the duality between H^{p_0} and H^{q_0}, the duality between H^1 and BMOA, and the Hadamand three-lines theorem (on which the method of complex interpolation is based) that the function

$$F(\zeta) = \int_{\mathbb{S}_n} f_\zeta \, \overline{g_\zeta} \, d\sigma$$

is analytic in $0 < \mathrm{Re}\,\zeta < 1$, continuous in $0 \leq \mathrm{Re}\,\zeta \leq 1$, and satisfies

$$|F(\theta)| \leq C'\|f\|_\theta \|g\|_q,$$

where C' is another positive constant (independent of f and g). This shows that

$$\left| \int_{\mathbb{S}_n} f \, \overline{g} \, d\sigma \right| \leq C'\|f\|_\theta \|g\|_q.$$

Since g is arbitrary, it follows from the duality between H^p and H^q that $f \in H^p$. □

We now discuss several issues related to the space VMOA.

Theorem 5.17. *The dual space of* VMOA *can be identified with* H^1 *under the integral pairing*

$$\langle f, g \rangle = \lim_{r \to 1^-} \int_{\mathbb{S}_n} f(r\zeta) \, \overline{g(\zeta)} \, d\sigma(\zeta),$$

where f is in VMOA *and g is in* H^1.

Proof. By Theorem 5.13, every function $g \in H^1$ induces a bounded linear functional on VMOA.

Conversely, if F is a bounded linear functional on VMOA, we define a holomorphic function g in $\mathbb{B}_n$ by

$$g(z) = \sum_m b_m z^m,$$

where

$$b_m = \frac{(n-1+|m|)!}{(n-1)! \, m!} \, \overline{F(z^m)}.$$

Since $\{z^m\}$ is a bounded sequence in VMOA, $\{F(z^m)\}$ must also be bounded, so the function g is indeed holomorphic in $\mathbb{B}_n$.

For $0 < r < 1$ and

$$f(z) = \sum_m a_m z^m$$

in VMOA we have

$$F(f_r) = \sum_m \frac{(n-1)! \, m!}{(n-1+|m|)!} \, a_m \overline{b_m} r^{|m|},$$

because the Taylor series of f_r converges in VMOA. Using Lemma 1.11 we can write

$$F(f_r) = \int_{\mathbb{S}_n} f \, \overline{g_r} \, d\sigma.$$

If g is in H^1, then by Lemma 1.11 again,

$$\int_{\mathbb{S}_n} f_r \overline{g} \, d\sigma = F(f_r),$$

and so

$$F(f) = \lim_{r \to 1^-} \int_{\mathbb{S}_n} f_r \overline{g} \, d\sigma$$

for all f in VMOA.

To show that g belongs to H^1, we use Theorem 5.13 to find a constant $M > 0$ (independent of r) such that

$$\|g_r\|_{H^1} \leq M \sup \left\{ \left| \int_{\mathbb{S}_n} g_r \overline{f} \, d\sigma \right| : \|f\|_* \leq 1 \right\}$$

for all $r \in (0, 1)$. Here we use the following norm on BMOA:

$$\|f\|_* = \inf\{\|h\|_\infty : f = C(h), h \in L^\infty(\mathbb{S}_n)\},$$

with $C(h)$ being the Cauchy transform of h. If $f = C(h)$ for $h \in L^\infty(\mathbb{S}_n)$, then it is easy to check that $f_r = C(h_r)$, where

$$h_r(\zeta) = \int_{\mathbb{S}_n} P(r\zeta, \eta) h(\eta) \, d\sigma(\eta), \qquad \zeta \in \mathbb{S}_n.$$

Since $\|h_r\|_\infty \le \|h\|_\infty$, it follows easily that $\|f_r\|_* \le \|f\|_*$.

Since f_r is in VMOA and since

$$\int_{\mathbb{S}_n} g_r \overline{f} \, d\sigma = \overline{F(f_r)},$$

we conclude that

$$\left| \int_{\mathbb{S}_n} g_r \overline{f} \, d\sigma \right| \le \|F\| \|f_r\|_* \le \|F\| \|f\|_*,$$

and so

$$\|g_r\|_{H^1} \le M \|F\|.$$

Since r is arbitrary, we must have $g \in H^1$ and $\|g\|_{H^1} \le M\|F\|$. This completes the proof of the theorem. $\qquad\square$

Theorem 5.18. *The Cauchy-Szegö projection maps the space $\mathbb{C}(\mathbb{S}_n)$ boundedly onto* VMOA .

Proof. It is clear that the Cauchy transform C maps every function of the form $z^m \overline{z}^l$ to a monomial, which belongs to VMOA. By Theorem 5.15 and the Stone-Weierstrass approximation theorem, C maps $\mathbb{C}(\mathbb{S}_n)$ boundedly into VMOA.

To show that the Cauchy transform C maps $\mathbb{C}(\mathbb{S}_n)$ *onto* VMOA, we fix a unit vector f in VMOA and use Theorem 5.15 to find a function $g \in L^\infty(\mathbb{S}_n)$ such that $f = C(g)$ and $\|g\|_\infty \le M$, where M is a positive constant independent of f. We extend the function g to $\mathbb{B}_n$ using the Poisson transform and still use g to denote the resulting extension. A use of Fubini's theorem shows that $f_r = C(g_r)$ for each $0 < r < 1$. It is clear that $\|g_r\|_\infty \le \|g\|_\infty \le M$.

Since f is in VMOA, there exists some $r \in (0, 1)$ such that

$$\|f - f_r\|_{\mathrm{BMO}} < \frac{1}{2}.$$

We then have the representation

$$f = f_1 + h^{(1)},$$

where $f_1 = f_r = C(g_r)$ with $g_r \in \mathbb{C}(\mathbb{S}_n)$ and $h^{(1)} = f - f_r$ satisfies $\|h^{(1)}\|_{\mathrm{BMO}} < 1/2$.

Choose a function $g^{(1)} \in L^\infty(\mathbb{S}_n)$ such that $h^{(1)} = C(g^{(1)})$ and $\|g^{(1)}\|_\infty \le M/2$, where M is the same constant from the previous paragraph. Since $h^{(1)}$ is still in VMOA, there exists some $r \in (0,1)$ (possibly different from the r in the previous paragraph) such that

$$\|h^{(1)} - h_r^{(1)}\|_{\mathrm{BMO}} < \frac{1}{4}.$$

We then have the representation

$$f = f_1 + f_2 + h^{(2)},$$

where $f_2 = h_r^{(1)} = C(g_r^{(1)})$ with $g_r^{(1)} \in C(\mathbb{S}_n)$ and $h^{(2)} = h^{(1)} - h_r^{(1)}$ satisfies $\|h^{(2)}\|_{\mathrm{BMO}} < 1/4$.

Continuing the above process infinitely, we obtain

$$f = C(g_1) + C(g_2) + \cdots + C(g_n) + \cdots,$$

where the convergence is in BMO norm, each g_n belongs to $C(\mathbb{S}_n)$, and

$$\|g_n\|_\infty \leq \frac{M}{2^{n-1}}.$$

Let

$$G = g_1 + g_2 + \cdots + g_n + \cdots,$$

then $G \in C(\mathbb{S}_n)$ and $f = C(G)$. This completes the proof of the theorem. $\quad\square$

Theorem 5.19. *Suppose f is holomorphic in $\mathbb{B}_n$. Then the following conditions are equivalent:*

(a) f belongs to VMOA.

(b) The measure $(1 - |z|^2)|\nabla f(z)|^2\, dv(z)$ is a vanishing Carleson measure.

(c) The measure $(1 - |z|^2)|Rf(z)|^2\, dv(z)$ is a vanishing Carleson measure.

Proof. If f is in VMOA, then by Theorem 5.12, the measure

$$(1 - |z|^2)^{-1}|\widetilde{\nabla} f(z)|^2\, dv(z)$$

is a vanishing Carleson measure. According to Lemma 2.14,

$$(1 - |z|^2)|\nabla f(z)|^2 \leq (1 - |z|^2)^{-1}|\widetilde{\nabla} f(z)|^2,$$

it is then clear that (a) implies (b). It also follows from Lemma 2.14 that (b) implies (c).

It follows from the open mapping theorem that the Garsia norm $\|f\|_G$ on BMOA is comparable to the Carleson norm $\|f\|_c$ defined by

$$\|f\|_c^2 = \sup\left\{ \int_{\mathbb{B}_n} |g(z)|^2\, d\mu_f(z) : \int_{\mathbb{S}_n} |g|^2\, d\sigma = 1 \right\},$$

where

$$d\mu_f(z) = (1 - |z|^2)|Rf(z)|^2\, dv(z).$$

If condition (c) holds, then f can be approximated by f_r in the norm $\|\ \|_c$, and so f can be approximated by f_r in the BMOA norm. This shows that (c) implies (a). $\quad\square$

The next result is the limit case of Theorem 4.39 when $p \to \infty$. The definitions of the operators R_k and E_k are given immediately prior to Theorem 4.39.

Theorem 5.20. *Suppose* $1 \leq k < n$. *Then the operator* R_k *maps* BMOA *and* VMOA *of* $\mathbb{B}_n$ *onto the Bloch space and the little Bloch space of* $\mathbb{B}_k$, *respectively.*

Proof. If $f \in \text{BMOA}(\mathbb{B}_n)$, then there exists a function $g \in L^\infty(\mathbb{S}_n)$ such that

$$f(z) = \int_{\mathbb{S}_n} \frac{g(\zeta)\, d\sigma(\zeta)}{(1 - \langle z, \zeta \rangle)^n}, \qquad z \in \mathbb{B}_n.$$

Write $z = (z', z'')$ and $\zeta = (\zeta', \zeta'')$ with z' and ζ' in $\mathbb{C}^k$. Then

$$(R_k f)(z_1, \cdots, z_k) = f(z', 0) = \int_{\mathbb{S}_n} \frac{g(\zeta)\, d\sigma(\zeta)}{(1 - \langle z', \zeta' \rangle)^n}.$$

Applying (1.15), we have

$$(R_k f)(z_1, \cdots, z_k) = \int_{\mathbb{B}_k} \frac{(1 - |w|^2)^{n-k-1}\, h(w)\, dv_k(w)}{(1 - \langle z', w \rangle)^n},$$

where

$$h(w) = \binom{n-1}{k} \int_{\mathbb{S}_{n-k}} g(w, \sqrt{1 - |w|^2}\, \eta)\, d\sigma_{n-k}(\eta)$$

is a bounded function on $\mathbb{B}_k$. According to part (d) of Theorem 3.4, the function $R_k f$ belongs to the Bloch space of $\mathbb{B}_k$.

A similar argument shows that R_k maps VMOA($\mathbb{B}_n$) boundedly onto the little Bloch space of $\mathbb{B}_k$. $\qquad\square$

We also have the following limit case of Corollary 4.40.

Theorem 5.21. *For each* $1 \leq k < n$ *the operator* E_k *maps the Bloch space and the little Bloch space of* $\mathbb{B}_k$ *into* BMOA *and* VMOA *of* $\mathbb{B}_n$, *respectively.*

Proof. Let f be a function in the Bloch space of $\mathbb{B}_k$. Then there exists a function $g \in L^\infty(\mathbb{B}_k)$ such that

$$f(z') = \binom{n-1}{k} \int_{\mathbb{B}_k} \frac{(1 - |w|^2)^{n-k-1} g(w)\, dv_k(w)}{(1 - \langle z', w \rangle)^n}.$$

Here again we write $z = (z', z'')$ for $z \in \mathbb{B}_n$ with $z' \in \mathbb{B}_k$. Define a function h (almost everywhere) on $\mathbb{S}_n$ by

$$h(\zeta', \zeta'') = g(\zeta'), \qquad \zeta = (\zeta', \zeta'') \in \mathbb{S}_n.$$

Then by (1.15),

$$\int_{\mathbb{S}_n} \frac{h(\zeta)\,d\sigma(\zeta)}{(1 - \langle z, \zeta \rangle)^n} = \binom{n-1}{k} \int_{\mathbb{B}_k} (1 - |w|^2)^{n-k-1} g(w)\,dv_k(w) \cdot$$

$$\cdot \int_{\mathbb{S}_{n-k}} \frac{d\sigma_{n-k}(\eta)}{(1 - \langle z', w \rangle - \langle z'', \sqrt{1 - |w|^2}\,\eta \rangle)^n}$$

$$= \binom{n-1}{k} \int_{\mathbb{B}_k} \frac{(1 - |w|^2)^{n-k-1}}{(1 - \langle z', w \rangle)^n} g(w)\,dv_k(w)$$

$$= f(z') = (E_k f)(z).$$

According to Theorem 5.15, we have $E_k f \in \mathrm{BMOA}(\mathbb{B}_n)$. A similar argument shows that E_k maps the little Bloch space of $\mathbb{B}_k$ boundedly into VMOA of $\mathbb{B}_n$. □

5.5 BMO in the Bergman Metric

Recall that

$$D(a, r) = \{z \in \mathbb{B}_n : \beta(z, a) < r\}, \qquad a \in \mathbb{B}_n,$$

where β is the Bergman metric. For a function f in $L^1(\mathbb{B}_n, dv_\alpha)$ we define

$$f_{\alpha, E} = \frac{1}{v_\alpha(E)} \int_E f(z)\,dv_\alpha(z), \tag{5.14}$$

where $\alpha > -1$ and E is any Lebesgue measurable set in $\mathbb{B}_n$. Two kinds of sets E will be of interest to us in this section, namely, Bergman metric balls and Carleson tubes. Our goal is to show that holomorphic functions in $\mathbb{B}_n$ that are of bounded mean oscillation with respect to the Bergman metric balls or Carleson tubes are exactly the Bloch functions.

Theorem 5.22. *Suppose $r > 0$, $\alpha > -1$, $p \geq 1$, and f is holomorphic in $\mathbb{B}_n$. Then the following conditions are equivalent:*

(a) $f \in \mathcal{B}$.
(b) There exists a constant $C > 0$ such that

$$\frac{1}{v_\alpha(D(a,r))} \int_{D(a,r)} |f(z) - f(a)|^p\,dv_\alpha(z) \leq C$$

for all $a \in \mathbb{B}_n$.
(c) There exists a constant $C > 0$ such that

$$\frac{1}{v_\alpha(D(a,r))} \int_{D(a,r)} |f(z) - f_{\alpha, D(a,r)}|^p\,dv_\alpha(z) \leq C$$

for all $a \in \mathbb{B}_n$.

(d) *There exists a constant $C > 0$ with the property that for every $a \in \mathbb{B}_n$ there is a complex number c_a such that*

$$\frac{1}{v_\alpha(D(a,r))} \int_{D(a,r)} |f(z) - c_a|^p \, dv_\alpha(z) \leq C.$$

Proof. Assume that $f \in \mathcal{B}$. By Corollary 3.8 there exists a constant $C_1 > 0$ such that

$$\int_{\mathbb{B}_n} |f(z) - f(a)|^p \frac{(1 - |a|^2)^{n+1+\alpha}}{|1 - \langle a, z \rangle|^{2(n+1+\alpha)}} \, dv_\alpha(z) \leq C_1$$

for all $a \in \mathbb{B}_n$. In particular,

$$\int_{D(a,r)} |f(z) - f(a)|^p \frac{(1 - |a|^2)^{n+1+\alpha}}{|1 - \langle z, a \rangle|^{2(n+1+\alpha)}} \, dv_\alpha(z) \leq C_1$$

for all $a \in \mathbb{B}_n$. By Lemma 2.20,

$$\int_{D(a,r)} |f(z) - f(a)|^p \frac{dv_\alpha(z)}{(1 - |a|^2)^{n+1+\alpha}} \leq C_2$$

for all $a \in \mathbb{B}_n$, where C_2 is another positive constant. Since

$$v_\alpha((D(a,r)) \sim (1 - |a|^2)^{n+1+\alpha}$$

by Lemma 1.24, we have

$$\frac{1}{v_\alpha(D(a,r))} \int_{D(a,r)} |f(z) - f(a)|^p \, dv_\alpha(z) \leq C_3$$

for all $a \in \mathbb{B}_n$, where C_3 is another positive constant. This proves that (a) implies (b).

To prove (b) implies (c), write

$$f(z) - f_{\alpha,D(a,r)} = f(z) - f(a) - (f_{\alpha,D(a,r)} - f(a))$$

and observe that

$$f_{\alpha,D(a,r)} - f(a) = \frac{1}{v_\alpha(D(a,r))} \int_{D(a,r)} (f(z) - f(a)) \, dv_\alpha(z).$$

The desired estimate then follows from the triangle inequality and Hölder's inequality.

That (c) implies (d) is trivial.

It remains to show that (d) implies (a). An examination of the proof of Lemma 2.4 reveals that there exists a constant $C_4 > 0$ such that

$$|\nabla g(0)|^p \leq C_4 \int_{D(0,r)} |g(z)|^p \, dv_\alpha(z)$$

for all holomorphic functions g in $\mathbb{B}_n$. For any $a \in \mathbb{B}_n$ we replace g by $f \circ \varphi_a - c_a$. Then

$$|\widetilde{\nabla} f(a)|^p \leq C_4 \int_{D(0,r)} |f \circ \varphi_a(z) - c_a|^p \, dv_\alpha(z).$$

Make an obvious change of variables according to Proposition 1.13 and then apply Lemmas 2.20 and 1.24. We obtain

$$|\widetilde{\nabla} f(a)|^p \leq \frac{C_5}{v_\alpha(D(a,r))} \int_{D(a,r)} |f(z) - c_a|^p \, dv_\alpha(z)$$

for all $a \in \mathbb{B}_n$, where C_5 is a new positive constant. This completes the proof of the theorem. $\square$

Since the first condition above is independent of p and α, it follows that the other three conditions are actually independent of p and α as well. However, if we allow non-holomorphic functions, then these conditions become dependent on the parameters p and α; see [131]. Also note that the assumption $p \geq 1$ was only used in the proof that (b) implies (c).

We shall also show that the Bloch space can be characterized by the boundedness of mean oscillation with respect to Carleson tubes

$$Q_r(\zeta) = \{z \in \mathbb{B}_n : d(z,\zeta) < r\}, \qquad r > 0, \zeta \in \mathbb{S}_n,$$

where

$$d(z,w) = |1 - \langle z, w \rangle|^{1/2}, \qquad z, w \in \overline{\mathbb{B}}_n.$$

First we show that a Carleson tube behaves much like a Bergman metric ball. To this end, we introduce the Euclidean tube

$$Q(\zeta, r) \times (s, 1) = \{z \in \mathbb{B}_n : s < |z| < 1, z/|z| \in Q(\zeta, r)\} \tag{5.15}$$

for any $\zeta \in \mathbb{S}_n, 0 < r < \sqrt{2}$, and $0 < s < 1$.

Lemma 5.23. *Suppose $0 < r < 1$ and $R > 0$. There exists a constant $\sigma \in (0,1)$ (depending on R but not on r) such that*

$$D(a, R) \subset Q_r(\zeta) \subset Q(\zeta, r') \times (1 - r^2, 1)$$

for all $\zeta \in \mathbb{S}_n$, where

$$a = (1 - \sigma r^2)\zeta, \qquad r' = \sqrt{\frac{2r^2}{1 - r^2}}.$$

Proof. First assume that $\zeta \in \mathbb{S}_n$ and $z \in Q_r(\zeta)$. Then $z \neq 0$ and we can write $z = |z|\eta$ for some $\eta \in \mathbb{S}_n$. Since

$$1 - \langle z, \zeta \rangle = 1 - |z| + |z|(1 - \langle \eta, \zeta \rangle),$$

an application of the triangle inequality gives

$$|z||1 - \langle \eta, \zeta \rangle| < r^2 + (1 - |z|).$$

On the other hand,

$$r^2 > |1 - \langle z, \zeta \rangle| \geq 1 - |z|.$$

Therefore, $1 - r^2 < |z| < 1$ and

$$|1 - \langle \eta, \zeta \rangle| < \frac{r^2 + (1 - |z|)}{|z|} < \frac{2r^2}{1 - r^2}.$$

This shows that

$$Q_r(\zeta) \subset Q(\zeta, r') \times (1 - r^2, 1).$$

Next assume that $\zeta \in \mathbb{S}_n$ and $a = (1 - \sigma r^2)\zeta$, where $\sigma \in (0, 1)$ is a constant to be specified later. If $z \in D(a, R)$, then $z = \varphi_a(w)$ for some w with $|w| < R'$, where $R' = \tanh(R) \in (0, 1)$. It follows that

$$1 - \langle z, \zeta \rangle = 1 - \left\langle \varphi_a(w), \frac{a}{1 - \sigma r^2} \right\rangle = \frac{1 - \langle \varphi_a(w), a \rangle - \sigma r^2}{1 - \sigma r^2}.$$

Since (see Lemma 1.3)

$$1 - \langle \varphi_a(w), a \rangle = \frac{1 - |a|^2}{1 - \langle w, a \rangle}, \qquad 1 - |a| = \sigma r^2,$$

we obtain

$$1 - \langle z, \zeta \rangle = \frac{\sigma r^2}{1 - \sigma r^2} \left[\frac{1 + |a|}{1 - \langle w, a \rangle} - 1 \right].$$

If we choose $\sigma \in (0, 1)$ so that

$$\frac{\sigma}{1 - \sigma r^2} \left| \frac{1 + |a|}{|1 - \langle w, a \rangle|} - 1 \right| < 1$$

for all $|w| < R'$, then $D(a, R) \subset Q_r(\zeta)$. $\square$

Corollary 5.24. *For any $\alpha > -1$ there exist positive constants c and C such that*

$$cr^{2(n+1+\alpha)} \leq v_\alpha(Q_r(\zeta)) \leq Cr^{2(n+1+\alpha)}$$

for all $\zeta \in \mathbb{S}_n$ and $0 \leq r \leq \sqrt{2}$.

Proof. With notation from the lemma above, we have

$$v_\alpha(D(a, R)) \sim (1 - |a|^2)^{n+1+\alpha} \sim r^{2(n+1+\alpha)}$$

as $r \to 0^+$. Also, it follows from polar coordinates and Lemma 4.6 that

$$v_\alpha\big(Q(\zeta, r') \times (1 - r^2, 1)\big) \sim r^{2(n+1+\alpha)}$$

as $r \to 0^+$. The desired result is obvious for r not near 0. $\square$

Theorem 5.25. *Suppose $\alpha > -1$, $p \geq 1$, and f is holomorphic in $\mathbb{B}_n$. Then the following conditions are equivalent:*

(a) $f \in \mathcal{B}$.
(b) There exists a constant $C > 0$ with the property that for each $r > 0$ and $\zeta \in \mathbb{S}_n$ there is a number c such that

$$\frac{1}{v_\alpha(Q_r(\zeta))} \int_{Q_r(\zeta)} |f - c|^p \, dv_\alpha \leq C.$$

(c) There exists a constant $C > 0$ such that

$$\frac{1}{v_\alpha(Q_r(\zeta))} \int_{Q_r(\zeta)} |f - f_{\alpha,Q_r(\zeta)}|^p \, dv_\alpha \leq C$$

for all $r > 0$ and all $\zeta \in \mathbb{S}_n$.

Proof. It is obvious that (c) implies (b). That (b) implies (c) follows from writing

$$f - f_{\alpha,Q_r(\zeta)} = f - c - \frac{1}{v_\alpha(Q_r(\zeta))} \int_{Q_r(\zeta)} (f - c) \, dv_\alpha$$

and applying the triangle and Hölder's inequalities.

To show that (a) implies (b), we fix $\zeta \in \mathbb{S}_n$, and without loss of generality, assume $0 < r < 1$. Define a point $a \in \mathbb{B}_n$ as in Lemma 5.23 so that $1 - |a|^2 \sim r^2$ and

$$D(a, R) \subset Q_r(\zeta),$$

where $R = 1$. Condition (a) implies that

$$\int_{\mathbb{B}_n} |f(z) - f(a)|^p \frac{(1 - |a|^2)^{n+1+\alpha}}{|1 - \langle z, a \rangle|^{2(n+1+\alpha)}} \, dv_\alpha(z) \leq M,$$

where $M > 0$ is a constant independent of a. In particular,

$$\int_{Q_r(\zeta)} |f(z) - f(a)|^p \frac{(1 - |a|^2)^{n+1+\alpha}}{|1 - \langle z, a \rangle|^{2(n+1+\alpha)}} \, dv_\alpha(z) \leq M.$$

For $z \in Q_r(\zeta)$, we have

$$|1 - \langle z, a \rangle| = |1 - |a|\langle z, \zeta \rangle|$$
$$= |1 - |a| + |a|(1 - \langle z, \zeta \rangle)|$$
$$\leq (1 - |a|) + |a||1 - \langle z, \zeta \rangle|$$
$$\leq \sigma r^2 + |a| r^2.$$

By Corollary 5.24, there exists another constant $M' > 0$ (independent of r and ζ) such that

$$\frac{1}{v_\alpha(Q_r(\zeta))} \int_{Q_r(\zeta)} |f(z) - f(a)|^p \, dv_\alpha(z) \leq M'$$

for all $\zeta \in \mathbb{S}_n$ and $0 < r < \sqrt{2}$.

It remains for us to prove (c) implies (a). So we fix $\zeta \in \mathbb{S}_n$ and $0 < r < 1$, and define $a \in \mathbb{B}_n$ according to Lemma 5.23. By condition (c), Lemma 5.23, and Corollary 5.24, there exists a constant $C > 0$ such that

$$\frac{1}{v_\alpha(D(a, R))} \int_{D(a,R)} |f(z) - c|^p \, dv_\alpha(z) \leq C,$$

where $c = f_{\alpha, Q_r(\zeta)}$. This easily implies that $f \in \mathcal{B}$. The proof of the theorem is now complete. $\square$

5.6 Atomic Decomposition

In this section we prove a decomposition theorem for the space BMOA. The decomposition and its proof are based on the Bergman kernel and related reproducing formulas.

Fix a parameter $b > n$ and fix a sequence $\{a_k\}$ satisfying the conditions in Theorem 2.23. The sequence $\{a_k\}$ induces a partition $\{D_k\}$ of $\mathbb{B}_n$. We also need the more dense sequence $\{a_{kj}\}$ and the associated finer partition $\{D_{kj}\}$ of $\mathbb{B}_n$ described in the preceding paragraphs of Lemma 2.29. Recall from Sections 2.5 and 3.6 that

$$Tf(z) = \int_{\mathbb{B}_n} \frac{(1 - |w|^2)^{b-n-1}}{|1 - \langle z, w \rangle|^b} f(w) \, dv(w),$$

and

$$Sf(z) = \sum_{k=1}^{\infty} \sum_{j=1}^{J} \frac{v_\alpha(D_{kj})f(a_{kj})}{(1 - \langle z, a_{kj} \rangle)^b},$$

where $\alpha = b - (n + 1)$.

For any $t > -1$ let QCM^t denote the space of Lebesgue measurable functions f in $\mathbb{B}_n$ for which

$$d\mu_f(z) = |f(z)|^2 (1 - |z|^2)^t \, dv(z)$$

is a Carleson measure. It is easy to check that QCM^t becomes a Banach space when equipped with the norm $\| \ \|_t$ defined by

$$\|f\|_t^2 = \sup \left\{ \frac{\mu_f(Q_s(\zeta))}{s^{2n}} : \zeta \in \mathbb{S}_n, 0 < s \leq 1 \right\}. \tag{5.16}$$

Theorem 5.26. *Suppose t and b satisfy*

$$0 < t + 1 < 2(b - n). \tag{5.17}$$

Then the operator T is bounded on QCM^t.

Proof. Suppose $f \in QCM^t$ and we must show that $g = Tf$ is also in QCM^t. Without loss of generality we may assume that $f \geq 0$.

Fix a Carleson tube $Q = Q_s(\zeta)$ and split f as $f = f_1 + f_2$, where $f_1 = f$ on $Q_{3s}(\zeta)$ and $f_1 = 0$ off $Q_{3s}(\zeta)$. We then have $g = g_1 + g_2$, where $g_k = Tf_k$ for $k = 1, 2$.

By Theorem 2.10, the assumptions in (5.17) tell us that the operator T is bounded on $L^2(\mathbb{B}_n, dv_t)$, so there exists a constant $C_1 > 0$ such that

$$\int_Q |g_1(z)|^2 (1 - |z|^2)^t \, dv(z) \leq \int_{\mathbb{B}_n} |g_1(z)|^2 (1 - |z|^2)^t \, dv(z)$$

$$\leq C_1 \int_{\mathbb{B}_n} |f_1(z)|^2 (1 - |z|^2)^t \, dv(z)$$

$$= C_1 \int_{Q_{3s}(\zeta)} |f(z)|^2 (1 - |z|^2)^t \, dv(z).$$

Since f is in QCM^t, there exists a constant $C_2 > 0$, independent of Q, such that

$$\int_Q |g_1(z)|^2 (1 - |z|^2)^t \, dv(z) \leq C_2 s^{2n}.$$

To estimate g_2, we consider the function h defined by

$$h(z) = \frac{1}{v(D(z))} \int_{D(z)} f_2(w) \, dv(w), \qquad z \in \mathbb{B}_n,$$

where $D(z) = D(z, \delta)$ is the Bergman metric ball about z with radius δ. Here δ is a positive constant so small that

$$D(z) \subset Q_{3s}(\zeta) \quad \text{whenever} \quad z \in Q_{2s}(\zeta). \tag{5.18}$$

To see that this is possible, choose δ so that

$$2\sqrt{\tanh \delta} < 5s^2.$$

Then $z \in Q_{2s}(\zeta)$ and $w \in D(z, \delta)$ imply that

$$|1 - \langle w, \zeta \rangle| \leq |1 - \langle z, \zeta \rangle| + |\langle z - w, \zeta \rangle| < 4s^2 + |z - w|.$$

Write $w = \varphi_z(u)$ with $|u| < \tanh \delta$ and use an identity from the proof of Lemma 3.3. We obtain

$$|z - w|^2 = |z - \varphi_z(u)|^2 = \frac{(1 - |z|^2)(|u|^2 - |\langle u, z \rangle|^2)}{|1 - \langle z, u \rangle|^2}$$

$$\leq \frac{(1 + |z|)(1 - |z|)(|u| - |\langle u, z \rangle|)(|u| + |\langle u, z \rangle|)}{(1 - |\langle z, u \rangle|)^2}$$

$$\leq \frac{2(1 - |z|)(1 - |\langle u, z \rangle|)(2|u|)}{(1 - |z|)(1 - |\langle z, u \rangle|)}$$

$$= 4|u|.$$

Therefore,

$$|1 - \langle w, \zeta \rangle| < 4s^2 + 2\sqrt{\tanh \delta} < 9s^2.$$

This shows that $D(z) \subset Q_{3s}(\zeta)$ whenever $z \in Q_{2s}(\zeta)$.

By Hölder's inequality,

$$|h(z)|^2 \le \frac{1}{v(D(z))} \int_{D(z)} |f_2(w)|^2 \, dv(w).$$

Since $v(D(z)) \sim (1 - |z|^2)^{n+1}$ and $1 - |w|^2 \sim 1 - |z|^2$ for $w \in D(z)$, we can find a constant $C_3 > 0$ such that

$$|h(z)|^2 \le \frac{C_3}{(1 - |z|^2)^{n+1+t}} \int_{D(z)} |f(w)|^2 (1 - |w|^2)^t \, dv(w)$$

for all $z \in \mathbb{B}_n$. Since f is in QCM^t, an application of Lemma 5.23 produces another constant $C_4 > 0$ such that

$$|h(z)|^2 \le \frac{C_4 (1 - |z|^2)^n}{(1 - |z|^2)^{n+1+t}} = C_4 (1 - |z|^2)^{-1-t}$$

for all $z \in \mathbb{B}_n$. Also, it follows from (5.18) that $h(z) = 0$ for all $z \in Q_{2s}(\zeta)$.

If $z \in Q$, then

$$|Th(z)| \le \sqrt{C_4} \int_{\mathbb{B}_n - Q_{2s}(\zeta)} \frac{(1 - |w|^2)^{b-(n+1)-\frac{1}{2}-\frac{t}{2}}}{|1 - \langle z, w \rangle|^b} \, dv(w).$$

Write

$$b = \epsilon + (b - \epsilon),$$

where $\epsilon > 0$ is small enough so that

$$b - (n+1) - \frac{t+1}{2} - \epsilon > -1, \qquad \frac{t+1}{2} > \epsilon.$$

Estimating $|1 - \langle z, w \rangle|^\epsilon$ by the triangle inequality

$$|1 - \langle z, w \rangle|^{\frac{1}{2}} = d(z, w) \ge d(\zeta, w) - d(\zeta, z) \ge 2s - s = s,$$

we obtain a constant $C_5 > 0$ such that

$$|Th(z)| \le \frac{C_5}{s^{2\epsilon}} \int_{\mathbb{B}_n} \frac{(1 - |w|^2)^{b-(n+1)-\frac{1}{2}-\frac{t}{2}}}{|1 - \langle z, w \rangle|^{b-\epsilon}} \, dv(w)$$

for $z \in Q$. Appealing to Theorem 1.12, we get another constant $C_6 > 0$ such that

$$|Th(z)| \le C_6 s^{-2\epsilon} (1 - |z|^2)^{\epsilon - \frac{t+1}{2}}$$

for all $z \in Q$. It follows that

$$\int_Q |Th(z)|^2 (1 - |z|^2)^t \, dv(z) \le C_6^2 s^{-4\epsilon} \int_Q (1 - |z|^2)^{2\epsilon - 1} \, dv(z).$$

The last integral is easily seen to be dominated by $s^{2n+4\epsilon}$. Therefore,

$$\int_Q |Th(z)|^2 (1 - |z|^2)^t \, dv(z) \le C_7 s^{2n},$$

where C_7 is a positive constant independent of Q.

It remains to show that $g_2 = Tf_2$ is dominated by Th. To this end, we use Fubini's theorem. First notice that $v(D(w))$ is comparable to $(1 - |w|^2)^{n+1}$, so there exists a constant $\delta_1 > 0$ such that

$$Th(z) \ge \delta_1 \int_{\mathbb{B}_n} \frac{(1 - |w|^2)^{b - 2(n+1)} \, dv(w)}{|1 - \langle z, w \rangle|^b} \int_{\mathbb{B}_n} f_2(u) \chi_{D(w)}(u) \, dv(u).$$

Since

$$\chi_{D(w)}(u) = \chi_{D(u)}(w),$$

a use of Fubini's theorem gives

$$Th(z) \ge \delta_1 \int_{\mathbb{B}_n} f_2(u) \, dv(u) \int_{D(u)} \frac{(1 - |w|^2)^{b - 2(n+1)} \, dv(w)}{|1 - \langle z, w \rangle|^b}.$$

By Lemmas 2.24 and 1.24, the inner integral above dominates

$$\frac{(1 - |u|^2)^{b - (n+1)}}{|1 - \langle z, u \rangle|^b}.$$

We conclude that there exists a constant $\delta_2 > 0$ such that

$$Th(z) \ge \delta_2 \int_{\mathbb{B}_n} \frac{(1 - |u|^2)^{b - (n+1)}}{|1 - \langle z, u \rangle|^b} f_2(u) \, dv(u) = \delta_2 Tf_2(z).$$

This completes the proof of the theorem.   □

For $a \in \mathbb{B}_n$ we use δ_a to denote the unit point-mass at the point a.

Lemma 5.27. *Suppose $R > 0$ and $\{a_k\}$ is any sequence in $\mathbb{B}_n$. Then the measure*

$$d\mu = \sum_k |c_k|^2 (1 - |a_k|^2)^n \delta_{a_k}$$

is Carleson if and only if the measure

$$d\lambda(z) = \sum_k \frac{|c_k|^2}{(1 - |a_k|^2)^2} (1 - |z|^2) \chi_k(z) \, dv(z)$$

is Carleson, where χ_k is the characteristic function of the Bergman ball $D(a_k, R)$.

Proof. For any $a \in \mathbb{B}_n$ we have

$$\int_{\mathbb{B}_n} \frac{(1 - |a|^2)^n \, d\lambda(z)}{|1 - \langle a, z \rangle|^{2n}} = \sum_k \frac{|c_k|^2}{(1 - |a_k|^2)^2} \int_{D(a_k, R)} \frac{(1 - |a|^2)^n (1 - |z|^2) \, dv(z)}{|1 - \langle a, z \rangle|^{2n}}.$$

It follows from Lemma 2.20 that

$$1 - |z|^2 \sim 1 - |a_k|^2 \qquad (k \to \infty)$$

for $z \in D(a_k, R)$, and it follows from (2.20) that

$$|1 - \langle a, z \rangle| \sim |1 - \langle a, a_k \rangle| \qquad (k \to \infty)$$

uniformly in a for $z \in D(a_k, R)$. Therefore,

$$\int_{\mathbb{B}_n} \frac{(1 - |a|^2)^n}{|1 - \langle a, z \rangle|^{2n}} \, d\lambda(z) \sim \sum_k \frac{|c_k|^2}{1 - |a_k|^2} \frac{(1 - |a|^2)^n}{|1 - \langle a, a_k \rangle|^{2n}} v(D(a_k, R)).$$

Since

$$v(D(a_k, R)) \sim (1 - |a_k|^2)^{n+1}$$

as $k \to \infty$ (see Lemma 1.24), we conclude that

$$\int_{\mathbb{B}_n} \frac{(1 - |a|^2)^n}{|1 - \langle a, z \rangle|^{2n}} \, d\lambda(z) \sim \sum_k |c_k|^2 (1 - |a_k|^2)^n \frac{(1 - |a|^2)^n}{|1 - \langle a, a_k \rangle|^{2n}}$$

$$= \int_{\mathbb{B}_n} \frac{(1 - |a|^2)^n}{|1 - \langle a, z \rangle|^{2n}} \, d\mu(z).$$

According to Theorem 5.4, μ is a Carleson measure if and only if λ is a Carleson measure. $\qquad\square$

Lemma 5.28. *Let $\{a_k\}$ be a sequence satisfying the conditions in Theorem 2.23. If a sequence $\{c_k\}$ has the property that*

$$\sum_k |c_k|^2 (1 - |a_k|^2)^n \delta_{a_k}$$

is a Carleson measure, then the function

$$f(z) = \sum_k c_k \left(\frac{1 - |a_k|^2}{1 - \langle z, a_k \rangle} \right)^b \tag{5.19}$$

belongs to BMOA whenever $b > n$.

Proof. For each k let $E_k = D(a_k, r/4)$ denote the Bergman metric ball about a_k with radius $r/4$. Consider the function

$$h(z) = \sum_{k=1}^{\infty} \frac{|c_k| \chi_k(z)}{1 - |a_k|^2},$$

where χ_k is the characteristic function of the set E_k. Since the sets E_k are disjoint, the measure

$$d\mu = |h(z)|^2 (1 - |z|^2) \, dv(z)$$

can be written as

$$d\mu = \sum_{k=1}^{\infty} \frac{|c_k|^2}{(1 - |a_k|^2)^2} (1 - |z|^2) \chi_k(z) \, dv(z).$$

It follows from Lemma 5.27 and the assumption on $\{c_k\}$ that μ is a Carleson measure, that is, h belongs to QCM^1.

Let T be the operator defined using the parameter $b + 1$. Then T is bounded on QCM^1 by Theorem 5.26. In particular, the function Th is in QCM^1.

Next consider

$$Th(z) = \sum_k \frac{|c_k|}{1 - |a_k|^2} \int_{E_k} \frac{(1 - |w|^2)^{b-n}}{|1 - \langle z, w \rangle|^{b+1}} \, dv(w).$$

Since $1 - |w|^2 \sim 1 - |a_k|^2$ for $w \in E_k$ and $v(E_k) \sim (1 - |a_k|^2)^{n+1}$, there exists a constant $\delta_1 > 0$ such that

$$Th(z) \geq \delta_1 \sum_k |c_k| (1 - |a_k|^2)^b \frac{1}{v(E_k)} \int_{E_k} \frac{dv(w)}{|1 - \langle z, w \rangle|^{b+1}}.$$

By Lemma 2.24, there exists a constant $\delta_2 > 0$ such that

$$Th(z) \geq \delta_2 \sum_k |c_k| (1 - |a_k|^2)^b \frac{1}{|1 - \langle z, a_k \rangle|^{b+1}}.$$

Since

$$Rf(z) = b \sum_k c_k \frac{\langle z, a_k \rangle (1 - |a_k|^2)^b}{(1 - \langle z, a_k \rangle)^{b+1}},$$

we have

$$Th(z) \geq \delta_2 \sum_k |c_k| \frac{(1 - |a_k|^2)^b}{|1 - \langle z, a_k \rangle|^{b+1}} \geq \frac{\delta_2}{b} |Rf(z)|.$$

Since Th is in QCM^1, the measure

$$|Rf(z)|^2 (1 - |z|^2) \, dv(z)$$

is Carleson, which, according to Theorem 5.14, shows that f is in BMOA . $\qquad \square$

Note that Lemma 5.28 remains true if we replace $\{a_k\}$ by the more dense sequence $\{a_{kj}\}$. In fact, if the measure

$$\sum_{k=1}^{\infty} \sum_{j=1}^{J} |c_{kj}|^2 (1 - |a_{kj}|^2)^n \delta_{a_{kj}}$$

is Carleson, then by Lemma 5.27, the measure

$$\sum_{k=1}^{\infty} \sum_{j=1}^{J} \frac{|c_{kj}|^2}{(1 - |a_{kj}|^2)^2} (1 - |z|^2) \chi_{D(a_{kj}, 2r)}(z) \, dv(z)$$

is Carleson. Since $1 - |a_{kj}|^2$ is comparable to $1 - |a_k|^2$ and $D(a_{kj}, 2r)$ contains $D(a_k, r)$, we see that the measure

$$\sum_{k=1}^{\infty} \frac{|d_k|^2}{(1 - |a_k|^2)^2} (1 - |z|^2) \chi_{D(a_k, r)}(z) \, dv(z)$$

is Carleson, where $|d_k|^2 = |c_{k1}|^2 + \cdots + |c_{kJ}|^2$. By Lemma 5.27 again, the measure

$$\sum_{k=1}^{\infty} |d_k|^2 (1 - |a_k|^2)^n \delta_{a_k}$$

is Carleson. Now if

$$f(z) = \sum_{k=1}^{\infty} \sum_{j=1}^{J} c_{kj} \left(\frac{1 - |a_{kj}|^2}{1 - \langle z, a_{kj} \rangle} \right)^b,$$

then an application (2.20) shows that

$$|Rf(z)| \leq C \sum_{k=1}^{\infty} \left(\sum_{j=1}^{J} |c_{kj}| \right) \frac{(1 - |a_k|^2)^b}{|1 - \langle z, a_k \rangle|^{b+1}}.$$

Since

$$\left(\sum_{j=1}^{J} |c_{kj}| \right)^2 \leq J |d_k|^2,$$

the desired estimate now follows from the proof of Lemma 5.28.

We can now prove the main result of this section.

Theorem 5.29. *For any $b > n$ there exists a sequence $\{a_k\}$ in $\mathbb{B}_n$ such that the space* BMOA *consists exactly of functions of the form*

$$f(z) = \sum_k c_k \left(\frac{1 - |a_k|^2}{1 - \langle z, a_k \rangle} \right)^b, \tag{5.20}$$

where the sequence $\{c_k\}$ has the property that

$$\sum_k |c_k|^2 (1 - |a_k|^2)^n \delta_{a_k}$$

is a Carleson measure, and the series in (5.20) converges to f in the weak-star topology of BMOA .

Proof. By Lemma 5.28 and the remark following it, every function f defined by (5.20) is in BMOA , as long as $\{a_k\}$ satisfies the conditions of Theorem 2.23, or when $\{a_k\}$ is replaced by the more dense sequence $\{a_{kj}\}$.

To show that every function in BMOA admits an atomic representation, we let

$$X = QCM^1 \cap H(\mathbb{B}_n),$$

let S and T be the operators on X defined with the parameter $b + 1$, and let the separation constants r for $\{a_k\}$ and η for $\{a_{kj}\}$ be chosen so that the constant $c = C\sigma$ from Lemma 3.22 satisfies $c\|T\| < 1$. By Lemma 3.22, the operator $I - S$ is bounded on X, and its norm on X satisfies $\|I - S\| < 1$, where I is the identity operator. In particular, S is invertible on X.

Fix f in BMOA and let $g = R^{\alpha,1} f$, where $\alpha = b - (n + 1)$. Since $R^{\alpha,1}$ is a linear partial differential operator of order 1 (see Proposition 1.15), it follows from Theorem 5.14 that $g \in X$. Since S is invertible on X, there exists a function $h \in X$ such that $g = Sh$. Thus g admits the representation

$$g(z) = \sum_{kj} \frac{v_\beta(D_{kj}) h(a_{kj})}{(1 - \langle z, a_{kj}\rangle)^{b+1}},$$

where

$$\beta = (b + 1) - (n + 1) = b - n.$$

Applying the inverse of $R^{\alpha,1}$ to both sides, we obtain

$$f(z) = \sum_{kj} \frac{v_\beta(D_{kj}) h(a_{kj})}{(1 - \langle z, a_{kj}\rangle)^b}.$$

Let

$$c_{kj} = \frac{v_\beta(D_{kj}) h(a_{kj})}{(1 - |a_{kj}|^2)^b}, \qquad k \geq 1, 1 \leq j \leq J.$$

Then

$$f(z) = \sum_{kj} c_{kj} \left(\frac{1 - |a_{kj}|^2}{1 - \langle z, a_{kj}\rangle} \right)^b.$$

It remains for us to show that the measure

$$\sum_{kj} |c_{kj}|^2 (1 - |a_{kj}|^2)^n \delta_{a_{kj}}$$

is Carleson. Since

$$v_\beta(D_{kj}) \le v_\beta(D_k) \sim (1 - |a_k|^2)^{n+1+\beta} = (1 - |a_k|^2)^{b+1} \sim (1 - |a_{kj}|^2)^{b+1},$$

it suffices for us to show that the measure

$$d\mu = \sum_{kj}(1 - |a_k|^2)^{n+2}|h(a_{kj})|^2\, \delta_{a_{kj}}$$

is Carleson.

For any $F \in H^2$, we have

$$\int_{\mathbb{B}_n} |F(z)|^2\, d\mu(z) = \sum_{kj}(1 - |a_k|^2)^{n+2}|h(a_{kj})|^2|F(a_{kj})|^2.$$

We use Lemma 2.24 to find a constant $C_1 > 0$ such that

$$|h(a_{kj})F(a_{kj})|^2 \le \frac{C_1}{v(D(a_{kj}, r))} \int_{D(a_{kj}, r)} |h(z)|^2|F(z)|^2\, dv(z)$$

for all k and j. We have

$$v(D(a_{kj}, r)) \sim (1 - |a_{kj}|^2)^{n+1} \sim (1 - |a_k|^2)^{n+1},$$

and

$$1 - |a_k|^2 \sim 1 - |z|^2, \qquad z \in D(a_k, R),$$

and $D(a_{kj}, r) \subset D(a_k, 2r)$. So we can find another constant $C_2 > 0$ such that

$$\int_{\mathbb{B}_n} |F(z)|^2\, d\mu(z) \le C_2 \sum_k \int_{D(a_k, 2r)} |F(z)|^2|h(z)|^2(1 - |z|^2)\, dv(z)$$

$$\le C_2 N \int_{\mathbb{B}_n} |F(z)|^2|h(z)|^2(1 - |z|^2)\, dv(z)$$

$$\le C_3 \int_{\mathbb{S}_n} |F(\zeta)|^2\, d\sigma(\zeta).$$

The last inequality is based on Theorem 5.9 and the assumption that

$$|h(z)|^2(1 - |z|^2)\, dv(z)$$

is Carleson. Using Theorem 5.9 one more time, we conclude that the measure μ is Carleson. This completes the proof of the theorem.   □

Using vanishing Carleson measures and the little oh version of the space X in the proof of the preceding theorem, we can also prove the following atomic decomposition for functions in VMOA . We leave the details to the interested reader.

Theorem 5.30. *For any $b > n$ there exists a sequence $\{a_k\}$ in $\mathbb{B}_n$ such that VMOA consists exactly of functions of the form*

$$f(z) = \sum_{k=1}^{\infty} c_k \frac{(1 - |a_k|^2)^b}{(1 - \langle z, a_k \rangle)^b},$$

where $\{c_k\}$ has the property that the measure

$$\sum_{k=1}^{\infty} |c_k|^2 (1 - |a_k|^2)^n \delta_{a_k}$$

is a vanishing Carleson measure.

Notes

The papers [35] and [36] of Fefferman and Stein are the original references for the theory of BMO. These papers discuss BMO in the context of $\mathbb{R}^n$. However, their ideas and techniques are readily applicable in the setting of the unit ball. In particular, Garnett's book [42] spells out the details for the open unit disk.

The proofs of Theorems 5.7, 5.8, and 5.9 are adapted from Hörmander's paper [54]. The key to all these results is the covering Lemma 5.6.

Sarason's paper [96] is probably the first one to study the space VMO systematically. In most situations, theorems concerning VMO are simply the little oh version of the corresponding results for BMO.

Duality between BMOA and H^1 is one of the highlights in the theory of BMO. The approach in Garnett's book does not work in higher dimensional cases, because the one-dimensional inner-outer factorization is used at a critical step. Our approach here includes a new ingredient, the use of Theorem 4.22.

The theory of BMO and VMO in the Bergman metric was first introduced in [121] in the case of the open unit disk and then fully developed in [19] in the context of bounded symmetric domains, of which the open unit ball is a special case.

There are several types of decomposition theorems for BMO. Theorem 5.29, based on Bergman type kernels, is due to Rochberg and Semmes [91].

Exercises

5.1. Show that the Cauchy transform maps BMO boundedly onto BMOA, and VMO boundedly onto VMOA. Here, BMO and VMO are subspaces of $L^2(\mathbb{S}_n, d\sigma)$ consisting of functions with bounded and vanishing mean oscillations, respectively.

5.2. Show that there exists a constant $C > 0$ such that

$$\int_{\mathbb{B}_n} |f(z)| |z|^{-2n} \log \frac{1}{|z|} \, dv(z) \leq C \int_{\mathbb{B}_n} |f(z)| (1 - |z|^2) \, dv(z)$$

for all holomorphic functions f in $\mathbb{B}_n$ with $f(0) = 0$.

5.3. Show that the Green function $G(z)$ for the invariant Laplacian satisfies the following estimates:

$$G(z) \geq \frac{n+1}{4n^2}(1-|z|^2)^n, \qquad z \in \mathbb{B}_n - \{0\},$$

and

$$G(z) \leq 16^n(1-|z|^2)^n, \qquad \frac{1}{4} < |z| < 1.$$

5.4. Show that the function $z \mapsto \log(1 - \langle z, \zeta \rangle)$ is in BMOA for any $\zeta \in \mathbb{S}_n$.

5.5. Show that VMOA contains the ball algebra.

5.6. Suppose f is holomorphic in $\mathbb{B}_n$ and $f(0) = 0$. Show that the measure

$$|f(z)|(1-|z|^2)\, dv(z)$$

is Carleson if and only if the measure

$$|f(z)||z|^{-2n}\log\frac{1}{|z|}\, dv(z)$$

is Carleson.

5.7. If $F : \mathbb{B}_n \to \mathbb{D}$ is Lipschitz, that is,

$$|F(z) - F(w)| \leq C|z - w|, \qquad z, w \in \mathbb{B}_n,$$

then $f \circ F$ belongs to BMOA of $\mathbb{B}_n$ whenever f is in the Bloch space of the unit disk $\mathbb{D}$. See [4].

5.8. Characterize the pointwise multipliers of BMOA and VMOA. See [103].

5.9. Develop the analagous theory for VMOA in the Bergman metric.

5.10. Suppose $f \in$ BMOA and $a = (a_1, \cdots, a_n) \in \mathbb{B}_n$. Show that there exist functions $f_k \in$ BMOA, $1 \leq k \leq n$, such that

$$f(z) - f(a) = \sum_{k=1}^{n}(z_k - a_k)f_k(z), \qquad z \in \mathbb{B}_n.$$

Do the same for VMOA.

5.11. Suppose $p \geq 1$, $\alpha > -1$, and $r > 0$. Define $\mathrm{BMO}_\partial(p, \alpha, r)$ to be the space of functions $f \in L^p(\mathbb{B}_n, dv_\alpha)$ such that

$$\sup_{z \in \mathbb{B}_n} \frac{1}{v_\alpha(D(z,r))} \int_{D(z,r)} |f(w) - c|^p\, dv_\alpha(w) < \infty,$$

where

$$c = \frac{1}{v_\alpha(D(z,r))} \int_{D(z,r)} f(u)\, dv_\alpha(u).$$

Show that the space $\mathrm{BMO}_\partial(p, \alpha, r)$ is independent of r. Therefore, we can write $\mathrm{BMO}_\partial(p, \alpha)$ for $\mathrm{BMO}_\partial(p, \alpha, r)$. For this and the next three problems, see [131].

5.12. Show that $\mathrm{BMO}_\partial(p, \alpha)$ consists exactly of functions f satisfying

$$\sup_{\zeta \in \mathbb{S}_n, r>0} \frac{1}{v_\alpha(Q_r(\zeta))} \int_{Q_r(\zeta)} |f(w) - c|^p \, dv_\alpha(w) < \infty,$$

where

$$c = \frac{1}{v_\alpha(Q_r(\zeta))} \int_{Q_r(\zeta)} f(w) \, dv_\alpha(w).$$

5.13. For $\alpha > -1$ and $f \in L^1(\mathbb{B}_n, dv_\alpha)$ define

$$B^\alpha f(z) = (1 - |z|^2)^{n+1+\alpha} \int_{\mathbb{B}_n} \frac{f(w) \, dv_\alpha(w)}{|1 - \langle z, w \rangle|^{2(n+1+\alpha)}}, \qquad z \in \mathbb{B}_n.$$

This is a weighted version of the Berezin transform. If $f \in \mathrm{BMO}_\partial(p, \alpha)$, show that there exists a constant $C > 0$ such that

$$|B^\alpha f(z) - B^\alpha f(w)| \le C\beta(z, w)$$

for all z and w in $\mathbb{B}_n$.

5.14. Show that a function f in $\mathbb{B}_n$ belongs to $\mathrm{BMO}_\partial(p, \alpha)$ if and only if $f = f_1 + f_2$, where f_1 satisfies

$$\sup_{a \in \mathbb{B}_n} \int_{\mathbb{B}_n} |f \circ \varphi_a(z)|^p \, dv_\alpha(w) < \infty,$$

and f_2 satisfies

$$\sup \left\{ \frac{|f_2(z) - f_2(w)|}{\beta(z, w)} : z, w \in \mathbb{B}_n, z \ne w \right\} < \infty.$$

5.15. Show that there exists a function f in VMOA such that f cannot be approximated by its Taylor polynomials in the norm topology of BMOA.

5.16. Show that QCM^t is a Banach space when equipped with the norm given in (5.16).

5.17. Suppose $b > n$ and $t > -1$ are parameters such that the operator T is bounded on QCM^t. If M is a positive constant greater than $\|T\|$ (the norm of T on QCM^t) and

$$\tilde{T} = \left(I - \frac{T}{M} \right)^{-1},$$

then for every holomorphic $f \in QCM^t$ the function

$$g(z) = \sum_k \tilde{T} f(a_k) \chi_k(z)$$

belongs to QCM^t, where χ_k is the characteristic function of the Bergman metric ball $D(a_k, R)$ and R is any fixed, sufficiently small radius.

5.18. For any $b > n$ and $R > 0$ there exists a positive constant C such that

$$T^k f(z) \le \frac{C}{v(D(z))} \int_{D(z)} T^k f(w)\, dv(w)$$

for all $f \ge 0$, $k \ge 1$, and all $z \in \mathbb{B}_n$. Here $D(z) = D(z, R)$ is the Bergman metric ball about z with radius R.

5.19. Show that there exist positive constants A and B such that

$$\int_{\mathbb{S}_n} e^{|f|}\, d\sigma \le A$$

for all f in BMOA with $\|f\|_{\mathrm{BMO}} \le B$. This is the John-Nirenberg theorem for the ball; see [6] and [42].

5.20. If $1 < p < \infty$, show that ultra-weak convergence in H^p is the same as weak convergence, which is also the same as weak-star convergence.

5.21. If $p = 1$, show that ultra-weak convergence in H^1 is the same as weak-star convergence but different from weak convergence.

5.22. Suppose $0 < p < \infty$ and

$$f(z) = \sum_{k=1}^{\infty} a_k z^{n_k}$$

is a lacunary series in d. Show that the following conditions are equivalent.

(a) f is in H^p.
(b) f is in BMOA .
(c) f is in VMOA .

See [43] and references there.

5.23. Show that BMOA is not separable.

5.24. Suppose φ is in H^2. Then the following conditions are equivalent.

(a) f is in BMOA .
(b) There exists a constant $C > 0$ such that

$$\int_{\mathbb{S}_n} |C(\varphi \bar{f})|^2\, d\sigma \le C \int_{\mathbb{S}_n} |f|^2\, d\sigma$$

for all $f \in H^2$, where $C(f)$ denotes the Cauchy-Szëgo projection of f.

5.25. Formulate and prove a little oh version of the above problem.

5.26. Suppose φ is in H^2. If there exists a constant $C > 0$ such that

$$\int_{\mathbb{S}_n} |\bar{\varphi} f - C(\bar{\varphi} f)|^2 \, d\sigma \leq C \int_{\mathbb{S}_n} |f|^2 \, d\sigma$$

for all $f \in H^2$, then φ must be in BMOA .

5.27. Formulate and prove a little oh version of the above problem.

5.28. Show that BMOA is contained in H^p for any $p > 0$.

5.29. Show that a function $f \in H^p$ is in BMOA if and only if there exists a constant $C > 0$ such that

$$\frac{1}{\sigma(Q)} \int_Q |f - f_Q|^p \, d\sigma \leq C$$

for all d-balls Q in $\mathbb{S}_n$.

6

Besov Spaces

In this chapter we study a class of holomorphic Besov spaces, B_p, for $0 < p \leq \infty$. When $p \geq 1$, the space B_p can be equipped with a (semi-)norm that is invariant under the action of the automorphism group. The space B_1 is the minimal Möbius invariant Banach space. The space B_2 plays the role of the Dirichlet space in higher dimensions. And the space B_∞ is just the Bloch space.

For each $0 < p < \infty$, the space B_p is the image of the Bergman space A_α^p under a suitable fractional integral operator. As a consequence, we obtain an atomic decomposition for functions in B_p. We also discuss complex interpolation and various duality issues for B_p.

6.1 The Spaces B_p

The Möbius invariant measure $d\tau$ plays a prominent role in this chapter. So recall that

$$d\tau(z) = \frac{dv(z)}{(1 - |z|^2)^{n+1}}.$$

The fractional radial differential operators $R^{\alpha,t}$ and the fractional radial integral operators $R_{\alpha,t}$ from Section 1.4 will also be used frequently in this chapter.

Theorem 6.1. *Suppose $0 < p < \infty$ and f is holomorphic in $\mathbb{B}_n$. Then the following two conditions are equivalent:*

(a) The functions

$$(1 - |z|^2)^N \frac{\partial^m f}{\partial z^m}(z), \qquad |m| = N,$$

are in $L^p(\mathbb{B}_n, d\tau)$ for some positive integer $N > n/p$.

(b) The functions

$$(1 - |z|^2)^N \frac{\partial^m f}{\partial z^m}(z), \qquad |m| = N,$$

are in $L^p(\mathbb{B}_n, d\tau)$ for every positive integer $N > n/p$.

Proof. It suffices to show that the conditions

$$(1 - |z|^2)^{|m|} \frac{\partial^m f}{\partial z^m}(z) \in L^p(\mathbb{B}_n, d\tau), \qquad |m| = N, \tag{6.1}$$

are equivalent to

$$(1 - |z|^2)^{|m|} \frac{\partial^m f}{\partial z^m}(z) \in L^p(\mathbb{B}_n, d\tau), \qquad |m| = N + 1. \tag{6.2}$$

Clearly, (6.1) holds if and only if

$$\frac{\partial^m f}{\partial z^m} \in L^p(\mathbb{B}_n, dv_\alpha), \qquad |m| = N, \tag{6.3}$$

where $\alpha = Np - (n + 1)$. By Theorem 2.17, (6.3) holds if and only if

$$(1 - |z|^2) \frac{\partial^m f}{\partial z^m} \in L^p(\mathbb{B}_n, dv_\alpha), \qquad |m| = N + 1. \tag{6.4}$$

This is obviously equivalent to (6.2). □

For any $0 < p < \infty$ we now define the Besov space B_p to be the space of holomorphic functions f in $\mathbb{B}_n$ such that the functions

$$(1 - |z|^2)^N \frac{\partial^m f}{\partial z^m}(z), \qquad |m| = N,$$

all belong to $L^p(\mathbb{B}_n, d\tau)$, where N is any fixed positive integer satisfying $pN > n$. According to Theorem 6.1, the definition of B_p is independent of the positive integer N used.

Proposition 6.2. *Suppose $0 < p < \infty$. Then B_p is complete with the "norm" defined by*

$$\|f\|_p^p = \sum_{|m| \leq N-1} \left| \frac{\partial^m f}{\partial z^m}(0) \right|^p + \sum_{|m|=N} \int_{\mathbb{B}_n} \left| (1 - |z|^2)^N \frac{\partial^m f}{\partial z^m}(z) \right|^p \, d\tau(z),$$

where N is any positive integer satisfying $pN > n$. Furthermore, the polynomials are dense in B_p.

Proof. If $\{f_n\}$ is a Cauchy sequence in B_p, then each of

$$\frac{\partial^m f}{\partial z^m}(0), \qquad |m| \leq N - 1,$$

is a numerical Cauchy sequence, and each of

$$\frac{\partial^m f}{\partial z^m}(z), \qquad |m| = N,$$

is a Cauchy sequence in A^p_α, where $\alpha = pN - (n+1)$. The completeness of B_p then follows from the completeness of A^p_α.

Every function f in B_p can approximated in norm by its dilations f_r, and each dilation f_r can be uniformly approximated by its Taylor polynomials in a neighborhood of the closed unit ball. In particular, every function in B_p can be approximated in norm by a sequence of polynomials. □

Lemma 6.3. *Suppose* $0 < p < \infty$, $n + \alpha$ *is not a negative integer,* N *is a positive integer satisfying* $Np > n$, *and* f *is holomorphic in* $\mathbb{S}_n$. *Then* $f \in B_p$ *if and only if the function*

$$F_N(z) = (1 - |z|^2)^N R^{\alpha,N} f(z)$$

belongs to $L^p(\mathbb{B}_n, d\tau)$.

Proof. If f is in B_p, then the functions $\partial^m f / \partial z^m$ belong to $A^p_{pN-(n+1)}$ for all $|m| = N$. It is then easy to see that the functions $\partial^m f / \partial z^m$ belong to $A^p_{pN-(n+1)}$ for all $|m| \leq N$. It follows from Proposition 1.15 that $R^{\alpha,N} f$ is in $A^p_{pN-(n+1)}$, or that F_N is in $L^p(\mathbb{B}_n, d\tau)$.

Conversely, we show that there exists a constant $C > 0$ such that

$$\left\| \frac{\partial^m f}{\partial z^m} \right\| \leq C \| R^{\alpha,N} f \|$$

for all holomorphic f and all $|m| = N$, where the norm is that of $A^p_{pN-(n+1)}$.

Fix a sufficiently large positive integer K and let $\beta = \alpha + K$. Then

$$R^{\alpha,N} f(z) = \int_{\mathbb{B}_n} \frac{R^{\alpha,N} f(w)\, dv_\beta(w)}{(1 - \langle z, w \rangle)^{n+1+\beta}}, \qquad z \in \mathbb{B}_n,$$

and

$$f(z) = R_{\alpha,N} \int_{\mathbb{B}_n} \frac{R^{\alpha,N} f(w)\, dv_\beta(w)}{(1 - \langle z, w \rangle)^{n+1+\beta}}, \qquad z \in \mathbb{B}_n.$$

By Lemma 2.18, there exists a polynomial $p(z, w)$ such that

$$f(z) = \int_{\mathbb{B}_n} \frac{p(z, w) R^{\alpha,N} f(w)\, dv_\beta(w)}{(1 - \langle z, w \rangle)^{n+1+\beta-N}}, \qquad z \in \mathbb{B}_n.$$

Differentiating inside the integral sign, we obtain

$$\left| \frac{\partial^m f}{\partial z^m}(z) \right| \leq C \int_{\mathbb{B}_n} \frac{|R^{\alpha,N} f(w)|\, dv_\beta(w)}{|1 - \langle z, w \rangle|^{n+1+\beta}}$$

for all $z \in \mathbb{B}_n$ and all m with $|m| = N$. If $1 \leq p < \infty$, this along with Theorem 2.10 shows that

$$\left\| \frac{\partial^m f}{\partial z^m} \right\| \leq C \| R^{\alpha,N} f \|$$

for $|m| = N$. If $0 < p < 1$, we write

$$\beta = \frac{n+1+\alpha'}{p} - (n+1),$$

where we assume that K is large enough so that $\alpha' > Np - (n+1)$. According to Lemma 2.15,

$$\left|\frac{\partial^m f}{\partial z^m}(z)\right|^p \leq C \int_{\mathbb{B}_n} \frac{|R^{\alpha,N} f(w)|^p}{|1 - \langle z, w\rangle|^{p(n+1+\beta)}} \, dv_{\alpha'}(w).$$

By Fubini's theorem, the integral

$$\int_{\mathbb{B}_n} \left|\frac{\partial^m f}{\partial z^m}(z)\right|^p (1 - |z|^2)^{Np-(n+1)} \, dv(z)$$

is dominated by

$$\int_{\mathbb{B}_n} |R^{\alpha,N} f(w)|^p \, dv_{\alpha'}(w) \int_{\mathbb{B}_n} \frac{(1 - |z|^2)^{Np-(n+1)} \, dv(z)}{|1 - \langle z, w\rangle|^{p(n+1+\beta)}}.$$

Apply part (3) of Theorem 1.12 to the inner integral and observe that

$$p(n + 1 + \beta) - [Np - (n+1)] - (n+1) = n + 1 + \alpha' - Np > 0.$$

We obtain

$$\left\|\frac{\partial^m f}{\partial z^m}\right\| \leq C\|R^{\alpha,N} f\|$$

for $|m| = N$. This completes the proof of the lemma. $\qquad\square$

We now show that the spaces B_p can be described in terms of more general fractional radial derivatives.

Theorem 6.4. *Suppose $0 < p < \infty$, $n + \alpha$ is not a negative integer, and f is holomorphic in $\mathbb{B}_n$. Then the following three conditions are equivalent:*

(a) $f \in B_p$.
(b) The function $(1 - |z|^2)^t R^{\alpha,t} f(z)$ belongs to $L^p(\mathbb{B}_n, d\tau)$ for some $t > n/p$, where $n + \alpha + t$ is not a negative integer.
(c) The function $(1 - |z|^2)^t R^{\alpha,t} f(z)$ belongs to $L^p(\mathbb{B}_n, d\tau)$ for all $t > n/p$, where $n + \alpha + t$ is not a negative integer.

Proof. It follows from Lemma 6.3 that (c) implies (a), and that (a) implies (b).

To prove that (b) and (c) are equivalent, it suffices to show that the norms of the two functions

$$(1 - |z|^2)^t R^{\alpha,t} f(z), \qquad (1 - |z|^2)^s R^{\alpha,s} f(z)$$

in $L^p(\mathbb{B}_n, d\tau)$ are comparable for all $f \in H^\infty(\mathbb{B}_n)$, where s and t are any two fixed constants greater than n/p, with $n + \alpha + t$ and $n + \alpha + s$ not a negative integer.

Without loss of generality, we assume that $s = t + \sigma$ for some $\sigma > 0$. It is easy to see from the definition of the fractional radial derivatives that

$$R^{\alpha+t,\sigma} R^{\alpha,t} = R^{\alpha,s}. \tag{6.5}$$

Now the norm of $(1 - |z|^2)^t R^{\alpha,t} f(z)$ in $L^p(\mathbb{B}_n, d\tau)$ is the same as the norm of $R^{\alpha,t} f$ in $A^p_{pt-(n+1)}$, which, according to Theorem 2.19, is comparable to the norm of

$$(1 - |z|^2)^\sigma R^{\alpha+t,\sigma} R^{\alpha,t} f(z)$$

in $L^p(\mathbb{B}_n, dv_{pt-(n+1)})$. This together with (6.5) shows that the norms of

$$(1 - |z|^2)^t R^{\alpha,t} f(z), \qquad (1 - |z|^2)^s R^{\alpha,s} f(z),$$

are comparable in $L^p(\mathbb{B}_n, d\tau)$. $\square$

The above result can be restated as follows.

Corollary 6.5. *Suppose $0 < p < \infty$ and $t > n/p$. If α is a real parameter such that the operator $R^{\alpha,t}$ is well defined, then $R^{\alpha,t}$ is a bounded invertible operator from B^p onto $A^p_{pt-(n+1)}$.*

As a consequence of the above corollary, we obtain the following atomic decomposition for functions in the Besov spaces B_p.

Theorem 6.6. *Given any $p \in (0, \infty)$ there exists a sequence $\{a_k\}$ in $\mathbb{B}_n$ such that for each $b > \max(0, n(p-1)/p)$ the space B_p consists exactly of functions of the form*

$$f(z) = \sum_k c_k \left(\frac{1 - |a_k|^2}{1 - \langle z, a_k \rangle} \right)^b, \tag{6.6}$$

where $\{c_k\} \in l^p$.

Proof. Fix any $t > n/p$. Let b and b' be two real parameters related by $b' = b + t$. Then the condition

$$b > \max\left(0, \frac{n(p-1)}{p}\right) \tag{6.7}$$

is equivalent to

$$b' > n \max\left(1, \frac{1}{p}\right) + \frac{pt - (n+1) + 1}{p}; \tag{6.8}$$

simply check this for $0 < p < 1$ and $p \geq 1$, respectively.

We write

$$b' = b + t = n + 1 + \alpha + t. \tag{6.9}$$

If b satisfies (6.7), then $n + 1 + \alpha = b > 0$. In particular, $n + \alpha$ is not a negative integer. Obviously, $n + \alpha + t = b' - 1$ is not a negative integer, so the operator $R^{\alpha,t}$ is well defined.

By Corollary 6.5, a holomorphic function f in $\mathbb{B}_n$ belongs to B_p if and only if $R^{\alpha,t} f \in A^p_{pt-(n+1)}$. Combining this with Theorem 2.30, we conclude that $f \in B_p$ if and only if

$$R^{\alpha,t} f(z) = \sum_k c_k \frac{(1-|a_k|^2)^{b'-t}}{(1-\langle z, a_k\rangle)^{b'}}, \tag{6.10}$$

where $\{c_k\} \in l^p$ and b' satisfies (6.8). Applying $R_{\alpha,t}$ to both sides of equation (6.10) gives

$$f(z) = \sum_k c_k \left(\frac{1-|a_k|^2}{1-\langle z, a_k\rangle}\right)^{b'-t} = \sum_k c_k \left(\frac{1-|a_k|^2}{1-\langle z, a_k\rangle}\right)^b.$$

This completes the proof of the theorem. □

The next result gives integral representations for the Besov spaces B_p when $p \geq 1$ and shows that such a B_p can be considered a quotient space of $L^p(\mathbb{B}_n, d\tau)$.

Theorem 6.7. *Suppose $1 \leq p < \infty$ and $\alpha > -1$. Then $B_p = P_\alpha L^p(\mathbb{B}_n, d\tau)$.*

Proof. Fix a parameter t such that $t > n/p$ and such that $R^{\alpha,t}$ is well defined. By Corollary 6.5, a holomorphic function f in $\mathbb{B}_n$ belongs to B_p if and only if its fractional radial derivative $R^{\alpha,t} f$ belongs to the weighted Bergman space A^p_γ, where $\gamma = pt - (n+1)$.

Let $\beta = t + \alpha$. Then $p(\beta+1) > \gamma + 1$, so by Theorem 2.11,

$$A^p_\gamma = P_\beta L^p(\mathbb{B}_n, dv_\gamma).$$

It follows that $f \in B_p$ if and only if $R^{\alpha,t} f = P_\beta g$, where $g \in L^p(\mathbb{B}_n, dv_\gamma)$, or

$$R^{\alpha,t} f(z) = c_\beta \int_{\mathbb{B}_n} \frac{g(w)(1-|w|^2)^\beta \, dv(w)}{(1-\langle z, w\rangle)^{n+1+\alpha+t}}.$$

Apply the fractional integral operator $R_{\alpha,t}$ to both sides and use Proposition 1.14. We obtain

$$f(z) = \frac{c_\beta}{c_\alpha} \int_{\mathbb{B}_n} \frac{(1-|w|^2)^t g(w) \, dv_\alpha(w)}{(1-\langle z, w\rangle)^{n+1+\alpha}}.$$

Since $g \in L^p(\mathbb{B}_n, dv_\gamma)$ if and only if the function $(1-|w|^2)^t g(w)$ is in $L^p(\mathbb{B}_n, d\tau)$, we conclude that $f \in B_p$ if and only if $f \in P_\alpha L^p(\mathbb{B}_n, d\tau)$. □

6.2 The Minimal Möbius Invariant Space

Recall that the Bloch space is maximal among Möbius invariant Banach spaces of holomorphic functions in $\mathbb{B}_n$. We show here that the Besov space B_1 is the minimal Möbius invariant Banach space of holomorphic functions in $\mathbb{B}_n$.

Theorem 6.8. *The space B_1 consists exactly of holomorphic functions of the form*

$$f(z) = c_0 + \sum_{k=1}^{\infty} c_k f_k(z), \tag{6.11}$$

where $\{c_k\} \in l^1$ and each f_k is a component of some automorphism of $\mathbb{B}_n$.

Proof. For the purpose of this proof, we write

$$\|f\|_1 = \sum_{|m|=n+1} \int_{\mathbb{B}_n} \left| \frac{\partial^m f}{\partial z^m}(z) \right| \, dv(z),$$

where f is any holomorphic function in $\mathbb{B}_n$. Clearly, f is in B_1 if and only if $\|f\|_1 < \infty$.

Let $\varphi \in \mathrm{Aut}(\mathbb{B}_n)$, and for $1 \le k \le n$ let f_k denote the kth component of φ. To show that the series (6.11) defines a function f in B_1, it suffices to show that there exists a constant $C > 0$, independent of φ and k, such that $\|f_k\|_1 \le C$.

We first consider the case in which $\varphi = \varphi_a$, where $a = (\lambda, 0, \cdots, 0)$. In this case, we have

$$f_1(z) = \frac{\lambda - z_1}{1 - \bar{\lambda} z_1}, \qquad z \in \mathbb{B}_n,$$

and for $2 \le k \le n$,

$$f_k(z) = -\frac{\sqrt{1 - |\lambda|^2}\, z_k}{1 - \bar{\lambda} z_1}, \qquad z \in \mathbb{B}_n.$$

It follows from the definition of the semi-norm $\| \ \|_1$ that

$$\|f_1\|_1 = (n+1)! |\lambda|^n (1 - |\lambda|^2) \int_{\mathbb{B}_n} \frac{dv(z)}{|1 - \bar{\lambda} z_1|^{n+2}}$$

$$= (n+1)! |a|^n (1 - |a|^2) \int_{\mathbb{B}_n} \frac{dv(z)}{|1 - \langle z, a \rangle|^{n+2}},$$

and for $2 \le k \le n$,

$$\|f_k\|_1 = (n+1)! |\lambda|^{n+1} \int_{\mathbb{B}_n} \frac{\sqrt{1 - |\lambda|^2}\, |z_k|}{|1 - \bar{\lambda} z_1|^{n+2}} \, dv(z) + n! |\lambda|^n \int_{\mathbb{B}_n} \frac{\sqrt{1 - |\lambda|^2}\, dv(z)}{|1 - \bar{\lambda} z_1|^{n+1}},$$

which is less than or equal to

$$\sqrt{2}(n+1)! |a|^{n+1} \int_{\mathbb{B}_n} \frac{\sqrt{1 - |a|^2}\, dv(z)}{|1 - \langle z, a \rangle|^{n+1+1/2}} + n! |a|^n \int_{\mathbb{B}_n} \frac{\sqrt{1 - |a|^2}\, dv(z)}{|1 - \langle z, a \rangle|^{n+1}}.$$

Here we used the fact that

$$|z_k| < (1 - |z_1|^2)^{1/2} \le \sqrt{2}(1 - |z_1|)^{1/2} \le \sqrt{2}|1 - \langle z, a \rangle|^{1/2}, \qquad 2 \le k \le n.$$

By Theorem 1.12, we can find a constant $C > 0$, depending on the dimension n alone, such that $\|f_k\|_1 \leq C$ for all $1 \leq k \leq n$.

Next we consider the case in which $\varphi = U\varphi_a$, where U is a unitary and $a = (\lambda, 0, \cdots, 0)$. If $U = (u_{ij})_{n \times n}$ and

$$\varphi_a(z) = (f_1(z), \cdots, f_n(z)),$$

then the kth component of φ is

$$g_k = u_{k1}f_1 + u_{k2}f_2 + \cdots + u_{kn}f_n.$$

It follows from what was proved in the previous paragraph that $\|g_k\|_1 \leq nC$ for $1 \leq k \leq n$.

For a more general point $a \in \mathbb{B}_n$, we can find a unitary transformation U of $\mathbb{C}^n$ such that $a = Ua'$, where $a' = (\lambda, 0, \cdots, 0)$. Since the automorphism

$$V = \varphi_{a'} \circ U^* \circ \varphi_a$$

fixes the origin, it must be a unitary, according to Lemma 1.1. It follows that $\varphi_a = U\varphi_{a'}V$. Combining this with Theorem 1.4, we see that every automorphism φ can be written as $\varphi = U\varphi_a V$, where $a = (\lambda, 0, \cdots, 0)$, U and V are unitary transformations of $\mathbb{C}^n$. If $f_1, \cdots, f_n$ are the components of $U\varphi_a$, then the kth component of φ is $g_k(z) = f_k(Vz)$. It is easy to see from the chain rule that $\|g_k\|_1 \leq C'\|f_k\|_1$ for $1 \leq k \leq n$, where C' is a positive constant depending only on n.

We have now shown that each series (6.11) defines a function in B_1.

Conversely, if we have a function $f \in B_1$, then by the atomic decomposition of B_1 (Theorem 6.6), there exists a sequence $\{c_k\} \in l^1$ and a sequence $\{a_k\} \in \mathbb{B}_n$ such that

$$f(z) = \sum_k c_k \frac{1 - |a_k|^2}{1 - \langle z, a_k \rangle}.$$

Recall from Lemma 1.3 that

$$\frac{1 - |a_k|^2}{1 - \langle z, a_k \rangle} = 1 - \langle \varphi_{a_k}(z), a_k \rangle.$$

Clearly, each $\langle \varphi_{a_k}(z), a_k \rangle$ is a linear combination of component functions of automorphisms of $\mathbb{B}_n$. This clearly shows that f admits a representation (6.11). □

Corollary 6.9. *The space B_1 is a Möbius invariant Banach space with the following norm:*

$$\|f\|_m = \inf \left\{ \sum_{k=0}^{\infty} |c_k| : f = c_0 + \sum_{k=1}^{\infty} c_k f_k \right\}, \tag{6.12}$$

where each f_k is a component of an element in $\mathrm{Aut}(\mathbb{B}_n)$.

Proof. It is easy to see that B_1 is a Banach space under the norm $\| \ \|_m$.

If f_k is the jth component of $\varphi \in \mathrm{Aut}(\mathbb{B}_n)$, and if $\psi \in \mathrm{Aut}(\mathbb{B}_n)$, then $f_k \circ \psi$ is the jth component of the automorphism $\varphi \circ \psi$. It follows easily that $\|f \circ \psi\|_m = \|f\|_m$ for all $f \in B_1$. This shows that B_1 is a Möbius invariant Banach space. □

We now prove that B_1 is the smallest among all Möbius invariant Banach spaces of holomorphic functions in $\mathbb{B}_n$.

Theorem 6.10. *If X is any Möbius invariant Banach space of holomorphic functions in $\mathbb{B}_n$ and if X contains a nonconstant function, then B_1 is continuously contained in X.*

Proof. By Lemma 3.18, X contains all the polynomials. Composing the coordinate functions z_k with $\varphi \in \mathrm{Aut}(\mathbb{B}_n)$, we see that X contains all n components of any $\varphi \in \mathrm{Aut}(\mathbb{B}_n)$. Now if

$$f(z) = c_0 + \sum_{k=1}^{\infty} c_k f_k \qquad (6.13)$$

is an arbitrary function in B_1, where each f_k is a component of an automorphism of $\mathbb{B}_n$, and $\{c_k\} \in l^1$, then

$$\|f\|_X \leq |c_0| \|1\|_X + \sum_{k=1}^{\infty} |c_k| \|f_k\|_X \leq M \sum_{k=0}^{\infty} |c_k|,$$

where

$$M = \max(\|1\|_X, \|z_1\|_X, \|z_2\|_X, \cdots, \|z_n\|_X).$$

So the series (6.13) converges to f in X and

$$\|f\|_X \leq M \sum_{k=0}^{\infty} |c_k|.$$

Taking the infimum of $\sum_k |c_k|$ over all representations of f in (6.13), and applying Theorem 6.8, we conclude that

$$\|f\|_X \leq C \|f\|_m,$$

where C is a positive constant independent of f. This completes the proof of the theorem. $\qquad \square$

6.3 Möbius Invariance of B_p

In this section we show that each space B_p can be equipped with a Möbius invariant semi-norm when $1 < p \leq \infty$. We established in the previous section that B_1 can be equipped with a Möbius invariant norm.

Let

$$\lambda_n = \begin{cases} 1, & n = 1 \\ 2n, & n > 1. \end{cases}$$

This dimensional constant will appear many times in the rest of this chapter. Sometimes λ_n is also referred to as the cut-off constant. In many situations it is easier to describe the behavior of the Besov spaces B_p when $p > \lambda_n$.

Theorem 6.11. *Suppose $p > \lambda_n$ and f is holomorphic in $\mathbb{B}_n$. Then $f \in B_p$ if and only if $|\widetilde{\nabla} f(z)|$ is in $L^p(\mathbb{B}_n, d\tau)$.*

Proof. When $n = 1$ and $p > 1$, it follows from the definition of B_p that a holomorphic function f is in B_p if and only if the function $(1 - |z|^2) f'(z)$ is in $L^p(\mathbb{D}, d\tau)$. The desired result then follows from the fact that

$$|\widetilde{\nabla} f(z)| = (1 - |z|^2)|f'(z)|,$$

where f is any analytic function in the unit disk $\mathbb{D}$.

For the rest of this proof we assume that $n > 1$. Recall from Lemma 2.14 that

$$(1 - |z|^2)|\nabla f(z)| \leq |\widetilde{\nabla} f(z)|.$$

Therefore, f is in B_p whenever the function $|\widetilde{\nabla} f(z)|$ is in $L^p(\mathbb{B}_n, d\tau)$.

Conversely, if $f \in B_p$, then by Theorem 6.7, there exists a function g in $L^p(\mathbb{B}_n, d\tau)$ such that

$$f(z) = \int_{\mathbb{B}_n} \frac{g(w)\, dv(w)}{(1 - \langle z, w \rangle)^{n+1}}, \qquad z \in \mathbb{B}_n.$$

According to Lemma 3.3,

$$|\widetilde{\nabla} f(z)| \leq (n+1)\sqrt{2}\,(1 - |z|^2)^{1/2} \int_{\mathbb{B}_n} \frac{|g(w)|\, dv(w)}{|1 - \langle z, w \rangle|^{n+1+1/2}}, \qquad z \in \mathbb{B}_n. \tag{6.14}$$

If $p > 2n$, we conclude from Theorem 2.10 that the operator

$$Tg(z) = (1 - |z|^2)^{1/2} \int_{\mathbb{B}_n} \frac{g(w)\, dv(w)}{|1 - \langle z, w \rangle|^{n+1+1/2}}$$

is bounded on $L^p(\mathbb{B}_n, d\tau)$. Therefore, the estimate in (6.14) shows that the function $|\widetilde{\nabla} f(z)|$ is in $L^p(\mathbb{B}_n, d\tau)$. $\qquad\square$

As a consequence of the above theorem, we see that for $p > \lambda_n$ the space B_p can be equipped with the following complete, Möbius invariant semi-norm.

$$\|f\|_{B_p} = \left[\int_{\mathbb{B}_n} |\widetilde{\nabla} f(z)|^p \, d\tau(z) \right]^{\frac{1}{p}}.$$

The remaining range of p is more difficult. We are going to use the technique of complex interpolation to settle this case.

Theorem 6.12. *Suppose $1 \leq p_0 < p_1 \leq \infty$. If $\theta \in (0, 1)$ and*

$$\frac{1}{p} = \frac{1 - \theta}{p_0} + \frac{\theta}{p_1},$$

then $[B_{p_0}, B_{p_1}]_\theta = B_p$ with equivalent norms.

Proof. For the purpose of this proof let $\|\ \|_p$ denote the quotient norm on B_p induced by the mapping $P : L^p(\mathbb{B}_n, d\tau) \to B_p$, where P is the Bergman projection. The basis for our analysis is Theorem 6.7.

If $f \in B_p$, then there exists a function $g \in L^p(\mathbb{B}_n, d\tau)$ such that $f = Pg$ with

$$\|g\|_{L^p(d\tau)} \le C_1 \|f\|_p,$$

where C_1 is a positive constant independent of f. For any complex number ζ with $0 \le \operatorname{Re}\zeta \le 1$ define

$$g_\zeta(z) = \frac{g(z)}{|g(z)|}|g(z)|^{p\left(\frac{1-\zeta}{p_0} + \frac{\zeta}{p_1}\right)}, \qquad z \in \mathbb{B}_n,$$

and set $f_\zeta = Pg_\zeta$. It is clear that $\|g_\zeta\|_{p_0}^{p_0} = \|g\|_p^p$ for $\operatorname{Re}\zeta = 0$ and $\|g_\zeta\|_{p_1}^{p_1} = \|g\|_p^p$ for $\operatorname{Re}\zeta = 1$. By Theorem 6.7, $\|f_\zeta\|_{p_0}^{p_0} \le C_2 \|f\|_p^p$ for all $\operatorname{Re}\zeta = 0$, and $\|f_\zeta\|_{p_1}^{p_1} \le C_2 \|f\|_p^p$ for all $\operatorname{Re}\zeta = 1$, where C_2 is a positive constant independent of f. Since $f_\theta = f$, we conclude that $f \in [B_{p_0}, B_{p_1}]_\theta$ with $\|f\|_\theta \le C_2 \|f\|_p$.

Conversely, if $f \in [B_{p_0}, B_{p_1}]_\theta$, then there exists f_ζ for $\operatorname{Re}\zeta \in [0,1]$ with the properties that $f_\theta = f$, $f_\zeta \in B_{p_0}$ when $\operatorname{Re}\zeta = 0$, and $f_\zeta \in B_{p_1}$ when $\operatorname{Re}\zeta = 1$. Define

$$g_\zeta(z) = c_{n+1}(1 - |z|^2)^{n+1} \int_{\mathbb{B}_n} \frac{f_\zeta(w)\, dv(w)}{(1 - \langle z, w\rangle)^{2(n+1)}}, \qquad 0 \le \operatorname{Re}\zeta \le 1, z \in \mathbb{B}_n,$$

or equivalently,

$$g_\zeta(z) = c_{n+1}(1 - |z|^2)^{n+1} R^{0,n+1} f_\zeta(z).$$

We see from Theorem 6.4 that $g_\zeta \in L^{p_0}(\mathbb{B}_n, d\tau)$ for $\operatorname{Re}\zeta = 0$, with the norm of g_ζ in $L^{p_0}(\mathbb{B}_n, d\tau)$ dominated by $\|f_\zeta\|_{p_0}$; and that $g_\zeta \in L^{p_1}(\mathbb{B}_n, d\tau)$ for $\operatorname{Re}\zeta = 1$, with the norm of g_ζ in $L^{p_1}(\mathbb{B}_n, d\tau)$ dominated by $\|f_\zeta\|_{p_1}$. By the complex interpolation of L^p spaces, we must have $g_\theta \in L^p(\mathbb{B}_n, d\tau)$, which, in light of Theorem 6.7, implies that $f = f_\theta = Pg_\theta$ is in B_p. This completes the proof of the theorem. $\square$

Theorem 6.13. *The space B_p is Möbius invariant for any $p \in [1, \infty]$.*

Proof. We only need to prove the theorem for $1 < p < \infty$, because we already know that B_1 admits a Möbius invariant norm $\|f\|_m$ and $B_\infty = \mathcal{B}$ admits a Möbius invariant semi-norm $\|f\|_\mathcal{B}$.

Fix $1 < p < \infty$ and let $p_0 = 1$, $p_1 = \infty$, and $\theta = (p-1)/p$. Then

$$\frac{1}{p} = \frac{1-\theta}{p_0} + \frac{\theta}{p_1}.$$

According to Theorem 6.12, we have $B_p = [B_1, B_\infty]_\theta$. It follows that for every function $f \in B_p$, there exists a function $F(z, \zeta)$, where $z \in \mathbb{B}_n$ and $0 \le \operatorname{Re}\zeta \le 1$, such that $F(z, \theta) = f(z)$ for all $z \in \mathbb{B}_n$, and

$$\|F\|_* = \max\left(\sup_{\operatorname{Re}\zeta = 0} \|F(\cdot, \zeta)\|_m, \sup_{\operatorname{Re}\zeta = 1} \|F(\cdot, \zeta)\|_\mathcal{B}\right) < \infty.$$

Define a semi-norm $\|\ \|_p$ on B_p by $\|f\|_p = \inf \|F\|_*$, where the infimum is taken over all F satisfying the conditions in the previous paragraph. It is then easy to check that $\|\ \|_p$ is a complete Möbius invariant semi-norm on B_p. $\square$

6.4 The Dirichlet Space B_2

In this section we focus on the space B_2 and show that it is the unique Möbius invariant Hilbert space in $\mathbb{B}_n$. We also obtain a characterization of B_2 in terms of Taylor coefficients.

Theorem 6.14. *Suppose f is holomorphic in $\mathbb{B}_n$ and*

$$f(z) = \sum_m a_m z^m$$

is its Taylor expansion. Then f belongs to B_2 if and only if

$$\sum_m |m| \frac{m!}{|m|!} |a_m|^2 < \infty.$$

Proof. Let $t = (n+1)/2$ and $\alpha = 0$. By Theorem 6.4, $f \in B_2$ if and only if the function $(1 - |z|^2)^t R^{\alpha,t} f(z)$ is in $L^2(\mathbb{B}_n, d\tau)$ if and only if $R^{\alpha,t} f$ is in $L^2(\mathbb{B}_n, dv)$. For any multi-index m of nonnegative integers let

$$b_m = \frac{\Gamma(n+1)\Gamma(n+1+|m|+t)}{\Gamma(n+1+t)\Gamma(n+1+|m|)}.$$

Then $|b_m| \sim |m|^t$ as $|m| \to \infty$, and by the definition of $R^{\alpha,t}$, we have

$$R^{\alpha,t} f(z) = \sum_m a_m b_m z^m.$$

Computing the norm of $R^{\alpha,t} f$ in $L^2(\mathbb{B}_n, dv)$, we conclude that $f \in B_2$ if and only if

$$\sum_m |a_m|^2 |m|^{2t} \int_{\mathbb{B}_n} |z^m|^2 \, dv(z) < \infty.$$

This, according to Lemma 1.11, is equivalent to

$$\sum_m |a_m|^2 |m|^{2t} \frac{m!}{(n+|m|)!} < \infty.$$

Since $2t = n+1$ and

$$(n+|m|)! = |m|!(|m|+1)\cdots(|m|+n),$$

we conclude that $f \in B_2$ if and only if

$$\sum_m |m| \frac{m!}{|m|!} |a_m|^2 < \infty.$$

This completes the proof of the theorem. $\square$

In the rest of this section, we are going to use the following notation.

$$\langle f, g \rangle = \sum_m |m| \frac{m!}{|m|!} a_m \bar{b}_m,$$

where

$$f(z) = \sum_m a_m z^m, \qquad g(z) = \sum_m b_m z^m.$$

It is clear that $\langle \, , \, \rangle$ is a semi-inner product, that is, $\langle f, f \rangle \geq 0$ for all f, $\overline{\langle f, g \rangle} = \langle g, f \rangle$ for all f and g, and $\langle f, g \rangle$ is linear in f. Throughout this section the semi-norm induced by this semi-inner product will simply be denoted by $\| \, \|$. By Theorem 6.14, B_2 is a semi-Hilbert space under the present semi-inner product.

We are going to show that the semi-inner product $\langle \, , \, \rangle$ is Möbius invariant, and up to a constant multiple, it is the only Möbius invariant (semi-)inner product that can be defined on a space of holomorphic functions in $\mathbb{B}_n$.

Theorem 6.15. *The semi-inner product $\langle \, , \, \rangle$ is Möbius invariant on the space B_2, that is,*

$$\langle f \circ \varphi, g \circ \varphi \rangle = \langle f, g \rangle$$

for all f and g in B_2 and all $\varphi \in \mathrm{Aut}(\mathbb{B}_n)$.

Proof. Let $f = \sum_k f_k$ and $g = \sum_k g_k$ be the homogeneous expansions of f and g. Also, let $f(z) = \sum_m a_m z^m$ and $g(z) = \sum_m b_m z^m$ be the Taylor expansions of f and g. It follows easily from Lemma 1.11 that

$$\int_{\mathbb{B}_n} f_k(z) \overline{g_k(z)} \, dv(z) = \sum_{|m|=k} \frac{n! \, m!}{(n+k)!} a_m \bar{b}_m.$$

From this we deduce that

$$\sum_{k=0}^{\infty} \frac{k(n+k)!}{n! \, k!} \int_{\mathbb{B}_n} f_k(z) \overline{g_k(z)} \, dv(z) = \langle f, g \rangle.$$

Since unitary transformations preserve homogeneous expansions, and since the volume measure dv is invariant under the action of the unitary group, we see that

$$\langle f \circ U, g \circ U \rangle = \langle f, g \rangle \tag{6.15}$$

for all f and g in B_2 and all unitary transformations U.

Next we consider the action on the semi-inner product $\langle \, , \, \rangle$ by an automorphism of the form

$$\varphi_a(z) = \left(\frac{r - z_1}{1 - r z_1}, \, -\frac{\sqrt{1 - r^2} \, z_2}{1 - r z_1}, \cdots, \, -\frac{\sqrt{1 - r^2} \, z_n}{1 - r z_1} \right),$$

where $r \in (0, 1)$ and $a = (r, 0, \cdots, 0)$.

If $m = (m_1, \cdots, m_n)$ is a multi-index of nonnegative integers, then $z^m \circ \varphi_a(z)$ is $z_2^{m_2} \cdots z_n^{m_n}$ times an analytic function of z_1. It follows that if $m' = (m'_1, \cdots, m'_n)$ is another multi-index of nonnegative integers with $m_j \neq m'_j$ for some $2 \leq j \leq n$, then no monomial in the Taylor expansion of $z^m \circ \varphi_a$ is equal to any monomial in the Taylor expansion of $z^{m'} \circ \varphi_a$, and so $\langle z^m \circ \varphi_a, z^{m'} \circ \varphi_a \rangle = 0$.

On the other hand, if

$$m = (m_1, m_2, \cdots, m_n), \qquad m' = (m'_1, m_2, \cdots, m_n),$$

that is, the only possible difference between m and m' is in their first component, then

$$z^m \circ \varphi_a(z) = (-1)^N (1 - r^2)^{N/2} \frac{(r - z_1)^{m_1}}{(1 - rz_1)^{m_1+N}} z_2^{m_2} \cdots z_n^{m_n},$$

and

$$z^{m'} \circ \varphi_a(z) = (-1)^N (1 - r^2)^{N/2} \frac{(r - z_1)^{m'_1}}{(1 - rz_1)^{m'_1+N}} z_2^{m_2} \cdots z_n^{m_n},$$

where $N = m_2 + \cdots + m_n$. Let

$$F(z_1) = \frac{(r - z_1)^{m_1}}{(1 - rz_1)^{m_1+N}} = \sum_{k=0}^{\infty} c_k z_1^k,$$

and

$$G(z_1) = \frac{(r - z_1)^{m'_1}}{(1 - rz_1)^{m'_1+N}} = \sum_{k=0}^{\infty} d_k z_1^k.$$

Then

$$\langle z^m \circ \varphi_a, z^{m'} \circ \varphi_a \rangle = (1 - r^2)^N \sum_{k=0}^{\infty} c_k \overline{d_k} (k + N) \frac{k! \, m_2! \cdots m_n!}{(k + N)!}.$$

We proceed to show that this is equal to $\langle z^m, z^{m'} \rangle$.

If $N = 0$, namely, if $m_2 = \cdots = m_n = 0$, we have

$$\langle z^m \circ \varphi_a, z^{m'} \circ \varphi_a \rangle = \sum_{k=0}^{\infty} k c_k \overline{d_k} = \int_{\mathbb{D}} F'(z_1) \overline{G'(z_1)} \, dA(z_1).$$

By a change of variables,

$$\int_{\mathbb{D}} F'(z_1) \overline{G'(z_1)} \, dA(z_1) = m_1 m'_1 \int_{\mathbb{D}} z_1^{m_1-1} \overline{z_1^{m'_1-1}} \, dA(z_1).$$

It is then clear that

$$\langle z^m \circ \varphi_a, z^{m'} \circ \varphi_a \rangle = \langle z^m, z^{m'} \rangle.$$

If $N = 1$, then

$$\langle z^m \circ \varphi_a, z^{m'} \circ \varphi_a \rangle = (1 - r^2) \sum_{k=0}^{\infty} c_k \overline{d_k} = \frac{1 - r^2}{2\pi} \int_0^{2\pi} F(e^{i\theta}) \overline{G(e^{i\theta})} \, d\theta$$

$$= \frac{1}{2\pi} \int_0^{2\pi} \left(\frac{r - e^{i\theta}}{1 - re^{i\theta}} \right)^{m_1} \overline{\left(\frac{r - e^{i\theta}}{1 - re^{i\theta}} \right)^{m_1'}} \frac{1 - r^2}{|1 - re^{i\theta}|^2} \, d\theta$$

$$= \frac{1}{2\pi} \int_0^{2\pi} e^{im_1\theta} \overline{e^{im_1'\theta}} \, d\theta$$

$$= \langle z^m, z^{m'} \rangle.$$

If $N \geq 2$, then we can write

$$\langle z^m \circ \varphi_a, z^{m'} \circ \varphi_a \rangle = (1 - r^2)^N \frac{m_2! \cdots m_n!}{(N-1)!} \sum_{k=0}^{\infty} c_k \overline{d_k} \frac{k!(N-1)!}{(k+N-1)!}$$

$$= (1 - r^2)^N \frac{m_2! \cdots m_n!}{(N-1)!} \sum_{k=0}^{\infty} c_k \overline{d_k} \int_{\mathbb{S}_N} |\zeta_1^k|^2 \, d\sigma_N(\zeta)$$

$$= (1 - r^2)^N \frac{m_2! \cdots m_n!}{(N-1)!} \int_{\mathbb{S}_N} F(\zeta_1) \overline{G(\zeta_1)} \, d\sigma_N(\zeta),$$

where $d\sigma_N$ is the normalized surface measure on the unit sphere $\mathbb{S}_N$ in $\mathbb{C}^N$. Evaluating the last integral according to (1.13) and then making an obvious change of variables on the unit disk, we obtain

$$\langle z^m \circ \varphi_a, z^{m'} \circ \varphi_a \rangle = (1 - r^2)^N \frac{m_2! \cdots m_n!}{(N-2)!} \int_{\mathbb{D}} (1 - |w|^2)^{N-2} F(w) \overline{G(w)} dA(w)$$

$$= \frac{m_2! \cdots m_n!}{(N-2)!} \int_{\mathbb{D}} w^{m_1} \overline{w^{m_1'}} (1 - |w|^2)^{N-2} \, dA(w)$$

$$= \langle z^m, z^{m'} \rangle.$$

We have now shown that $\langle z^m \circ \varphi_a, z^{m'} \circ \varphi_a \rangle = \langle z^m, z^{m'} \rangle$ for all m, m', and r. Expanding f and g into Taylor series, we conclude that $\langle f \circ \varphi_a, g \circ \varphi_a \rangle = \langle f, g \rangle$ for all f, g in B_2 and $r \in (0, 1)$. Recall from (6.15) that this is also true when φ_a is replaced by any unitary transformation of $\mathbb{C}^n$. Since (see the proof of Theorem 6.8) every automorphism $\varphi \in \mathrm{Aut}(\mathbb{B}_n)$ can be written as $\varphi = U \circ \varphi_a \circ V$, where $a = (r, 0, \cdots, 0)$ with $r \in (0, 1)$ and U and V are unitaries, the proof of the theorem is complete. □

Finally in this section we show that there is only one natural Möbius invariant Hilbert space of holomorphic functions in $\mathbb{B}_n$. We say that H is a Möbius invariant Hilbert space if H is a Möbius invariant Banach space according to the conventions set forth in Section 3.4, and the norm on H is induced by a semi-inner product:

$$\|f\|_H^2 = \langle f, f \rangle_H, \qquad f \in H.$$

A polarization argument then shows that

$$\langle f \circ \varphi, g \circ \varphi \rangle_H = \langle f, g \rangle_H$$

for all f and g in H and all $\varphi \in \mathrm{Aut}(\mathbb{B}_n)$.

Theorem 6.16. *Suppose H is a Möbius invariant Hilbert space with semi-inner product $\langle \ , \ \rangle_H$. If H contains a nonconstant function, then $H = B_2$ and there exists a constant $c > 0$ such that $\langle f, g \rangle_H = c \langle f, g \rangle$ for all f and g in H.*

Proof. By Lemma 3.18, H contains all the polynomials. It suffices for us to show that there exists a constant $c > 0$ such that

$$\langle z^m, z^{m'} \rangle_H = c \langle z^m, z^{m'} \rangle \tag{6.16}$$

for all multi-indexes $m = (m_1, \cdots, m_n)$ and $m' = (m'_1, \cdots, m'_n)$ of nonnegative integers.

If $m \neq m'$, then there exists some k, $1 \leq k \leq n$, such that $m_k \neq m'_k$. For any real θ let $U = U_\theta$ be the unitary defined by $Uz = w$, where $w_j = z_j$ for $j \neq k$ and $w_k = z_k e^{i\theta}$. We have

$$\langle z^m, z^{m'} \rangle_H = \langle z^m \circ U, z^{m'} \circ U \rangle_H = e^{i(m_k - m'_k)\theta} \langle z^m, z^{m'} \rangle_H.$$

Since θ is arbitrary and $m_k - m'_k \neq 0$, we must have

$$\langle z^m, z^{m'} \rangle_H = 0 = \langle z^m, z^{m'} \rangle.$$

To compute $\langle z^m, z^m \rangle_H$, we consider the following special automorphisms:

$$\varphi_a(z) = \left(\frac{r - z_1}{1 - rz_1}, -\frac{\sqrt{1 - r^2}\, z_2}{1 - rz_1}, \cdots, -\frac{\sqrt{1 - r^2}\, z_n}{1 - rz_1} \right), \qquad z \in \mathbb{B}_n,$$

where $a = (r, 0, \cdots, 0)$ with $r \in (0, 1)$. For the function $f(z) = 1 - rz_1$ in H, we have

$$\langle f, f \rangle_H = \langle 1, 1 \rangle_H + r^2 \langle z_1, z_1 \rangle_H.$$

Since $\langle f \circ \varphi_a, f \circ \varphi_a \rangle_H = \langle f, f \rangle_H$ and

$$f \circ \varphi_a(z) = 1 - r \frac{r - z_1}{1 - rz_1} = \frac{1 - r^2}{1 - rz_1},$$

we obtain

$$\langle 1, 1 \rangle_H + r^2 \langle z_1, z_1 \rangle_H = \left\langle \frac{1 - r^2}{1 - rz_1}, \frac{1 - r^2}{1 - rz_1} \right\rangle_H$$

$$= (1 - r^2)^2 \sum_{k=0}^{\infty} r^{2k} \langle z_1^k, z_1^k \rangle_H$$

$$= \langle 1, 1 \rangle_H + r^2 \langle z_1, z_1 \rangle_H - 2r^2 \langle 1, 1 \rangle_H + \sum_{k=2}^{\infty} c_k r^{2k},$$

where

$$c_k = \langle z_1^k, z_1^k \rangle_H - 2\langle z_1^{k-1}, z_1^{k-1} \rangle_H + \langle z_1^{k-2}, z_1^{k-2} \rangle_H.$$

Since r is arbitrary, we must have $\langle 1, 1 \rangle_H = 0$ and $c_k = 0$ for all $k \geq 2$. It follows from the recursive relation $c_k = 0$ that

$$\langle z_1^k, z_1^k \rangle_H = k\langle z_1, z_1 \rangle_H \qquad (6.17)$$

for all $k \geq 2$. Note that this identity also holds when $k = 0$ and $k = 1$.

Let U be an arbitrary unitary transformation on $\mathbb{C}^n$ and we assume that U acts on $\mathbb{C}^n$ as matrix multiplication Uz, where $z \in \mathbb{C}^n$ is thought of as a column vector. Then

$$z_1^k \circ U(z) = (u_1 z_1 + u_2 z_2 + \cdots + u_n z_n)^k = \sum_{|m|=k} \frac{k!}{m!} u^m z^m,$$

where u is the transpose of the first row of U. By the Möbius invariance of the semi-inner product in H, we have

$$\langle z_1^k, z_1^k \rangle_H = \sum_{|m|=k} t_m \frac{k!}{m!} \langle z^m, z^m \rangle_H,$$

where each

$$t_m = \frac{k!}{m!} |u^m|^2 = \frac{k!}{m!} |u_1|^{2m_1} \cdots |u_n|^{2m_n}$$

is nonnegative and

$$\sum_{|m|=k} t_m = (|u_1|^2 + \cdots + |u_n|^2)^k = 1.$$

As U runs over the whole unitary group, the sum

$$\sum_{|m|=k} t_m \frac{k!}{m!} \langle z^m, z^m \rangle_H$$

runs over the closed convex hull of the points $(k!/m!)\langle z^m, z^m \rangle_H$. Since the above sum is always $\langle z_1^k, z_1^k \rangle_H$, we conclude that

$$\frac{k!}{m!} \langle z^m, z^m \rangle_H = \langle z_1^k, z_1^k \rangle_H \qquad (6.18)$$

for every m with $|m| = k$. Combining (6.17) and (6.18), we obtain

$$\langle z^m, z^m \rangle_H = |m| \frac{m!}{|m|!} \langle z_1, z_1 \rangle_H.$$

This completes the proof of the theorem with $c = \langle z_1, z_1 \rangle_H$.   □

Theorem 6.17. *The reproducing kernel of the space B_2 (when equipped with the Möbius invariant inner product) is given by*

$$K(z,w) = \log \frac{1}{1 - \langle z, w \rangle}.$$

Proof. For each multi-index m of nonnegative integers, we have

$$\|z^m\|^2 = |m| \frac{m!}{|m|!}.$$

It follows that

$$K(z,w) = \sum_{|m|>0} \frac{z^m}{\|z^m\|} \frac{\overline{w}^m}{\|w^m\|} = \sum_{k=1}^{\infty} \frac{1}{k} \sum_{|m|=k} \frac{k!}{m!} z^m \overline{w}^m$$

$$= \sum_{k=1}^{\infty} \frac{1}{k} \langle z, w \rangle^k = \log \frac{1}{1 - \langle z, w \rangle}.$$

$\square$

6.5 Duality of Besov Spaces

In this section we consider the dual space of B_p for $0 < p < \infty$. Identification of the dual space depends on the duality pairing being used, and for the Besov space B_p, we introduce a pairing based on the operator V defined by

$$V f(z) = c_{n+1}(1 - |z|^2)^{n+1} R^{0,n+1} f(z), \qquad f \in H(\mathbb{B}_n). \qquad (6.19)$$

Lemma 6.18. *If f and g are bounded holomorphic functions in $\mathbb{B}_n$, then*

$$\int_{\mathbb{B}_n} f(z) \overline{g(z)} \, dv(z) = \int_{\mathbb{B}_n} f(z) \overline{V g(z)} \, dv(z).$$

Proof. Represent $R^{0,n+1}g$ by Corollary 2.3, use Fubini's theorem, and then apply the reproducing formula in Theorem 2.2. The integral on the right-hand side is then reduced to the one on the left-hand side. $\square$

Obviously, by using approximation arguments, the assumptions that f and g be in $H^\infty(\mathbb{B}_n)$ can be relaxed in many different ways.

Theorem 6.19. *Suppose $1 < p < \infty$ and $1/p + 1/q = 1$. Then the dual space of B_p can be identified with B_q (with equivalent norms) under the pairing*

$$\langle f, g \rangle = \int_{\mathbb{B}_n} V f(z) \, \overline{V g(z)} \, d\tau(z), \qquad (6.20)$$

where $f \in B_p$ and $g \in B_q$.

Proof. By Theorem 6.4 and Hölder's inequality, every $g \in B_q$ induces a bounded linear functional on B_p via the integral pairing (6.20).

To show that every bounded linear functional on B_p arises this way, recall from Corollay 6.5 that $R^{0,n+1}$ is a bounded invertible operator from B_p onto $A^p_{(p-1)(n+1)}$. Therefore, $F \in (B_p)^*$ if and only if $F \circ R_{0,n+1} \in (A^p_{(p-1)(n+1)})^*$, which, according to Theorem 2.12, is equivalent to the existence of some $\varphi \in A^q_{(p-1)(n+1)}$ such that

$$F \circ R_{0,n+1}(f) = \int_{\mathbb{B}_n} f(z)\overline{\varphi(z)}(1 - |z|^2)^{(p-1)(n+1)} \, dv(z),$$

for all $f \in A^p_{(p-1)(n+1)}$. Let

$$g(z) = \frac{1}{c_{n+1}} \int_{\mathbb{B}_n} \frac{\varphi(w)(1 - |w|^2)^{(p-1)(n+1)} \, dv(w)}{(1 - \langle z, w \rangle)^{n+1}}.$$

By Theorem 6.7, we have $g \in B_q$, because the numerator of the integrand above is a function in $L^q(\mathbb{B}_n, d\tau)$. By Fubini's theorem,

$$F \circ R_{0,n+1}(f) = c_{n+1} \int_{\mathbb{B}_n} f(z)\overline{g(z)} \, dv(z), \qquad f \in A^p_{(p-1)(n+1)}.$$

Replacing f by $R^{0,n+1}f$, we obtain

$$F(f) = c_{n+1} \int_{\mathbb{B}_n} R^{0,n+1}f(z)\overline{g(z)} \, dv(z), \qquad f \in B_p.$$

By Lemma 6.18 and the remark following it, we have

$$F(f) = c_{n+1} \int_{\mathbb{B}_n} R^{0,n+1}f(z)\overline{Vg(z)} \, dv(z),$$

or

$$F(f) = \int_{\mathbb{B}_n} Vf(z)\overline{Vg(z)} \, d\tau(z), \qquad f \in B_p.$$

This completes the proof of the theorem. $\qquad\square$

Theorem 6.20. *Under the integral pairing in (6.20) we can identify (with equivalent norms) the dual space of B_1 as $\mathcal{B}$, and the dual space of $\mathcal{B}_0$ with B_1.*

Proof. It follows from Theorem 6.4 that, via the integral pairing in (6.20), every function $g \in \mathcal{B}$ induces a bounded linear functional on B_1, and every function $g \in B_1$ induces a bounded linear functional on $\mathcal{B}_0$.

If F is a bounded linear functional on B_1, then $F \circ R_{0,n+1}$ is a bounded linear functional on A^1 (the unweighted Bergman space), because $R_{0,n+1}$ is a bounded invertible operator from A^1 onto B_1. It follows from Theorem 3.17 that there exists a function $g \in \mathcal{B}$ such that

$$F \circ R_{0,n+1} f = c_{n+1} \int_{\mathbb{B}_n} f(z) \overline{g(z)} \, dv(z), \qquad f \in A^1.$$

By Lemma 6.18,

$$F \circ R_{0,n+1} f = c_{n+1} \int_{\mathbb{B}_n} f(z) \overline{Vg(z)} \, dv(z), \qquad f \in A^1,$$

or equivalently,

$$F(f) = c_{n+1} \int_{\mathbb{B}_n} R^{0,n+1} f(z) \, \overline{Vg(z)} \, dv(z) = \int_{\mathbb{B}_n} Vf(z) \overline{Vg(z)} \, d\tau(z)$$

for all $f \in B_1$.

If F is a bounded linear functional on $\mathcal{B}_0$, we use Theorem 3.16 to find a function $h \in A^1$ such that

$$F(f) = \int_{\mathbb{B}_n} f(z) \overline{h(z)} \, dv(z), \qquad f \in \mathcal{B}_0.$$

According to Corollary 6.5, there exists $g \in B_1$ such that $h = c_{n+1} R^{0,n+1} g$. Thus for $f \in \mathcal{B}_0$,

$$\begin{aligned} F(f) &= c_{n+1} \int_{\mathbb{B}_n} f(z) \overline{R^{0,n+1} g(z)} \, dv(z) \\ &= c_{n+1} \int_{\mathbb{B}_n} Vf(z) \overline{R^{0,n+1} g(z)} \, dv(z) \\ &= \int_{\mathbb{B}_n} Vf(z) \overline{Vg(z)} \, d\tau(z). \end{aligned}$$

Here we used Lemma 6.18 to justify the second equality above.   □

The case $0 < p < 1$ is similar to the case $p = 1$.

Theorem 6.21. *Suppose $0 < p \le 1$ and $t > n/p$. Then the dual space of B_p can be identified with $\mathcal{B}$ under the following duality pairing:*

$$\langle f, g \rangle = \int_{\mathbb{B}_n} (1 - |z|^2)^t R^{\alpha,t} f(z) \overline{g(z)} \, d\tau(z), \qquad f \in B_p, g \in \mathcal{B},$$

where α is any real number such that neither $n + \alpha$ nor $n + \alpha + t$ is a negative integer.

Proof. Recall from Corollary 6.5 that $R^{\alpha,t}$ is a bounded invertible operator from B_p onto $A^p_{pt-(n+1)}$. So F is a bounded linear functional on B_p if and only if $F \circ R_{\alpha,t}$ is a bounded linear functional on $A^p_{pt-(n+1)}$. Combining this with Theorem 3.17, we see that F is a bounded linear functional on B_p if and only if there exists a function $g \in \mathcal{B}$ such that

$$(F \circ R_{\alpha,t})(f) = \int_{\mathbb{B}_n} f(z)\overline{g(z)} \, dv_\beta(z),$$

where

$$\beta = \frac{n+1+pt-(n+1)}{p} - (n+1) = t-(n+1).$$

Equivalently, F is a bounded linear functional on B_p if and only if

$$F(f) = \int_{\mathbb{B}_n} (1-|z|^2)^t R^{\alpha,t} f(z) \overline{g(z)} \, d\tau(z)$$

for all $f \in B_p$, where $g \in \mathcal{B}$. This completes the proof of the theorem.   □

When $p = 1$, we can choose $\alpha = 0$ and $t = n+1$. In this case, Lemma 6.18 shows that Theorem 6.21 includes Theorem 6.20 as a special case.

It is interesting that the integral pairing in Theorem 6.21 is "almost" independent of p; the only requirement is that $pt > n$. In particular, for any two different p_1 and p_2 in $(0, 1]$, there exists a common integral pairing under which the dual spaces of B_{p_1} and B_{p_2} are both the Bloch space $\mathcal{B}$.

Our next goal is to identify the dual space B_p using the more natural Möbius invariant pairing. To this end, we need to introduce a partial differential operator D. Thus for f holomorphic in $\mathbb{B}_n$ with homogeneous expansion

$$f(z) = \sum_{k=0}^{\infty} f_k(z),$$

we define

$$Df(z) = \sum_{k=1}^{\infty} \frac{\Gamma(n+k)}{\Gamma(n)\Gamma(k)} f_k(z). \tag{6.21}$$

It is clear that D annihilates constant functions.

Lemma 6.22. *We have*

$$D = RR^{-n,n-1} = n(R^{-n,n} - R^{-n,n-1}),$$

and D acts on $H(\mathbb{B}_n)$ as a linear partial differential operator of order n with polynomial coefficients.

Proof. For any fixed $w \in \mathbb{B}_n$ we have

$$D\frac{1}{1-\langle z,w \rangle} = \sum_{k=1}^{\infty} \frac{\Gamma(n+k)}{\Gamma(n)\Gamma(k)} \langle z,w \rangle^k = R \sum_{k=1}^{\infty} \frac{\Gamma(n+k)}{k!\,\Gamma(n)} \langle z,w \rangle^k$$

$$= R\frac{1}{(1-\langle z,w \rangle)^n} = RR^{-n,n-1}\frac{1}{1-\langle z,w \rangle}.$$

This shows that $D = RR^{-n,n-1}$.

On the other hand,

$$R\frac{1}{(1 - \langle z, w \rangle)^n} = \frac{n \langle z, w \rangle}{(1 - \langle z, w \rangle)^{n+1}} = \frac{n}{(1 - \langle z, w \rangle)^{n+1}} - \frac{n}{(1 - \langle z, w \rangle)^n}$$

$$= n(R^{-n,n} - R^{-n,n-1})\frac{1}{1 - \langle z, w \rangle}.$$

Combining this with the previous paragraph, we see that

$$D = n(R^{-n,n} - R^{-n,n-1}).$$

That D is an nth order partial differential operator on $H(\mathbb{B}_n)$ with polynomial coefficients now follows from Proposition 1.15. □

Proposition 6.23. *For every holomorphic function f in $\mathbb{B}_n$ we have*

$$\int_{\mathbb{B}_n} |Df(z)|^2 \frac{(1 - |z|^2)^{n-1}}{|z|^{2n}} \, dv(z) = n \sum_m |m| \frac{m!}{|m|!} |a_m|^2, \qquad (6.22)$$

where a_m are the Taylor coefficients of f.

Proof. It suffices to prove the result for polynomials (to avoid issues of convergence); the general case then follows from an approximation argument.

If f is a polynomial, then the integral

$$I(f) = \int_{\mathbb{B}_n} |Df(z)|^2 \frac{(1 - |z|^2)^{n-1} \, dv(z)}{|z|^{2n}}$$

is equal to

$$\sum_{|m|>0} |a_m|^2 \frac{\Gamma(n + |m|)^2}{\Gamma(n)^2 \Gamma(|m|)^2} \int_{\mathbb{B}_n} (1 - |z|^2)^{n-1} |z|^{-2n} |z^m|^2 \, dv(z).$$

Integrating in polar coordinates and using Lemma 1.11, we obtain

$$I(f) = \sum_{|m|>0} |a_m|^2 \frac{\Gamma(n + |m|)^2}{\Gamma(n)^2 \Gamma(|m|)^2} \frac{(n-1)!m!}{(n-1+|m|)!} \frac{n\Gamma(n)\Gamma(|m|)}{\Gamma(n+|m|)},$$

which, after simplification, proves (6.22). □

Lemma 6.24. *Suppose $p > 1$ and f is holomorphic in $\mathbb{B}_n$. Then $f \in B_p$ if and only if*

$$\int_{\mathbb{B}_n} (1 - |z|^2)^{pn} |Df(z)|^p |z|^{-2n} \, d\tau(z) < \infty. \qquad (6.23)$$

Proof. Since Df vanishes at the origin, the integral in (6.23) is always convergent near $z = 0$, and so (6.23) holds if and only if

$$\int_{\mathbb{B}_n} (1 - |z|^2)^{pn} |Df(z)|^p \, d\tau(z) < \infty, \tag{6.24}$$

which is equivalent to the function $Df(z)$ being in $A^p_{pn-(n+1)}$.

If $f \in B_p$, then the function

$$(1 - |z|^2)^n \frac{\partial^m f}{\partial z^m}(z)$$

belongs to $L^p(\mathbb{B}_n, d\tau)$ for every $|m| \leq n$. Since D is a differential operator of order n with polynomial coefficients, inequality (6.24) holds.

Conversely, if the function $Df = RR^{-n,n-1} f$ belongs to $A^p_{pn-(n+1)}$, then its anti-derivative $R^{-n,n-1} f$ is also in $A^p_{pn-(n+1)}$. It follows that the function

$$nR^{-n,n} f = Df + nR^{-n,n-1} f$$

belongs to $A^p_{pn-(n+1)}$. By Theorem 6.4, we have $f \in B_p$. $\qquad\square$

We now identify the dual space of B_p, $1 \leq p < \infty$, under the Möbius invariant pairing.

Theorem 6.25. *Suppose $1 < p < \infty$ and $1/p + 1/q = 1$. Then the dual space of B_p can be identified with B_q (with equivalent norms) under the Möbius invariant pairing*

$$\langle f, g \rangle = \sum_m |m| \frac{m!}{|m|!} a_m \overline{b_m}, \tag{6.25}$$

where

$$f(z) = \sum_m a_m z^m \in B_p, \qquad g(z) = \sum_m b_m z^m \in B_q.$$

Proof. By polarizing the identity in Proposition 6.23 we can write the Möbius invariant semi-inner product as

$$\langle f, g \rangle = \frac{1}{n} \int_{\mathbb{B}_n} (1 - |z|^2)^n Df(z) \overline{(1 - |z|^2)^n Dg(z)} \, |z|^{-2n} \, d\tau(z).$$

By Hölder's inequality and Lemma 6.24,

$$|\langle f, g \rangle| \leq C \|f\|_p \|g\|_q,$$

so every function $g \in B_q$ induces a bounded linear functional on B_p via the Möbius invariant pairing (6.25).

Conversely, if F is a bounded linear functional on B_p, then by Theorem 6.19, there exists a function $h \in B_q$ such that

$$F(f) = \int_{\mathbb{B}_n} Vf(z)\overline{Vh(z)}\, d\tau(z), \qquad f \in B_p,$$

where

$$Vf(z) = c_{n+1}(1 - |z|^2)^{n+1} R^{0,n+1} f(z).$$

A computation with Taylor series shows that

$$F(f) = C_n \sum_m a_m \bar{b}_m \frac{m!\,\Gamma(2n+2+|m|)}{\Gamma(n+1+|m|)^2},$$

where C_n is a positive constant, and $\{a_m\}$ and $\{b_m\}$ are the Taylor coefficients of f and h, respectively. Define $g \in H(\mathbb{B}_n)$ by

$$g(z) = C_n \sum_{|m|>0} \frac{(|m|-1)!\,\Gamma(2n+2+|m|)}{\Gamma(n+1+|m|)^2} b_m z^m,$$

then for all $f \in B_p$ with $f(0) = 0$ we have $F(f) = \langle f, g \rangle$ in the Möbius invariant pairing. It remains to show that $g \in B_q$, which we isolate as the next lemma. $\square$

Lemma 6.26. *Suppose $1 < p < \infty$ and $f(z) = \sum_m a_m z^m$ is holomorphic in $\mathbb{B}_n$. Then f is in B_p if and only if the function*

$$F(z) = \sum_{|m|>0} \frac{|m|!\,\Gamma(2n+2+|m|)}{|m|\Gamma(n+1+|m|)^2} a_m z^m$$

is in B_p.

Proof. It is easy to see that there exists a constant $c > 0$ such that

$$DF = cR_{-1,1}R^{0,n+1}f,$$

provided that $f(0) = 0$. By Lemma 6.24, $F \in B_p$ if and only if $DF \in A^p_{pn-(n+1)}$, which, according to Theorem 2.19, is equivalent to

$$(1 - |z|^2)R^{-1,1}DF(z) \in L^p(\mathbb{B}_n, dv_{pn-(n+1)}).$$

Therefore, $F \in B_p$ if and only if

$$(1 - |z|^2)R^{0,n+1}f(z) \in L^p(\mathbb{B}_n, dv_{pn-(n+1)}),$$

which is obviously the same as

$$(1 - |z|^2)^{n+1}R^{0,n+1}f(z) \in L^p(\mathbb{B}_n, d\tau).$$

Combining this with Theorem 6.4, we see that $F \in B_p$ if and only if $f \in B_p$. $\square$

A more precise duality theorem can be formulated for the minimal invariant space B_1 and the maximal invariant space $B_\infty = B$. We use the invariant norm (not just a semi-norm) $\| \; \|_m$ (see Section 6.2) on B_1, and we use the following norm (not just a semi-norm) on the Banach space $B/\mathbb{C}$:

$$\|f\|_B = \sup\{|\widetilde{\nabla} f(z)| : z \in \mathbb{B}_n\}.$$

Theorem 6.27. *Under the Möbius invariant pairing $\langle \, , \, \rangle$ defined in (6.25) we have*

$$B_1^* = B/\mathbb{C}, \qquad (B_0/\mathbb{C})^* = B_1,$$

with equal (not just equivalent) norms.

Proof. Suppose

$$\varphi = (\varphi_1, \cdots, \varphi_n) \in \mathrm{Aut}(\mathbb{B}_n).$$

For each $g \in B$ and $1 \le k \le n$ we have

$$\langle g, \varphi_k \rangle = \langle g \circ \varphi^{-1}, z_k \rangle = \frac{\partial(g \circ \varphi^{-1})}{\partial z_k}(0).$$

Therefore,

$$|\langle g, \varphi_k \rangle| \le |\nabla(g \circ \varphi^{-1})(0)| = |\widetilde{\nabla} g(\varphi^{-1}(0))|. \tag{6.26}$$

Let F denote the set consisting of all coordinate functions of all automorphisms of $\mathbb{B}_n$. It follows from (6.26) that

$$\sup_{f \in F} |\langle f, g \rangle| \le \|g\|_B.$$

We claim that equality holds here. To see this, choose a unitary U that maps the vector

$$v_a = (\langle \psi_1, g \rangle, \cdots, \langle \psi_n, g \rangle)$$

to the vector $|v_a| e_1$, where ψ_k are the coordinate functions of φ_a. Then by (6.26),

$$|\widetilde{\nabla} g(a)| = |v_a| = |U v_a| = \left| \sum_{j=1}^n u_{1j} \langle \psi_k, g \rangle \right| = |\langle \phi_1, g \rangle|,$$

where ϕ_1 is the first coordinate function of $U \circ \varphi_a$. This shows that

$$\sup_{f \in F} |\langle f, g \rangle| = \|g\|_B \tag{6.27}$$

for every $g \in B$ with $g(0) = 0$.

Now if $g \in B$, $g(0) = 0$, and

$$f = c_0 + \sum_{k=1}^\infty c_k f_k, \qquad f_k \in F,$$

is a function in B_1, then

$$\langle f, g \rangle = c_0 + \sum_{k=1}^{\infty} c_k \langle f_k, g \rangle,$$

and so

$$|\langle f, g \rangle| \le \|g\|_{\mathcal{B}} \sum_{k=0}^{\infty} |c_k|.$$

Taking the infimum over all such representations of f, we obtain

$$|\langle f, g \rangle| \le \|f\|_m \|g\|_{\mathcal{B}}.$$

This shows that every function g in $\mathcal{B}/\mathbb{C}$ induces a bounded linear functional on B_1 via the Möbius invariant pairing, and the norm of this linear functional is equal to $\|g\|_{\mathcal{B}}$; this also shows that every function $f \in B_1$ induces a bounded linear functional on $\mathcal{B}_0/\mathbb{C}$, and the norm of this linear functional is no more than $\|f\|_m$.

Let L be a bounded linear functional on B_1. For each multi-index m of nonnegative integers with $|m| > 0$ let

$$a_m = \frac{|m|!}{|m|m!} \overline{L(z^m)},$$

and define

$$f(z) = \sum_{|m|>0} a_m z^m.$$

If

$$g(z) = \sum_{m} b_m z^m$$

is any polynomial, then

$$\langle g, f \rangle = \sum_{m} b_m |m| \frac{m!}{|m|!} \overline{a_m} = \sum_{m} b_m L(z^m) = L(g).$$

Since the polynomials are dense in B_1, we have

$$L(g) = \langle g, f \rangle, \qquad g \in B_1.$$

Furthermore, by (6.27),

$$\|f\|_{\mathcal{B}} = \sup_{g \in F} |\langle g, f \rangle| = \sup_{g \in F} |L(g)| \le \|L\|.$$

This completes the proof that $B_1^* = \mathcal{B}/\mathbb{C}$ with equal norms.

Finally, if L is a bounded linear functional on $\mathcal{B}_0/\mathbb{C}$, then by Theorem 6.20 and the proof of Theorem 6.25, there exists a function $g \in H(\mathbb{B}_n)$ such that

$$L(f) = \langle f, g \rangle, \qquad f \in \mathcal{B}_0/\mathbb{C}.$$

From the already established duality $B_1^* = \mathcal{B}/\mathbb{C}$ (with equal norms) we have

$$\|g\|_m = \sup\{|\langle f, g \rangle| : \|f\|_\mathcal{B} \leq 1\}.$$

Since $\mathcal{B}_0$ is dense in $\mathcal{B}$ in the weak-star topology, we also have

$$\|g\|_m = \sup\{|\langle f, g \rangle| : f \in \mathcal{B}_0, \|f\|_\mathcal{B} \leq 1\},$$

or

$$\|g\|_m = \sup\{|L(f)| : f \in \mathcal{B}_0, \|f\|_\mathcal{B} \leq 1\} = \|L\|.$$

This completes the proof of the duality $(\mathcal{B}_0/\mathbb{C})^* = B_1$ with equal norms. $\square$

Note that the duality between B_p and B_q under the Möbius invariant pairing enables us to define a canonical Möbius invariant semi-norm on B_p for $1 < p < 2n/(2n-1)$ when $n > 1$. In fact, when $p > 2n$, Theorem 6.11 shows that B_p has a canonical invariant semi-norm given by

$$\|f\|_p = \left[\int_{\mathbb{B}_n} |\widetilde{\nabla} f(z)|^p \, d\tau(z) \right]^{1/p}.$$

Now if $1 < p < 2n/(2n-1)$, then the conjugate exponent q, defined by $1/p + 1/q = 1$, satisfies $2n < q < \infty$. Therefore, we can define a canonical Möbius invariant semi-norm on B_p as follows:

$$\|f\| = \sup\{|\langle f, g \rangle| : \|g\|_q \leq 1\},$$

where

$$\|g\|_q = \left[\int_{\mathbb{B}_n} |\widetilde{\nabla} g(z)|^q \, d\tau(z) \right]^{1/q}.$$

It is an interesting problem to find explicit realizations for Möbius invariant semi-norms on B_p for $1 < p \leq 2n$ when $n > 1$; the case $p = 2$ was settled by Theorem 6.14.

6.6 Other Characterizations

In this section we obtain several other characterizations for the spaces B_p. Recall that

$$\lambda_n = \begin{cases} 1, & n = 1 \\ 2n, & n > 1 \end{cases}$$

is a dimensional constant.

Theorem 6.28. *Suppose $p > \lambda_n$, $\alpha > -1$, and f is holomorphic in $\mathbb{B}_n$. Then $f \in B_p$ if and only if*

$$\int_{\mathbb{B}_n} \int_{\mathbb{B}_n} \frac{|f(z) - f(w)|^p \, dv_\alpha(z) \, dv_\alpha(w)}{|1 - \langle z, w \rangle|^{2(n+1+\alpha)}} < \infty.$$

Proof. For a holomorphic function f in $\mathbb{B}_n$ let

$$I_\alpha(f) = \int_{\mathbb{B}_n} \int_{\mathbb{B}_n} \frac{|f(z) - f(w)|^p \, dv_\alpha(z) \, dv_\alpha(w)}{|1 - \langle z, w \rangle|^{2(n+1+\alpha)}}.$$

We can rewrite

$$I_\alpha(f) = c_\alpha \int_{\mathbb{B}_n} d\tau(z) \int_{\mathbb{B}_n} |f(w) - f(z)|^p \frac{(1 - |z|^2)^{n+1+\alpha}}{|1 - \langle z, w \rangle|^{2(n+1+\alpha)}} \, dv_\alpha(w),$$

where

$$d\tau(z) = \frac{dv(z)}{(1 - |z|^2)^{n+1}}$$

is the Möbius invariant measure on $\mathbb{B}_n$. A change of variables (see Proposition 1.13) gives

$$I_\alpha(f) = c_\alpha \int_{\mathbb{B}_n} d\tau(z) \int_{\mathbb{B}_n} |f \circ \varphi_z(w) - f \circ \varphi_z(0)|^p \, dv_\alpha(w).$$

By Theorem 2.16, the integral $I_\alpha(f)$ is comparable to the integral

$$J_\alpha(f) = \int_{\mathbb{B}_n} d\tau(z) \int_{\mathbb{B}_n} |\widetilde{\nabla}(f \circ \varphi_z)(w)|^p \, dv_\alpha(w).$$

Since

$$|\widetilde{\nabla}(f \circ \varphi_z)(w)| = |\widetilde{\nabla} f(\varphi_z(w))|,$$

changing variables again leads to

$$J_\alpha(f) = \int_{\mathbb{B}_n} d\tau(z) \int_{\mathbb{B}_n} |\widetilde{\nabla} f(w)|^p \frac{(1 - |z|^2)^{n+1+\alpha}}{|1 - \langle z, w \rangle|^{2(n+1+\alpha)}} \, dv_\alpha(w),$$

and an application of Fubini's theorem gives

$$J_\alpha(f) = \int_{\mathbb{B}_n} |\widetilde{\nabla} f(w)|^p \, dv_\alpha(w) \int_{\mathbb{B}_n} \frac{(1 - |z|^2)^\alpha \, dv(z)}{|1 - \langle z, w \rangle|^{2(n+1+\alpha)}}.$$

The inner integral above is equal to $(1 - |w|^2)^{-(n+1+\alpha)}/c_\alpha$, so

$$J_\alpha(f) = \int_{\mathbb{B}_n} |\widetilde{\nabla} f(w)|^p \, d\tau(w).$$

In view of Theorem 6.11, the proof of the theorem is complete. □

Theorem 6.29. *Suppose $\alpha > -1$, $r > 0$, and $p > \lambda_n$. Then the following conditions are equivalent for a holomorphic function f in $\mathbb{B}_n$:*

(a) $f \in B_p$.
(b) *The function $MO(f)$ belongs to $L^p(\mathbb{B}_n, d\tau)$, where*

$$MO(f)(z) = \int_{\mathbb{B}_n} |f \circ \varphi_z(w) - f(z)| \, dv_\alpha(w). \tag{6.28}$$

(c) The function $MO_r(f)$ belongs to $L^p(\mathbb{B}_n, d\tau)$, where

$$MO_r(f)(z) = \frac{1}{v_\alpha(D(z,r))} \int_{D(z,r)} |f(w) - f_{\alpha,D(z,r)}| \, dv_\alpha(w). \qquad (6.29)$$

Proof. By Theorem 2.16, there exists a positive constant C_1 such that

$$MO(f)(z) \leq C_1 \int_{\mathbb{B}_n} |\tilde{\nabla}(f \circ \varphi_z)(w)| \, dv_\alpha(w)$$

for all $f \in H(\mathbb{B}_n)$ and all $z \in \mathbb{B}_n$. By Hölder's inequality and the fact that

$$|\tilde{\nabla}(f \circ \varphi_z)(w)| = |\tilde{\nabla}f(\varphi_z(w))|,$$

we have

$$MO(f)(z)^p \leq C_1^p \int_{\mathbb{B}_n} |\tilde{\nabla}f(\varphi_z(w))|^p \, dv_\alpha(w).$$

Changing variables according to Proposition 1.13, we obtain

$$MO(f)(z)^p \leq C_1^p \int_{\mathbb{B}_n} |\tilde{\nabla}f(w)|^p \frac{(1-|z|^2)^{n+1+\alpha}}{|1-\langle z,w\rangle|^{2(n+1+\alpha)}} \, dv_\alpha(w).$$

Fubini's theorem then gives

$$\int_{\mathbb{B}_n} MO(f)(z)^p \, d\tau(z) \leq C_1^p \int_{\mathbb{B}_n} |\tilde{\nabla}f(w)|^p \, dv_\alpha(w) \int_{\mathbb{B}_n} \frac{(1-|z|^2)^\alpha \, dv(z)}{|1-\langle z,w\rangle|^{2(n+1+\alpha)}}.$$

Since

$$\int_{\mathbb{B}_n} \frac{dv_\alpha(z)}{|1-\langle z,w\rangle|^{2(n+1+\alpha)}} = c_\alpha \int_{\mathbb{B}_n} \frac{(1-|z|^2)^\alpha \, dv(z)}{|1-\langle z,w\rangle|^{2(n+1+\alpha)}} = \frac{1}{(1-|w|^2)^{n+1+\alpha}},$$

we must have

$$\int_{\mathbb{B}_n} MO(f)(z)^p \, d\tau(z) \leq C_1^p \int_{\mathbb{B}_n} |\tilde{\nabla}f(w)|^p \, d\tau(w).$$

This together with Theorem 6.11 shows that (a) implies (b).

We write

$$f(w) - f_{\alpha,D(z,r)} = f(w) - f(z) - \left[f_{\alpha,D(z,r)} - f(z) \right],$$

and

$$f_{\alpha,D(z,r)} - f(z) = \frac{1}{v_\alpha(D(z,r))} \int_{D(z,r)} (f(w) - f(z)) \, dv_\alpha(w).$$

It follows that

$$MO_r(f)(z) \leq \frac{2}{v_\alpha(D(z,r))} \int_{D(z,r)} |f(w) - f(z)| \, dv_\alpha(w).$$

By Lemmas 1.24 and 2.20, there exists a positive constant $C_2 > 0$ such that

$$\frac{2}{v_\alpha(D(z,r))} \leq \frac{C_2(1 - |z|^2)^{n+1+\alpha}}{|1 - \langle z, w \rangle|^{2(n+1+\alpha)}}$$

for all $z \in \mathbb{B}_n$ and $w \in D(z,r)$. Therefore,

$$MO_r(f)(z) \leq C_2 \int_{\mathbb{B}_n} |f(w) - f(z)| \frac{(1 - |z|^2)^{n+1+\alpha}}{|1 - \langle z, w \rangle|^{2(n+1+\alpha)}} \, dv_\alpha(w).$$

Changing variables according to Proposition 1.13, we see that

$$MO_r(f)(z) \leq C_2 MO(f)(z),$$

which proves that (b) implies (c).

By Lemma 2.4, there exists a positive constant C_3 such that

$$|\nabla g(0)| \leq C_3 \int_{D(0,r)} |g(w) - c| \, dv_\alpha(w)$$

for all holomorphic g in $\mathbb{B}_n$ and all complex constants c. Replace g by $f \circ \varphi_z$ and c by $f_{\alpha, D(z,r)}$. We obtain

$$|\widetilde{\nabla} f(z)| \leq C_3 \int_{D(0,r)} |f \circ \varphi_z(w) - f_{\alpha, D(z,r)}| \, dv_\alpha(w).$$

Changing variables according to Proposition 1.13, we obtain

$$|\widetilde{\nabla} f(z)| \leq C_3 \int_{D(z,r)} |f(w) - f_{\alpha, D(z,r)}| \frac{(1 - |z|^2)^{n+1+\alpha}}{|1 - \langle z, w \rangle|^{2(n+1+\alpha)}} \, dv_\alpha(w).$$

Since

$$\frac{(1 - |z|^2)^{n+1+\alpha}}{|1 - \langle z, w \rangle|^{2(n+1+\alpha)}}$$

is comparable to $1/v_\alpha(D(z,r))$ when $w \in D(z,r)$ (see Lemmas 1.24 and 2.20), we can find a constant $C_4 > 0$ such that

$$|\widetilde{\nabla} f(z)| \leq C_4 MO_r(f)(z), \qquad z \in \mathbb{B}_n.$$

This together with Theorem 6.11 shows that (c) implies (a).   □

For any $r > 0$ and f holomorphic in $\mathbb{B}_n$ we define

$$\omega_r(f)(z) = \sup\{|f(z) - f(w)| : w \in D(z,r)\}, \qquad z \in \mathbb{B}_n. \tag{6.30}$$

This is the oscillation of f in the Bergman metric at the point z.

Theorem 6.30. *Suppose $r > 0$ and $p > \lambda_n$. Then a holomorphic function f in $\mathbb{B}_n$ belongs to B_p if and only if $\omega_r(f)$ belongs to $L^p(\mathbb{B}_n, d\tau)$.*

Proof. Fix any $\alpha > -1$ and recall that

$$MO_r(f)(z) = \frac{1}{v_\alpha(D(z,r))} \int_{D(z,r)} |f(w) - f_{\alpha,D(z,r)}| \, dv_\alpha(w), \qquad z \in \mathbb{B}_n.$$

Since

$$f(w) - f_{\alpha,D(z,r)} = \frac{1}{v_\alpha(D(z,r))} \int_{D(z,r)} (f(w) - f(u)) \, dv_\alpha(u),$$

the triangle inequality

$$|f(u) - f(w)| \le |f(z) - f(w)| + |f(z) - f(u)|$$

shows that

$$MO_r(f)(z) \le 2\,\omega_r(f)(z), \qquad z \in \mathbb{B}_n.$$

If $\omega_r(f)$ is in $L^p(\mathbb{B}_n, d\tau)$, then $MO_r(f)$ is in $L^p(\mathbb{B}_n, d\tau)$, so $f \in B_p$ by Theorem 6.29.

On the other hand, there exists a positive constant C_1 such that

$$|f(w)| \le \frac{C_1}{v_\alpha(D(w,r))} \int_{D(w,r)} |f(u)| \, dv_\alpha(u)$$

for all $f \in H(\mathbb{B}_n)$ and $w \in \mathbb{B}_n$; see Lemma 2.24. Replacing f by $f - f(z)$, we obtain

$$|f(z) - f(w)| \le \frac{C_1}{v_\alpha(D(w,r))} \int_{D(w,r)} |f(u) - f(z)| \, dv_\alpha(u)$$

for all $z \in \mathbb{B}_n$ and $w \in \mathbb{B}_n$. By Lemma 2.24 and (2.20) there exists a constant $C_2 > 0$ such that for $w \in D(z,r)$,

$$|f(z) - f(w)| \le C_2 \int_{D(w,r)} |f(u) - f(z)| \frac{(1 - |z|^2)^{n+1+\alpha} \, dv_\alpha(u)}{|1 - \langle z, u \rangle|^{2(n+1+\alpha)}}$$

$$\le C_2 \int_{\mathbb{B}_n} |f(u) - f(z)| \frac{(1 - |z|^2)^{n+1+\alpha} \, dv_\alpha(u)}{|1 - \langle z, u \rangle|^{2(n+1+\alpha)}}$$

$$= C_2 \int_{\mathbb{B}_n} |f \circ \varphi_z(u) - f(z)| \, dv_\alpha(u)$$

$$= C_2 MO(f)(z).$$

Taking the supremum over all $w \in D(z,r)$, we obtain

$$\omega_r(f)(z) \le C_2 MO(f)(z).$$

It follows from Theorem 6.29 that $f \in B_p$ implies $\omega_r(f) \in L^p(\mathbb{B}_n, d\tau)$. $\square$

Notes

The theory of Besov spaces is a classical topic in analysis. Our coverage here is only on a very special class of such spaces, the so-called diagonal Besov spaces.

What makes the diagonal Besov spaces especially interesting is the fact that they are natually invariant under the action of automorphisms. The minimality of B_1 among Möbius invariant Banach spaces was first studied by Arazy and Fisher in [10] in the case of the unit disk, and then by Peloso in [83] in the case of the unit ball. The uniqueness of B_2 among Möbius invariant Hilbert spaces was first proved by Arazy and Fisher in [11] for the unit disk, and then generalized to the unit ball by Zhu in [126] and by Peetre in an unpublished manuscript.

The Möbius invariance of Besov spaces have been explored by numerous authors in various situations. Our Sections 6.2, 6.3, and 6.5 are mainly based on the paper [83].

Once the Besov spaces B_p are realized as the image of weighted Bergman spaces under the action of fractional integral operators, atomic decomposition and duality (with weighted volume-integral pairings) follow easily. However, duality using the Möbius invariant pairing is more subtle, and probably more interesting and more natural.

Theorems 6.7 and 6.28 were proved in [127] for the unit disk, and the proofs there essentially carry over to higher dimensions. Similarly, Theorems 6.29 and 6.30 are rooted in [127].

Exercises

6.1. Suppose $0 < p \le \lambda_n$ and f is holomorphic in $\mathbb{B}_n$. Then the function $|\widetilde{\nabla} f(z)|$ belongs to $L^p(\mathbb{B}_n, d\tau)$ if and only if f is constant.

6.2. Show that Theorems 6.11, 6.28, 6.29, and 6.30 are false for $0 < p \le \lambda_n$.

6.3. Does Theorem 6.7 remain true when $0 < p < 1$?

6.4. Find sharp pointwise estimates for functions in B_p.

6.5. Find sharp growth estimates for the Taylor coefficients of $f \in B_p$.

6.6. Show that $f \in B_p$ if and only if $z^m f \in B_p$, where f is holomorphic in $\mathbb{B}_n$ and m is any multi-index of nonnegative integers.

6.7. Show that for $n/(n+1) < p \le 1$ we have $(B_p)^* = \mathcal{B}$ (with equivalent norms) under the integral pairing

$$\langle f, g \rangle = \int_{\mathbb{B}_n} Vf(z)\overline{Vg(z)} \, d\tau(z).$$

6.8. Suppose $n < p < \infty$ and f is holomorphic in $\mathbb{B}_n$. Show that $f \in B_p$ if and only if $(1 - |z|^2)Rf(z)$ is in $L^p(\mathbb{B}_n, d\tau)$.

6.9. Suppose $n < p < \infty$ and f is holomorphic in $\mathbb{B}_n$. Show that $f \in B_p$ if and only if $(1 - |z|^2)|\nabla f(z)|$ is in $L^p(\mathbb{B}_n, d\tau)$.

6.10. Show that the operator V defined in (6.19) has the following properties:

(a) $V(Vf) = Vf$ whenever f is in B_p.
(b) $V(Ph) = Vh$ whenever $h \in L^q(\mathbb{B}_n, d\tau)$.
(c) V is self-adjoint with respect to the integral pairing induced by $d\tau$.

6.11. Suppose $0 < p < \infty$ and $f \in B_p$. If $f(0) = 0$, show that there exist functions $f_k \in B_p$, $1 \le k \le n$, such that

$$f(z) = \sum_{k=1}^{n} z_k f_k(z)$$

for $z \in \mathbb{B}_n$.

6.12. Suppose $1 < p < \infty$, $1/p + 1/q = 1$, and $\alpha > -1$. Find an integral pairing under which $(B_p)^*$ can be identified with A_α^q.

6.13. Show that $B_{p_1} \subset B_{p_2}$ whenever $0 < p_1 \le p_2 \le \infty$.

6.14. Show that Theorem 6.29 remains true if $MO(f)$ is defined by

$$MO(f)(z) = \left[\int_{\mathbb{B}_n} |f \circ \varphi_z(w) - f(z)|^q \, dv_\alpha(w) \right]^{1/q}$$

and $MO_r(f)$ is defined by

$$MO_r(f)(z) = \left[\frac{1}{v_\alpha(D(z,r))} \int_{D(z,r)} |f(w) - f_{\alpha,D(z,r)}|^q \, dv_\alpha(w) \right]^{1/q},$$

where q is any fixed positive exponent.

6.15. If $n = 1$, show that a holomorphic function f in $\mathbb{B}_n$ belongs to B_2 if and only if

$$\int_{\mathbb{S}_n} \int_{\mathbb{S}_n} \frac{|f(\zeta) - f(\eta)|^2}{|1 - \langle \zeta, \eta \rangle|^{2n}} \, d\sigma(\zeta) \, d\sigma(\eta) < \infty. \tag{6.31}$$

Show that this is not true when $n > 1$. In fact, if $n > 1$, then (6.31) holds only when f is constant.

6.16. Suppose f is analytic in the unit disk $\mathbb{D}$ and F is a continuous function from $[0, \infty)$ into $(0, \infty)$. Show that the integral

$$\int_{\mathbb{D}} \int_{\mathbb{D}} \frac{|f(z) - f(w)|^2}{|1 - \bar{z}w|^4} F\left(\left|\frac{z - w}{1 - \bar{z}w}\right|\right) \, dA(z) \, dA(w)$$

is equal to

$$\int_{\mathbb{D}} |f'(z)|^2 \, dA(z) \int_0^1 F(\sqrt{r}) \log \frac{1}{1 - r} \, dr.$$

See [126].

6.17. Suppose $n > 1$, f is holomorphic in $\mathbb{B}_n$, and F is continuous from $[0, \infty)$ into $[0, \infty)$. Show that

$$\int_{\mathbb{B}_n} \int_{\mathbb{B}_n} \frac{|f(z) - f(w)|^2}{|1 - \langle z, w \rangle|^{2(n+1)}} F\left(|\varphi_z(w)|\right) \, dv(z) \, dv(w) < \infty$$

if and only if f is constant or F is identically 0. See [126].

6.18. Suppose $0 < p < \infty$ and $pt > n$. Show that a holomorphic function f in $\mathbb{B}_n$ belongs to B_p if and only if the function $(1 - |z|^2)^t R^t f(z)$ is in $L^p(\mathbb{B}_n, d\tau)$.

6.19. Suppose α is a real parameter such that the fractional differential operator $R^{\alpha, n/2}$ is well defined. Show that $R^{\alpha, n/2}$ is bounded linear operator from B_2 onto H^2.

6.20. Show that a holomorphic function f in $\mathbb{B}_n$ belongs to B_2 if and only if $R^{n/2} f$ belongs to H^2.

6.21. For any $1 \leq p \leq \infty$ there exists a constant $C_p > 0$ such that

$$|f(z) - f(w)| \leq C_p \beta(z, w)^{1/q}$$

for all $f \in B_p$ and z and w in $\mathbb{B}_n$, where $1/p + 1/q = 1$.

6.22. Show that every function in B_p can be approximated in norm by its Taylor polynomials if and only if $1 < p < \infty$.

6.23. Study the behavior of the integral pairing in Theorem 6.21 as $t \to \infty$.

6.24. Suppose $1 < p < \infty$ and $1/p + 1/q = 1$. Show that $(B_p)^* = B_q$ under any of the following integral pairings:

$$\langle f, g \rangle_{\alpha, t} = \int_{\mathbb{B}_n} (1 - |z|^2)^t R^{\alpha, t} f(z) \, \overline{(1 - |z|^2)^t R^{\alpha, t} g(z)} \, d\tau(z),$$

where $2t > n$, and neither $n + \alpha$ nor $n + \alpha + t$ is a negative integer. Similar assertions can be made about the dual spaces of B_0 and B_p for $0 < p \leq 1$.

6.25. Show that B_1 is an algebra. Moreover, there exists a constant $C > 0$ such that

$$\|fg\|_{B_1} \le C\|f\|_{B_1}\|g\|_{B_1}$$

for all f and g in B_1. See [30].

6.26. For an analytic function f in the unit disk $\mathbb{D}$ define

$$r_k(f) = \inf \|f - g\|_B, \qquad = 1, 2, \cdots,$$

and

$$s_k(f) = \inf \|f - g\|_{\mathrm{BMO}}, \qquad k = 1, 2, \cdots,$$

where both infima are taken over all rational functions g of degree k with poles outside the closed unit disk. Show that the following conditions are equivalent for $0 < p < \infty$.

(a) $f \in B_p$.
(b) $\{r_k\} \in l^p$.
(c) $\{s_k\} \in l^p$.

See [30] and references there.

6.27. If H is a Hilbert space of holomorphic functions in $\mathbb{B}_n$ and $K(z, w)$ is the reproducing kernel of H, show that

$$\sup\{|f(z) - f(w)|^2 : \|f\| \le 1\} = K(z, z) + K(w, w) - K(z, w) - K(w, z)$$

for all a and w in $\mathbb{B}_n$.

6.28. Fix any radius $r > 0$. For any holomorphic function f in $\mathbb{B}_n$ define

$$I_r f(z) = \int_{D(z,r)} |\widetilde{\nabla} f(w)| \, d\tau(w), \qquad z \in \mathbb{B}_n.$$

If $1 < p < \infty$, show that $f \in B_p$ if and only if $I_r f \in L^p(\mathbb{B}_n, d\tau)$. See [72][73].

6.29. If f is holomorphic in $\mathbb{B}_n$ and $2n < p < \infty$, show that $f \in B_p$ if and only if

$$\int_{\mathbb{B}_n} \int_{\mathbb{B}_n} \left[\frac{|f(z) - f(w)|}{|w - P_w(z) - s_w Q_w(z)|} \right]^p (1 - |z|^2)^{p/2} (1 - |w|^2)^{p/2} \, d\tau(z) \, d\tau(w) < \infty,$$

where $s_w = (1 - |w|^2)^{1/2}$, P_w is the orthogonal projection from $\mathbb{C}^n$ onto the one-dimensional subspace $[w]$ spanned by w, and Q_w is the orthogonal projection from $\mathbb{C}^n$ onto $\mathbb{C}^n \ominus [w]$. See [72][73].

6.30. Let H_2 denote the space of holomorphic functions

$$f(z) = \sum_m a_m z^m$$

in the unit ball $\mathbb{B}_n$ such that

$$\|f\|^2 = \sum_m |a_m|^2 < \infty.$$

Show that H_2 is a Hilbert space with inner product

$$\langle f, g \rangle = \sum_m a_m \bar{b}_m, \qquad f(z) = \sum_m a_m z^m, g(z) = \sum_m b_m z^m.$$

Show that the reproducing kernel of H_2 is given by

$$K(z, w) = \frac{1}{1 - \langle z, w \rangle}.$$

This space has attracted much attention lately in multi-variable operator theory and is sometimes referred to as the Arveson space.

6.31. Characterize H_2 in terms of higher order derivatives and membership in A_α^2.

6.32. Characterize H_2 in terms of fractional derivatives and membership in A_α^2.

6.33. Characterize H_2 in terms of higher order derivatives and membership in H^2.

6.34. Characterize H_2 in terms of fractional derivatives and membership in H^2.

7

Lipschitz Spaces

In this chapter we study two classes of holomorphic functions in the unit ball, both depending on a positive parameter α, and are denoted by Λ_α and $\mathcal{B}_\alpha$, respectively. As α increases, the spaces $\mathcal{B}_\alpha$ get larger and larger, while the spaces Λ_α get smaller and smaller. The two classes have an intersection; more specifically, we have $\mathcal{B}_\alpha = \Lambda_{1-\alpha}$ for $0 < \alpha < 1$.

Our emphasis is on the holomorphic Lipschitz spaces Λ_α, and the most interesting cases are when $0 < \alpha \leq 1$. Our approach here is to treat the Lipschitz spaces as close relatives of the Bloch space. In fact, Λ_α is simply the image of the Bloch space under a certain fractional integral operator. Consequently, we will obtain integral representations, estimates in terms various derivatives, atomic decomposition, and duality theorems for the Lipschitz spaces Λ_α. We will also discuss the tangential growth of functions in Λ_α, which is an important feature of the classical theory of Lipschitz spaces.

7.1 $\mathcal{B}_\alpha$ Spaces

In this section we introduce a class of spaces similar to the Bloch space, which later will be shown to include the holomorphic Lipschitz spaces. Thus for any $\alpha > 0$ we let $\mathcal{B}_\alpha$ denote the space of holomorphic functions f in $\mathbb{B}_n$ such that the functions

$$(1 - |z|^2)^\alpha \frac{\partial f}{\partial z_k}(z), \qquad 1 \leq k \leq n,$$

are all bounded in $\mathbb{B}_n$. Clearly, a holomorphic function f in $\mathbb{B}_n$ belongs to $\mathcal{B}_\alpha$ if and only if

$$\|f\|_\alpha = |f(0)| + \sup_{z \in \mathbb{B}_n} (1 - |z|^2)^\alpha |\nabla f(z)| < \infty. \tag{7.1}$$

It is an easy exercise to show that $\mathcal{B}_\alpha$ is a Banach space when equipped with the above norm; see the proof of Proposition 3.2.

Theorem 7.1. *Suppose $\alpha > 0$, $\beta > -1$, and f is holomorphic in $\mathbb{B}_n$. Then the following conditions are equivalent:*

(a) $f \in \mathcal{B}_\alpha$.
(b) The function $(1 - |z|^2)^\alpha |Rf(z)|$ is bounded in $\mathbb{B}_n$.
(c) There exists a function $g \in L^\infty(\mathbb{B}_n)$ such that

$$f(z) = \int_{\mathbb{B}_n} \frac{g(w)\, dv_\beta(w)}{(1 - \langle z, w \rangle)^{n+\beta+\alpha}}, \qquad z \in \mathbb{B}_n.$$

Proof. It is obvious that (a) implies (b).

If (b) holds, then the function

$$g(z) = \frac{c_{\alpha+\beta}}{c_\beta}(1 - |z|^2)^\alpha \left(f(z) + \frac{Rf(z)}{n + \alpha + \beta} \right)$$

is bounded in $\mathbb{B}_n$. Consider the holomorphic function

$$F(z) = \int_{\mathbb{B}_n} \frac{g(w)\, dv_\beta(w)}{(1 - \langle z, w \rangle)^{n+\alpha+\beta}}, \qquad z \in \mathbb{B}_n,$$

or

$$F(z) = \int_{\mathbb{B}_n} \frac{1}{(1 - \langle z, w \rangle)^{n+\alpha+\beta}} \left(f(w) + \frac{Rf(w)}{n + \alpha + \beta} \right) dv_{\alpha+\beta}(w), \qquad z \in \mathbb{B}_n.$$

Applying the fractional differential operator $T = R^{\alpha+\beta-1,1}$ inside the integral, then using Proposition 1.14 and Theorem 2.2, we obtain

$$TF(z) = f(z) + \frac{Rf(z)}{n + \alpha + \beta}.$$

Since $T^{-1} = R_{\alpha+\beta-1,1}$, a calculation using (1.34) shows that

$$F = R_{\alpha+\beta-1,1} \left(f + \frac{Rf}{n + \alpha + \beta} \right) = f.$$

This shows that (b) implies (c).

That (c) implies (a) follows from differentiating under the integral sign and then applying Theorem 1.12.   □

Theorem 7.2. *Suppose $n > 1$ and f is holomorphic in $\mathbb{B}_n$.*

(a) *If $\alpha > \frac{1}{2}$, then $f \in \mathcal{B}_\alpha$ if and only if the function*

$$(1 - |z|^2)^{\alpha-1} |\widetilde{\nabla} f(z)|$$

is bounded in $\mathbb{B}_n$.
(b) *If $\alpha = \frac{1}{2}$ and $f \in \mathcal{B}_\alpha$, then there exists a constant $C > 0$ such that*

$$|\widetilde{\nabla} f(z)| \le C(1 - |z|^2)^{\frac{1}{2}} \log \frac{2}{1 - |z|^2}$$

for all $z \in \mathbb{B}_n$.

(c) If $0 < \alpha < \frac{1}{2}$ and $f \in \mathcal{B}_\alpha$, then there exists a constant $C > 0$ such that

$$|\widetilde{\nabla} f(z)| \le C(1 - |z|^2)^{\frac{1}{2}}$$

for all $z \in \mathbb{B}_n$.

Proof. Recall from Lemma 2.14 that

$$(1 - |z|^2)|\nabla f(z)| \le |\widetilde{\nabla} f(z)|, \qquad z \in \mathbb{B}_n.$$

So the boundedness of $(1 - |z|^2)^{\alpha-1}|\widetilde{\nabla} f(z)|$ implies that of $(1 - |z|^2)^\alpha |\nabla f(z)|$.
On the other hand, if $f \in \mathcal{B}_\alpha$, then by Theorem 7.1,

$$f(z) = \int_{\mathbb{B}_n} \frac{g(w)\, dv(w)}{(1 - \langle z, w \rangle)^{n+\alpha}}, \qquad z \in \mathbb{B}_n,$$

where g is a function in $L^\infty(\mathbb{B}_n)$. An application of Lemma 3.3 gives

$$|\widetilde{\nabla} f(z)| \le \sqrt{2}\,(n + \alpha)(1 - |z|^2)^{\frac{1}{2}} \int_{\mathbb{B}_n} \frac{|g(w)|\, dv(w)}{|1 - \langle z, w \rangle|^{n+\alpha+\frac{1}{2}}}$$

for all $z \in \mathbb{B}_n$. Since g is bounded, the rest of the proof follows from Theorem 1.12. $\qquad\square$

Note that when $n = 1$, we have

$$|\widetilde{\nabla} f(z)| = (1 - |z|^2)|f'(z)|,$$

so part (a) of the above theorem holds for all $\alpha > 0$. When $n > 1$, the condition $\alpha > \frac{1}{2}$ is indispensible.

Let $\mathcal{B}_{\alpha,0}$ denote the closure of the set of polynomials in $\mathcal{B}_\alpha$. It is easy to see that $\mathcal{B}_{\alpha,0}$ consists exactly of holomorphic functions f such that the function

$$h(z) = (1 - |z|^2)^\alpha |\nabla f(z)|$$

is in $C_0(\mathbb{B}_n)$. Note that here and below, the space $C_0(\mathbb{B}_n)$ can be replaced by $C(\overline{\mathbb{B}}_n)$.

Theorem 7.3. *Suppose* $\alpha > 0$, $\beta > -1$, *and* f *is holomorphic in* $\mathbb{B}_n$. *Then the following conditions are equivalent:*

(a) $f \in \mathcal{B}_{\alpha,0}$.
(b) *The function* $(1 - |z|^2)^\alpha |Rf(z)|$ *is in* $C_0(\mathbb{B}_n)$.
(c) *There exists a function* $g \in C_0(\mathbb{B}_n)$ *such that*

$$f(z) = \int_{\mathbb{B}_n} \frac{g(w)\, dv_\beta(w)}{(1 - \langle z, w \rangle)^{n+\beta+\alpha}}, \qquad z \in \mathbb{B}_n.$$

We leave the proof of this theorem as well as the next one to the interested reader.

Theorem 7.4. *Suppose $n > 1$, $\alpha > \frac{1}{2}$, and f is holomorphic in $\mathbb{B}_n$. Then $f \in \mathcal{B}_{\alpha,0}$ if and only if the function*

$$h(z) = (1 - |z|^2)^{\alpha-1} |\widetilde{\nabla} f(z)|$$

belongs to $\mathbb{C}_0(\mathbb{B}_n)$.

For any $s > -1$ we write

$$\langle f, g \rangle_s = \lim_{r \to 1^-} \int_{\mathbb{B}_n} f(rz) \overline{g(rz)} \, dv_s(z), \tag{7.2}$$

whenever the limit exists, where f and g are holomorphic in $\mathbb{B}_n$.

Theorem 7.5. *Suppose $\alpha > 0$, $\beta > -1$, and $\alpha + \beta > 0$. If $s = \alpha + \beta - 1$, then $(\mathcal{B}_{\alpha,0})^* = A_\beta^1$ under the integral pairing $\langle \, , \, \rangle_s$, where the equality holds with equivalent norms.*

Proof. If $f \in \mathcal{B}_\alpha$, then by Theorem 7.1 there exists a function $h \in L^\infty(\mathbb{B}_n)$ such that

$$f(z) = \int_{\mathbb{B}_n} \frac{h(w) \, dv_\beta(w)}{(1 - \langle z, w \rangle)^{n+\beta+\alpha}}, \qquad z \in \mathbb{B}_n.$$

Moreover, we can choose h so that $\|h\|_\infty \le C\|f\|_\alpha$, where C is a positive constant independent of f. For $g \in A_\beta^1$ and $0 < r < 1$ we write $g_r(z) = g(rz)$, $z \in \mathbb{B}_n$. Then

$$\langle f, g_r \rangle_s = \int_{\mathbb{B}_n} h(w) \overline{g_r(w)} \, dv_\beta(w)$$

by Fubini's theorem and the reproducing formula in Theorem 2.2. It follows that

$$|\langle f, g_r \rangle_s| \le \|h\|_\infty \|g\|_{\beta,1} \le C\|f\|_\alpha \|g\|_{\beta,1}.$$

In particular, every function $g \in A_\beta^1$ induces a bounded linear functional on $\mathcal{B}_{\alpha,0}$ via the integral pairing $\langle \, , \, \rangle_s$.

To show that every bounded linear functional on $\mathcal{B}_{\alpha,0}$ arises from a function in A_β^1 via the integral pairing $\langle \, , \, \rangle_s$, we fix a sufficiently large positive parameter a and consider the operator T defined by

$$Tf(z) = \frac{c_{a+\beta}}{c_\beta} (1 - |z|^2)^a \int_{\mathbb{B}_n} \frac{f(w) \, dv_s(w)}{(1 - \langle z, w \rangle)^{n+1+\beta+a}}.$$

If $f \in \mathcal{B}_\alpha$, then by the previous paragraph, we have

$$|Tf(z)| \le C(1 - |z|^2)^a \|f\|_\alpha \int_{\mathbb{B}_n} \frac{dv_\beta(w)}{|1 - \langle z, w \rangle|^{n+1+\beta+a}}.$$

An application of Theorem 1.12 then shows that $Tf \in L^\infty(\mathbb{B}_n)$, and T is actually a bounded operator from $\mathcal{B}_\alpha$ into $L^\infty(\mathbb{B}_n)$. On the other hand, using Fubini's theorem and the reproducing formula in Theorem 2.2, we can verify that

$$\int_{\mathbb{B}_n} \frac{Tf(w)\,dv_\beta(w)}{(1-\langle z,w\rangle)^{n+\beta+\alpha}} = f(z)$$

for all $f \in \mathcal{B}_\alpha$ and $z \in \mathbb{B}_n$. This implies that

$$\|f\|_\alpha \leq C\|Tf\|_\infty,$$

where C is some positive constant independent of f. We conclude that T is an embedding of $\mathcal{B}_\alpha$ into $L^\infty(\mathbb{B}_n)$.

If f is a polynomial, then Tf is $(1-|z|^2)^a$ times a polynomial, which is a function in $C_0(\mathbb{B}_n)$. Since $C_0(\mathbb{B}_n)$ is closed in $L^\infty(\mathbb{B}_n)$, we see that T is an embedding of $\mathcal{B}_{\alpha,0}$ into $C_0(\mathbb{B}_n)$. Let X be the image of $\mathcal{B}_{\alpha,0}$ in $C_0(\mathbb{B}_n)$ under the mapping T. Then X is a closed subspace of $C_0(\mathbb{B}_n)$.

Now if F is a bounded linear functional on $\mathcal{B}_{\alpha,0}$, then $F \circ T^{-1}$ is a bounded linear functional on X. Continuously extend $F \circ T^{-1}$ to the whole space $C_0(\mathbb{B}_n)$ via the Hahn-Banach extension theorem, and then apply the classical Riesz representation theorem for $C_0(\mathbb{B}_n)$. We obtain a finite complex Borel measure μ on $\mathbb{B}_n$ such that

$$F \circ T^{-1}(f) = \int_{\mathbb{B}_n} f(z)\,d\mu(z), \qquad f \in X,$$

or equivalently,

$$F(f) = \int_{\mathbb{B}_n} Tf(z)\,d\mu(z), \qquad f \in \mathcal{B}_{\alpha,0}.$$

If f is a polynomial (recall that the polynomials are dense in $\mathcal{B}_{\alpha,0}$), we use Fubini's theorem to rewrite the above integral as

$$F(f) = \int_{\mathbb{B}_n} f(z)\overline{g(z)}\,dv_s(z),$$

where

$$g(z) = \frac{c_{a+\beta}}{c_\beta} \int_{\mathbb{B}_n} \frac{(1-|w|^2)^a\,d\bar\mu(w)}{(1-\langle z,w\rangle)^{n+1+\beta+a}}.$$

By Fubini's theorem and Theorem 1.12, we easily check that $g \in A^1_\beta$. This completes the proof of the theorem. □

Theorem 7.6. *Suppose $\alpha > 0$, $\beta > -1$, $0 < p \leq 1$, and*

$$s = \frac{n+1+\beta}{p} + \alpha - (n+2).$$

If $s > -1$, then under the integral pairing $\langle\,,\,\rangle_s$ we have the duality

$$(A^p_\beta)^* = \mathcal{B}_\alpha,$$

where the equality holds with equivalent norms.

Proof. It is easy to see that $1 - \alpha + s > -1$. If $g \in \mathcal{B}_\alpha$, then by Theorem 7.1, there exists a function $h \in L^\infty(\mathbb{B}_n)$ such that

$$g(z) = \int_{\mathbb{B}_n} \frac{h(w)\, dv_{1-\alpha+s}(w)}{(1 - \langle z, w \rangle)^{n+1+s}}, \qquad z \in \mathbb{B}_n,$$

and $\|h\|_\infty \le C\|g\|_\alpha$, where C is a positive constant independent of g. By Fubini's theorem,

$$\langle f, g \rangle_s = \int_{\mathbb{B}_n} f(z)\overline{h(z)}\, dv_{1-\alpha+s}(z)$$

$$= c_{1-\alpha+s} \int_{\mathbb{B}_n} f(z)\overline{h(z)}(1-|z|^2)^{\frac{n+1+\beta}{p}-(n+1)}\, dv(z).$$

Combining this with Lemma 2.15, we see that g induces a bounded linear functional on A^p_β under the integral pairing $\langle\,,\,\rangle_s$.

Conversely, if F is a bounded linear functional on A^p_β and $f \in A^p_\beta$, then

$$f_r(z) = \int_{\mathbb{B}_n} \frac{f_r(w)\, dv_s(w)}{(1 - \langle z, w \rangle)^{n+1+s}}$$

for $0 < r < 1$, and it is easy to verify (using the homogeneous expansion of the kernel function) that

$$F(f_r) = \int_{\mathbb{B}_n} f_r(w) F_z \left[\frac{1}{1 - \langle z, w \rangle)^{n+1+s}} \right] dv_s(w).$$

Define a function g on $\mathbb{B}_n$ by

$$\overline{g(w)} = F_z \left[\frac{1}{(1 - \langle z, w \rangle)^{n+1+s}} \right].$$

Then

$$F(f_r) = \int_{\mathbb{B}_n} f(rw)\overline{g(w)}\, dv_s(w) = \langle f_r, g \rangle_s.$$

It remains for us to show that $g \in \mathcal{B}_\alpha$.

We interchange differentiation and the application of F, which can be justified by using the homogeneous expansion of the kernel function. The result is

$$\overline{Rg(w)} = (n+1+s) F_z \left[\frac{\langle z, w \rangle}{(1 - \langle z, w \rangle)^{n+2+s}} \right].$$

Since F is bounded on A^p_β, we have

$$|Rg(w)| \le (n+1+s)\|F\| \left[\int_{\mathbb{B}_n} \frac{dv_\beta(z)}{|1 - \langle z, w \rangle|^{p(n+2+s)}} \right]^{1/p}.$$

An application of Theorem 1.12 then shows that

$$|Rg(w)| \le \frac{C\|F\|}{(1 - |w|^2)^\alpha}, \qquad w \in \mathbb{B}_n.$$

This shows that $g \in \mathcal{B}_\alpha$ and completes the proof of the theorem. □

An atomic decomposition theorem similar to that for the Bloch space also holds for the spaces $\mathcal{B}_\alpha$.

Theorem 7.7. *Suppose $\alpha > 0$ and*

$$b > \max(n, n + \alpha - 1).$$

There exists a sequence $\{a_k\}$ in $\mathbb{B}_n$ such that $\mathcal{B}_\alpha$ consists exactly of functions of the form

$$f(z) = \sum_{k=1}^{\infty} c_k \frac{(1 - |a_k|^2)^{b+1-\alpha}}{(1 - \langle z, a_k \rangle)^b},$$

where $\{c_k\} \in l^\infty$.

Proof. The proof is similar to that of Theorem 3.23. We leave the details to the interested reader. □

A little oh version of this result also holds, with $\mathcal{B}_\alpha$ replaced by $\mathcal{B}_{\alpha,0}$, and l^∞ replaced by c_0.

7.2 The Lipschitz Spaces Λ_α for $0 < \alpha < 1$

For $0 < \alpha < 1$ we define Λ_α to be the space of holomorphic functions f in $\mathbb{B}_n$ such that

$$\|f\|_\alpha = \sup \left\{ \frac{|f(z) - f(w)|}{|z - w|^\alpha} : z, w \in \mathbb{B}_n, z \ne w \right\} < \infty. \tag{7.3}$$

The space Λ_α will be called the holomorphic Lipschitz space of order α. It is clear that each space Λ_α contains the polynomials, and is contained in the ball algebra.

Proposition 7.8. *For each $\alpha \in (0,1)$ the holomorphic Lipschitz space Λ_α is a Banach space with the norm*

$$\|f\| = |f(0)| + \|f\|_\alpha.$$

Proof. If $\{f_k\}$ is a Cauchy sequence in Λ_α, then $\{f_k\}$ is uniformly Cauchy in $\mathbb{B}_n$, so $\{f_k\}$ converges uniformly to some holomorphic function f in $\mathbb{B}_n$.

Given any $\epsilon > 0$, choose a positive integer N such that

$$|f_k(0) - f_l(0)| + \|f_k - f_l\|_\alpha < \epsilon$$

for all $l > N$ and $k > N$. Then

$$|(f_l(z) - f_k(z)) - (f_l(w) - f_k(w))| \leq \epsilon |z - w|^\alpha$$

for all z and w in $\mathbb{B}_n$, and all l and k greater than N. Let $l \to \infty$. We obtain

$$|(f_k(z) - f(z)) - (f_k(w) - f(w))| \leq \epsilon |z - w|^\alpha$$

for all $k > N$ and all z and w in $\mathbb{B}_n$. It follows that $f_k - f$ is in Λ_α and

$$\lim_{k \to \infty} \|f_k - f\|_\alpha = 0.$$

Since $f_k(0) \to f(0)$, we have $f_k \to f$ in Λ_α, so the space Λ_α is complete.   □

The space Λ_α is not separable. We let $\Lambda_{\alpha,0}$ denote the closure in Λ_α of the set of polynomials. So $\Lambda_{\alpha,0}$ is a separable Banach space by itself.

The space X of holomorphic functions f in $\mathbb{B}_n$ such that

$$\sup \left\{ \frac{|f(z) - f(w)|}{|z - w|} : z, w \in \mathbb{B}_n, z \neq w \right\} < \infty,$$

and the separable subspace X_0 of X generated by the polynomials, are fundamentally different from the Lipschitz spaces Λ_α and $\Lambda_{\alpha,0}$ when $0 < \alpha < 1$. In fact, a moment of thought reveals that X consists of holomorphic functions whose complex gradient is bounded; and X_0 contains exactly the holomorphic functions whose first order partial derivatives are continuous up to the boundary. In particular, the radial differential operator

$$f(z) \mapsto f(0) + \sum_{k=1}^{n} z_k \frac{\partial f}{\partial z_k}(z)$$

maps the space X (respectively, X_0) boundedly onto H^∞ (respectively, the ball algebra A). Because of this observation, we will not discuss the spaces X and X_0 in this chapter. The appropriate definition for the Lipschitz space Λ_1 will be introduced in the next section.

Theorem 7.9. *Suppose $0 < \alpha < 1$, $\beta > -1$, and f is holomorphic in $\mathbb{B}_n$. Then the following conditions are equivalent:*

(a) f is in Λ_α.
(b) f is in the ball algebra and its boundary values satisfy

$$\sup \left\{ \frac{|f(\zeta) - f(\xi)|}{|\zeta - \xi|^\alpha} : \zeta, \xi \in \mathbb{S}_n, \zeta \neq \xi \right\} < \infty.$$

(c) $(1 - |z|^2)^{1-\alpha} |Rf(z)|$ is bounded in $\mathbb{B}_n$.
(d) There exists a function $g \in L^\infty(\mathbb{B}_n)$ such that

$$f(z) = \int_{\mathbb{B}_n} \frac{g(w) \, dv_\beta(w)}{(1 - \langle z, w \rangle)^{n+1+\beta-\alpha}}$$

for all $z \in \mathbb{B}_n$.

(e) $(1 - |z|^2)^{1-\alpha} |\nabla f(z)|$ *is bounded in* $\mathbb{B}_n$.

Proof. The equivalence of (c), (d), and (e) follows from Theorem 7.1.

It is obvious that (a) implies (b).

Suppose f is in the ball algebra and its boundary function satisfies

$$|f(\zeta_1) - f(\zeta_2)| \le C|\zeta_1 - \zeta_2|^\alpha$$

for all ζ_1 and ζ_2 in $\mathbb{S}_n$. In particular,

$$|f(\zeta e^{i\theta}) - f(\zeta e^{it})| \le C|e^{i\theta} - e^{it}|^\alpha$$

for all $\zeta \in \mathbb{S}_n$ and all real θ and t. Since

$$f(z) = \int_{\mathbb{S}_n} \frac{f(\zeta)\, d\sigma(\zeta)}{(1 - \langle z, \zeta \rangle)^n}, \qquad z \in \mathbb{B}_n,$$

a calculation using (1.14) shows that

$$Rf(z) = n \int_{\mathbb{S}_n} d\sigma(\zeta) \frac{1}{2\pi} \int_0^{2\pi} \frac{\langle z, \zeta \rangle e^{-i\theta} f(e^{i\theta}\zeta)}{(1 - \langle z, \zeta \rangle e^{-i\theta})^{n+1}}\, d\theta$$

for every $z \in \mathbb{B}_n$. Fix z and ζ for the moment and denote the inner integral above by J. Since the integral J is zero when f is constant, we can write

$$J = \frac{1}{2\pi} \int_0^{2\pi} \frac{\langle z, \zeta \rangle e^{-i\theta}(f(e^{i\theta}\zeta) - f(e^{it}\zeta))}{(1 - \langle z, \zeta \rangle e^{-i\theta})^{n+1}}\, d\theta,$$

where $\langle z, \zeta \rangle = re^{it}$ with $0 \le r < 1$ and t real. By the Lipschitz condition of f on the boundary,

$$|J| \le \frac{r}{2\pi} \int_0^{2\pi} \frac{|e^{i\theta} - e^{it}|^\alpha}{|1 - re^{i(t-\theta)}|^{n+1}}\, d\theta.$$

Since

$$\frac{|1 - \lambda|}{|1 - r\lambda|} \le \frac{1-r}{|1 - r\lambda|} + \frac{|r - \lambda|}{|1 - r\lambda|} \le 2$$

for all $r \in [0, 1)$ and all $\lambda \in \mathbb{C}$ with $|\lambda| = 1$,

$$|J| \le \frac{2^\alpha}{2\pi} \int_0^{2\pi} \frac{d\theta}{|1 - \langle z, \zeta \rangle e^{-i\theta}|^{n+1-\alpha}}.$$

Therefore,

$$\begin{aligned}
|Rf(z)| &\le \frac{2^\alpha n}{2\pi} \int_{\mathbb{S}_n} d\sigma(\zeta) \int_0^{2\pi} \frac{d\theta}{|1 - \langle z, \zeta \rangle e^{-i\theta}|^{n+1-\alpha}} \\
&= 2^\alpha n \int_{\mathbb{S}_n} \frac{d\sigma(\zeta)}{|1 - \langle z, \zeta \rangle|^{n+1-\alpha}} \\
&\le \frac{C}{(1 - |z|^2)^{1-\alpha}},
\end{aligned}$$

where C is a positive constant whose existence follows from Theorem 1.12. This shows that (b) implies (c).

To finish the proof, we assume that there is a positive constant C such that

$$(1 - |z|)^{1-\alpha}|\nabla f(z)| \leq (1 - |z|^2)^{1-\alpha}|\nabla f(z)| \leq C$$

for all $z \in \mathbb{B}_n$, and proceed to show that $f \in \Lambda_\alpha$. Fix two points a and b in $\mathbb{B}_n$ such that

$$0 < |a| \leq |b| < 1;$$

the case $a = 0$ will then follow from an obvious limit argument. Let

$$a' = \frac{1 - \delta}{|a|} a, \qquad b' = \frac{1 - \delta}{|b|} b,$$

where $\delta = |a - b| \in [0, 2)$, so that $1 - \delta \in (-1, 1]$. There are three cases to consider.

Case I. If $\delta \leq 1 - |b|$, then

$$|(\nabla f)(\gamma(t))| \leq C(1 - |b|)^{\alpha-1} \leq C\delta^{\alpha-1}$$

for all $0 \leq t \leq 1$, where

$$\gamma(t) = ta + (1 - t)b, \qquad 0 \leq t \leq 1,$$

is the line segment from a to b. It follows that

$$|f(b) - f(a)| = \left| \int_0^1 \frac{df(\gamma(t))}{dt} dt \right|$$

$$= \left| \int_0^1 \sum_{k=1}^n \frac{\partial f}{\partial z_k}(\gamma(t))\gamma'_k(t) dt \right|$$

$$\leq \int_0^1 |\nabla f(\gamma(t))| |\gamma'(t)| dt$$

$$\leq C\delta^{\alpha-1} \int_0^1 |\gamma'(t)| dt$$

$$= C|a - b|^\alpha.$$

Case II. If $1 - |b| < \delta \leq 1 - |a|$, then $|a - b'| \leq |a - b|$ and

$$|f(a) - f(b)| \leq |f(a) - f(b')| + |f(b') - f(b)|.$$

The first term on the right-hand side above is estimated as in Case I, while the second term is estimated using the line segment

$$\gamma(t) = tb + (1 - t)b', \qquad 0 \leq t \leq 1,$$

as follows:

$$|f(b') - f(b)| \le \int_0^1 |\nabla f(\gamma(t))| \, |\gamma'(t)| \, dt$$

$$\le C|b - b'| \int_0^1 \left(1 - [|b|t + (1 - \delta)(1 - t)]\right)^{\alpha-1} dt$$

$$= C \int_{1-\delta}^{|b|} (1 - x)^{\alpha-1} \, dx$$

$$< C\alpha^{-1}|a - b|^\alpha.$$

Case III. If $1 - |a| < \delta$, then $|a' - b'| < |a - b|$ and

$$|f(a) - f(b)| \le |f(a) - f(a')| + |f(a') - f(b')| + |f(b') - f(b)|.$$

The first and third terms on the right-hand side above are estimated directly using line segments as in Case II, while the second term is estimated as in Case I.

The proof of the theorem is now complete. □

Note that in the proof that (b) implies (c) we only need to assume that for each $\zeta \in \mathbb{S}_n$ the slice function

$$f_\zeta(z) = f(\zeta z)$$

satisfies the Lipschitz α-condition on the unit circle in $\mathbb{C}$.

As a consequence of Theorem 7.9, we see that $\Lambda_\alpha = \mathcal{B}_{1-\alpha}$ for any $\alpha \in (0, 1)$.

Theorem 7.10. *Suppose $0 < \alpha < 1$, $\beta > -1$, and f is holomorphic in $\mathbb{B}_n$. Then the following conditions are equivalent:*

(a) $f \in \Lambda_{\alpha,0}$.
(b) The restriction of f on $\mathbb{S}_n$ can be approximated in the Lipschitz α-norm of $\mathbb{S}_n$ by polynomials.
(c) $(1 - |z|^2)^{1-\alpha} Rf(z)$ is in $C_0(\mathbb{B}_n)$.
(d) There exists a function $g \in C_0(\mathbb{B}_n)$ such that

$$f(z) = \int_{\mathbb{B}_n} \frac{g(w) \, dv_\beta(w)}{(1 - \langle z, w \rangle)^{n+1+\beta-\alpha}}$$

for all z in $\mathbb{B}_n$.
(e) $(1 - |z|^2)^{1-\alpha}|\nabla f(z)|$ is in $C_0(\mathbb{B}_n)$.

Proof. The proof is similar to that of Theorem 7.9. We omit the details. □

Note that Theorem 7.10 still holds if $C_0(\mathbb{B}_n)$ is replaced by $C(\overline{\mathbb{B}}_n)$.

7.3 The Zygmund Class

In this section we introduce the appropriate limit case of Λ_α when $\alpha \to 1^-$. Instead of defining it to be the space of holomorphic functions f satisfying

$$\sup\left\{ \frac{|f(z) - f(w)|}{|z - w|} : z, w \in \mathbb{B}_n, z \neq w \right\} < \infty,$$

we define Λ_1 to be the space of holomorphic functions in $\mathbb{B}_n$ whose first order partial derivatives are in the Bloch space. The space Λ_1 so defined is also called the Zygmund class.

Theorem 7.11. *Let $\beta > -1$ and f be holomorphic in $\mathbb{B}_n$. The following conditions are equivalent:*

(a) $f \in \Lambda_1$, that is, $\partial f / \partial z_k$ is in the Bloch space for each $1 \leq k \leq n$.
(b) Rf is in the Bloch space.
(c) There exists $g \in L^\infty(\mathbb{B}_n)$ such that

$$f(z) = \int_{\mathbb{B}_n} \frac{g(w) \, dv_\beta(w)}{(1 - \langle z, w \rangle)^{n+\beta}}$$

for all $z \in \mathbb{B}_n$.

Proof. Since each coordinate function z_k, $1 \leq k \leq n$, is a pointwise multiplier of the Bloch space $\mathcal{B}$, it is clear that (a) implies (b).

If Rf is in the Bloch space, then so is f. By Theorem 3.4, there exists a function $g \in L^\infty(\mathbb{B}_n)$ such that

$$f(z) + \frac{Rf(z)}{n + \beta} = \int_{\mathbb{B}_n} \frac{g(w) \, dv_\beta(w)}{(1 - \langle z, w \rangle)^{n+1+\beta}} \tag{7.4}$$

for $z \in \mathbb{B}_n$. Consider the fractional integral operator $R_{\beta-1,1}$ from Section 1.4. It is easy to check that

$$R_{\beta-1,1}\left(f + \frac{Rf}{n + \beta} \right) = f$$

for every holomorphic function f in $\mathbb{B}_n$. Also recall from Proposition 1.14 that

$$R_{\beta-1,1}\left(\frac{1}{(1 - \langle z, w \rangle)^{n+1+\beta}} \right) = \frac{1}{(1 - \langle z, w \rangle)^{n+\beta}}$$

for every $w \in \mathbb{B}_n$. If we apply the operator $R_{\beta-1,1}$ to both sides of (7.4), the result is

$$f(z) = \int_{\mathbb{B}_n} \frac{g(w) \, dv_\beta(w)}{(1 - \langle z, w \rangle)^{n+\beta}}$$

for $z \in \mathbb{B}_n$. This shows that (b) implies (c).

If f admits the integral representation in (c), then for each $1 \leq k \leq n$ we can differentiate under the integral sign to get

$$\frac{\partial f}{\partial z_k}(z) = \int_{\mathbb{B}_n} \frac{g_k(w)\,dv_\beta(w)}{(1 - \langle z, w \rangle)^{n+1+\beta}}, \qquad z \in \mathbb{B}_n,$$

where

$$g_k(w) = (n + \beta)\,\overline{w}_k g(w)$$

is still bounded in $\mathbb{B}_n$. By Theorem 3.4, each partial derivative $\partial f / \partial z_k$ is in the Bloch space. This shows that (c) implies (a), and the proof of the theorem is complete. □

Define

$$\|f\|_1 = |f(0)| + \|Rf\|_\mathcal{B} \tag{7.5}$$

for $f \in \Lambda_1$. Then it is easy to see that Λ_1 is a Banach space with this norm. We let $\Lambda_{1,0}$ denote the closure in Λ_1 of the set of all polynomials.

Theorem 7.12. *For $\beta > -1$ and f holomorphic in $\mathbb{B}_n$ the following conditions are equivalent:*

(a) $f \in \Lambda_{1,0}$.
(b) Each $\partial f / \partial z_k$, $1 \leq k \leq n$, is in the little Bloch space $\mathcal{B}_0$.
(c) Rf is in $\mathcal{B}_0$.
(d) There exists $g \in C_0(\mathbb{B}_n)$ such that

$$f(z) = \int_{\mathbb{B}_n} \frac{g(w)\,dv_\beta(w)}{(1 - \langle z, w \rangle)^{n+\beta}}$$

for all $z \in \mathbb{B}_n$.
(e) There exists $g \in C(\overline{\mathbb{B}}_n)$ such that

$$f(z) = \int_{\mathbb{B}_n} \frac{g(w)\,dv_\beta(w)}{(1 - \langle z, w \rangle)^{n+\beta}}$$

for all $z \in \mathbb{B}_n$.

Proof. The proof is similar to that of Theorem 7.11. We leave the details to the interested reader. □

7.4 The case $\alpha > 1$

In this section we generalize the definition of holomorphic Lipschitz spaces Λ_α to the case $\alpha > 1$. Thus for $\alpha > 1$ we define Λ_α to be the space of holomorphic functions f in $\mathbb{B}_n$ whose k-th order partial derivatives all belong to $\Lambda_{\alpha-k}$, where k is the positive integer such that

$$k < \alpha \leq k + 1. \tag{7.6}$$

The following theorem characterizes the membership of f in Λ_α, $\alpha > 1$, in terms of various derivatives of f. Notice that the conditions in (7.6) above and (7.7) below are meant to be different; there is no misprint here.

Theorem 7.13. *Suppose $\alpha > 1$ and k is the positive integer such that*

$$k \le \alpha < k + 1. \tag{7.7}$$

Then the following conditions are equivalent for a holomorphic function f in $\mathbb{B}_n$:

(a) The function f belongs to Λ_α.

(b) The function $(1 - |z|^2)^{k+1-\alpha} R^{k+1} f(z)$ is bounded in $\mathbb{B}_n$, where R^{k+1} is the $(k+1)$th power of the radial differential operator R.

(c) The function $(1 - |z|^2)^{k+1-\alpha} \partial^m f / \partial z^m$ is bounded in $\mathbb{B}_n$ for each multi-index m of nonnegative integers with $|m| = k + 1$.

Proof. If α is not an integer, then $f \in \Lambda_\alpha$ if and only if all k-th order partial derivatives of f belong to $\Lambda_{\alpha-k}$, which, in view of Theorem 7.9 and the identity

$$1 - (\alpha - k) = k + 1 - \alpha,$$

shows that conditions (a) and (c) are equivalent in this case.

If α is an integer, then $\alpha = k$, and so $f \in \Lambda_\alpha$ if and only if all partial derivatives of f of order $k - 1$ belongs to Λ_1. Since Λ_1 consists of holomorphic functions whose first order partial derivatives belong to the Bloch space, we see that conditions (a) and (c) are also equivalent in this case. Thus (a) and (c) are equivalent.

It is easy to see that if condition (c) holds, then the functions

$$(1 - |z|^2)^{k+1-\alpha} \frac{\partial^m f}{\partial z^m}$$

are bounded for all m with $|m| \le k + 1$. This observation, together with the fact that R^{k+1} is a linear differential operator of order $k + 1$ with polynomial coefficients, shows that (c) implies (b).

Next assume that the function

$$g(z) = c_{k+1-\alpha}(1 - |z|^2)^{k+1-\alpha} R^{k+1} f(z)$$

is bounded in $\mathbb{B}_n$. Then by Theorem 2.2,

$$R^{k+1} f(z) = \int_{\mathbb{B}_n} \frac{g(w)\, dv(w)}{(1 - \langle z, w \rangle)^{n+1+k+1-\alpha}}, \qquad z \in \mathbb{B}_n.$$

From the identity

$$\int_0^1 \frac{Rh(tz)}{t}\, dt = h(z) - h(0), \qquad z \in \mathbb{B}_n,$$

where h is any holomorphic function in $\mathbb{B}_n$, we deduce that

$$f(z) - f(0) = \int_0^1 \frac{dt_1}{t_1} \cdots \int_0^1 \frac{dt_{k+1}}{t_{k+1}} \int_{\mathbb{B}_n} \frac{g(w)\, dv(w)}{(1 - t_1 \cdots t_{k+1} \langle z, w \rangle)^{n+1+k+1-\alpha}}$$

for all $z \in \mathbb{B}_n$. Now if m is any multi-index of nonnegative integers with $|m| = k+1$, then differentiation under the integral signs shows that there exists a constant $C > 0$ such that

$$\left| \frac{\partial^m f}{\partial z^m}(z) \right| \leq C \int_0^1 dt_1 \cdots \int_0^1 dt_{k+1} \int_{\mathbb{B}_n} \frac{(t_1 \cdots t_{k+1})^{k+1} \|g\|_\infty \, dv(w)}{|1 - t_1 \cdots t_{k+1}\langle z, w\rangle|^{n+1+2(k+1)-\alpha}}$$

for all $z \in \mathbb{B}_n$. Applying Theorem 1.12, we obtain a constant $C' > 0$ such that

$$\left| \frac{\partial^m f}{\partial z^m}(z) \right| \leq C' \|g\|_\infty \int_0^1 dt_1 \cdots \int_0^1 \frac{(t_1 \cdots t_{k+1})^{k+1} dt_{k+1}}{(1 - t_1 \cdots t_{k+1}|z|)^{2(k+1)-\alpha}}$$

for all $z \in \mathbb{B}_n$. This iterated integral can be estimated elementarily, and we obtain another constant $C'' > 0$ such that

$$\left| \frac{\partial^m f}{\partial z^m}(z) \right| \leq \frac{C''}{(1 - |z|)^{k+1-\alpha}}, \qquad z \in \mathbb{B}_n.$$

This shows that (b) implies (c), and completes the proof of the theorem. $\square$

The integral representation for functions in Λ_α is slightly more complicated when α is large.

Theorem 7.14. *Suppose $\alpha > 1$, $\beta > -1$, and $n + \beta - \alpha$ is not a negative integer. Then a holomorphic function f in $\mathbb{B}_n$ belongs to Λ_α if and only if there exists a function $g \in L^\infty(\mathbb{B}_n)$ such that*

$$f(z) = \int_{\mathbb{B}_n} \frac{g(w) \, dv_\beta(w)}{(1 - \langle z, w\rangle)^{n+1+\beta-\alpha}}$$

for all $z \in \mathbb{B}_n$.

Proof. First assume that f admits the integral representation. Differentiating under the integral sign and applying Theorem 1.12, we see that condition (c) in Theorem 7.13 holds, so $f \in \Lambda_\alpha$.

Next we assume that $f \in \Lambda_\alpha$. Let k be the positive integer satisfying $k < \alpha \leq k + 1$. By definition, all k-th order partial derivatives of f belong to $\Lambda_{\alpha-k}$. It follows that all partial derivatives of f of order less than or equal to k also belong to $\Lambda_{\alpha-k}$. Since $\Lambda_{\alpha-k}$ is invariant under multiplication by polynomials, and since the operator $R^{\beta-\alpha,k}$ is a k-th order linear differential operator with polynomial coefficients (see Proposition 1.15), the function $R^{\beta-\alpha,k} f$ belongs to $\Lambda_{\alpha-k}$ as well. By Theorems 7.9 and 7.11, there exists a function $g \in L^\infty(\mathbb{B}_n)$ such that

$$R^{\beta-\alpha,k} f(z) = \int_{\mathbb{B}_n} \frac{g(w) \, dv_\beta(w)}{(1 - \langle z, w\rangle)^{n+1+\beta+k-\alpha}}, \qquad z \in \mathbb{B}_n.$$

Apply $R_{\beta-\alpha,k}$ and use Proposition 1.14. We obtain

$$f(z) = \int_{\mathbb{B}_n} \frac{g(w) \, dv_\beta(w)}{(1 - \langle z, w\rangle)^{n+1+\beta-\alpha}}$$

for all $z \in \mathbb{B}_n$. $\square$

It is easy to see that

$$\|f\| = |f(0)| + \sup\{(1 - |z|^2)^{k+1-\alpha}|R^{k+1}f(z)| : z \in \mathbb{B}_n\}$$

is a norm on Λ_α, where k is the positive integer satisfying $k \leq \alpha < k + 1$. Furthermore, Λ_α is a Banach space in this norm.

Let $\Lambda_{\alpha,0}$ be the closure of the set of polynomials in Λ_α. It is easy to check that $\Lambda_{\alpha,0}$ consists of holomorphic functions in $\mathbb{B}_n$ whose k-th order partial derivatives all belong to $\Lambda_{\alpha-k,0}$, where $k < \alpha \leq k + 1$.

Theorem 7.15. *Suppose $\alpha > 1$, $\beta > -1$, and k is the positive integer such that $k \leq \alpha < k + 1$. If $n + \beta - \alpha$ is not a negative integer, then the following conditions are equivalent for a holomorphic function f in $\mathbb{B}_n$:*

(a) The function f belongs to $\Lambda_{\alpha,0}$.

(b) The function $(1 - |z|^2)^{k+1-\alpha}R^{k+1}f(z)$ is in $C_0(\mathbb{B}_n)$.

(c) The function $(1 - |z|^2)^{k+1-\alpha}\partial^m f/\partial z^m$ is in $C_0(\mathbb{B}_n)$ for every multi-index m of nonnegative integers with $|m| = k + 1$.

(d) There exists a function $g \in C_0(\mathbb{B}_n)$ such that

$$f(z) = \int_{\mathbb{B}_n} \frac{g(w)\,dv_\beta(w)}{(1 - \langle z, w \rangle)^{n+1+\beta-\alpha}}$$

for all $z \in \mathbb{B}_n$.

Proof. The proof is similar to those of Theorems 7.13 and 7.14. We omit the details.
□

As usual, the space $C_0(\mathbb{B}_n)$ in the above theorem can be replaced by the space $C(\overline{\mathbb{B}}_n)$.

7.5 A Unified Treatment

In this section we unify the treatment of Λ_α for all $\alpha > 0$ and reveal the close relationship between Lipschitz spaces Λ_α and the Bloch space $\mathcal{B}$. This will be done via fractional differentiation and integration. As a consequence, we shall obtain atomic decomposition for the Lipschitz spaces.

Theorem 7.16. *Suppose $\alpha > 0$, $\beta > -1$, and f is holomorphic in $\mathbb{B}_n$. If $n + \beta - \alpha$ is not a negative integer, then $f \in \Lambda_\alpha$ if and only if there exists a function $g \in L^\infty(\mathbb{B}_n)$ such that*

$$f(z) = \int_{\mathbb{B}_n} \frac{g(w)\,dv_\beta(w)}{(1 - \langle z, w \rangle)^{n+1+\beta-\alpha}}, \qquad z \in \mathbb{B}_n.$$

Similar characterizations hold for $\Lambda_{\alpha,0}$ with $L^\infty(\mathbb{B}_n)$ replaced by $C_0(\mathbb{B}_n)$ or $C(\overline{\mathbb{B}}_n)$.

Proof. This follows from Theorems 7.9, 7.11, and 7.14.   □

Theorem 7.17. *Suppose $t > \alpha > 0$. If s is a real parameter such that neither $n + s$ nor $n + s + t$ is a negative integer, then a holomorphic function f in $\mathbb{B}_n$ belongs to Λ_α if and only if the function*

$$\varphi(z) = (1 - |z|^2)^{t-\alpha} R^{s,t} f(z)$$

is bounded in $\mathbb{B}_n$.

Proof. Let $\beta = s + \alpha + N$, where N is a sufficiently large positive integer.

If $f \in \Lambda_\alpha$, then by Theorem 7.16, we can then find a function $g \in L^\infty(\mathbb{B}_n)$ such that

$$f(z) = \int_{\mathbb{B}_n} \frac{g(w)\, dv_\beta(w)}{(1 - \langle z, w \rangle)^{n+1+s+N}}, \qquad z \in \mathbb{B}_n.$$

By Lemma 2.18, there exists a one-variable polynomial h such that

$$R^{s,t} f(z) = \int_{\mathbb{B}_n} \frac{h(\langle z, w \rangle) g(w)\, dv_\beta(w)}{(1 - \langle z, w \rangle)^{n+1+s+N+t}}.$$

Since

$$n + 1 + s + N + t = n + 1 + \beta + (t - \alpha),$$

an application of Theorem 1.12 shows that the function φ is bounded in $\mathbb{B}_n$.

On the other hand, if the function φ is bounded in $\mathbb{B}_n$, then by the reproducing formula in Theorem 2.2,

$$R^{s,t} f(z) = \frac{c_{s+N+t}}{c_\beta} \int_{\mathbb{B}_n} \frac{\varphi(w)\, dv_\beta(w)}{(1 - \langle z, w \rangle)^{n+1+s+N+t}}.$$

Apply the operator $R_{s,t}$ inside the integral sign and use Lemma 2.18. We obtain a polynomial $h(z, w)$ such that

$$f(z) = \int_{\mathbb{B}_n} \frac{h(z, w) \varphi(w)\, dv_\beta(w)}{(1 - \langle z, w \rangle)^{n+1+\beta-\alpha}}$$

$$= \sum_m p_m(z) \int_{\mathbb{B}_n} \frac{\overline{w^m}\, \varphi(w)\, dv_\beta(w)}{(1 - \langle z, w \rangle)^{n+1+\beta-\alpha}},$$

where the summation is over a finite number of terms and each $p_m(z)$ is a polynomial of z. By Theorem 7.16, each integral in the above sum defines a function in Λ_α. Since Λ_α is closed under multiplication by polynomials, we see that f is in Λ_α.   □

We state the little oh version of the preceding theorem without proof.

Theorem 7.18. *Suppose $t > \alpha > 0$. If s is a real parameter such that neither $n + s$ nor $n + s + t$ is a negative integer, then a holomorphic function f in $\mathbb{B}_n$ belongs to $\Lambda_{\alpha,0}$ if and only if the function*

$$h(z) = (1 - |z|^2)^{t-\alpha} R^{s,t} f(z)$$

belongs to $\mathbb{C}_0(\mathbb{B}_n)$.

As usual, the space $\mathbb{C}_0(\mathbb{B}_n)$ here can be replaced by the space $\mathbb{C}(\overline{\mathbb{B}}_n)$.

The next result shows that all the Lipschitz spaces Λ_α are isomorphic to the Bloch space as Banach spaces.

Theorem 7.19. *Suppose $\alpha > 0$ and s is a real parameter such that neither $n + s$ nor $n + s + \alpha$ is a negative integer. Then the fractional differential operator $R^{s,\alpha}$ maps the Lipschitz space Λ_α onto the Bloch space $\mathcal{B}$. Equivalently, a holomorphic function f in $\mathbb{B}_n$ belongs to Λ_α if and only if $R^{s,\alpha} f$ belongs to $\mathcal{B}$.*

Proof. Let t be a positive number large enough so that $t > \alpha$ and $n + s - t + \alpha$ is not a negative integer. Then the operators

$$R^{s-t+\alpha,t}, \quad R^{s-t+\alpha,t-\alpha}, \quad R^{s,\alpha},$$

are all well defined. By Theorem 7.17, the assumption $f \in \Lambda_\alpha$ is equivalent to the condition that the function

$$(1 - |z|^2)^{t-\alpha} R^{s-t+\alpha,t} f(z)$$

is bounded in $\mathbb{B}_n$. Since

$$R^{s-t+\alpha,t} = R^{s-t+\alpha,t-\alpha} R^{s,\alpha},$$

we conclude that $f \in \Lambda_\alpha$ if and only if the function

$$(1 - |z|^2)^{t-\alpha} R^{s-t+\alpha,t-\alpha} R^{s,\alpha} f(z)$$

is bounded in $\mathbb{B}_n$, which, according to Theorem 3.5, is equivalent to $R^{s,\alpha} f \in \mathcal{B}$. $\qquad\square$

Once again, we state the little oh version of the above theorem without proof.

Theorem 7.20. *Suppose $\alpha > 0$ and s is a real parameter such that neither $n + s$ nor $n + s + \alpha$ is a negative integer. Then the fractional differential operator $R^{s,\alpha}$ maps the space $\Lambda_{\alpha,0}$ onto the little Bloch space $\mathcal{B}_0$. Equivalently, a holomorphic function f in $\mathbb{B}_n$ belongs to $\Lambda_{\alpha,0}$ if and only if $R^{s,\alpha} f$ belongs to $\mathcal{B}_0$.*

We now obtain an atomic decomposition for functions in Λ_α.

Theorem 7.21. *Suppose $\alpha > 0$, $b > n$, and $b - \alpha - 1$ is not a negative integer. Then there exists a sequence $\{a_k\}$ in $\mathbb{B}_n$ such that Λ_α consists exactly of functions of the form*

$$f(z) = \sum_{k=1}^\infty c_k \frac{(1 - |a_k|^2)^b}{(1 - \langle z, a_k \rangle)^{b-\alpha}}, \tag{7.8}$$

where $\{c_k\} \in l^\infty$. A similar representation holds for $\Lambda_{\alpha,0}$ with l^∞ replaced by c_0.

Proof. Write $b - \alpha = n + 1 + s$. The assumptions on b imply that neither $n + s$ nor $n + s + \alpha$ is a negative integer, so that the operator $R^{s,\alpha}$ is well defined. Now a holomorphic function f in $\mathbb{B}_n$ admits a representation as in (7.8) if and only if

$$R^{s,\alpha} f(z) = \sum_{k=1}^\infty c_k \frac{(1 - |a_k|^2)^b}{(1 - \langle z, a_k \rangle)^{n+1+s+\alpha}} = \sum_{k=1}^\infty c_k \frac{(1 - |a_k|^2)^b}{(1 - \langle z, a_k \rangle)^b},$$

which, according to Theorem 3.23, is equivalent to $R^{s,\alpha} f \in \mathcal{B}$. This combined with Theorem 7.19 completes the proof. $\qquad\square$

7.6 Growth in Tangential Directions

For a holomorphic function f in $\mathbb{B}_n$ and a complex direction u (a unit vector in $\mathbb{C}^n$) we define the (complex) directional derivative

$$\frac{\partial f}{\partial u}(z) = \lim_{\lambda \to 0} \frac{f(z + \lambda u) - f(z)}{\lambda}, \tag{7.9}$$

where $\lambda \in \mathbb{C}$ and $z \in \mathbb{B}_n$. A simple application of the chain rule gives

$$\frac{\partial f}{\partial u}(z) = u_1 \frac{\partial f}{\partial z_1}(z) + \cdots + u_n \frac{\partial f}{\partial z_n}(z), \tag{7.10}$$

where $u = (u_1, \cdots, u_n)$.

By the Cauchy-Schwarz inequality for vectors in $\mathbb{C}^n$, the maximum modulus of all directional derivatives of f at z is given by

$$|\nabla f(z)| = \left(\left| \frac{\partial f}{\partial z_1}(z) \right|^2 + \cdots + \left| \frac{\partial f}{\partial z_n}(z) \right|^2 \right)^{1/2}. \tag{7.11}$$

So the quantity $|\nabla f(z)|$ is called the full complex gradient of f at z, or simply the gradient of f at z.

More generally, if X is any subspace of $\mathbb{C}^n$, we define $|\nabla_X f(z)|$ to be the maximum modulus of all the directional derivatives of f at z in directions $u \in X$. Two special situations will be considered.

First, if $z \neq 0$, we consider the case $X = [z]$, the one-dimensional subspace of $\mathbb{C}^n$ generated by z. There is only one (complex) direction in X and the correponding directional derivative is

$$\frac{1}{|z|} \left(z_1 \frac{\partial f}{\partial z_1}(z) + \cdots + z_n \frac{\partial f}{\partial z_n} \right) = \frac{Rf(z)}{|z|}.$$

Secondly, if $z \neq 0$, we consider the case where $X = \mathbb{C}^n \ominus [z]$, the subspace of $\mathbb{C}^n$ that is orthogonal to z. Directions determined by vectors in X are usually called tangential directions at z and the gradient corresponding to this $(n-1)$-dimensional subspace is denoted by $|\nabla_T f(z)|$ and will be called the tangential gradient of f at z. Thus

$$|\nabla_T f(z)| = \max \left\{ \left| \frac{\partial f}{\partial u}(z) \right| : u \perp z \right\}. \tag{7.12}$$

The following result exhibits the relationship among the radial derivative, the tangential gradient, and the invariant gradient of a holomorphic function in $\mathbb{B}_n$.

Theorem 7.22. *If $n > 1$ and $z \in \mathbb{B}_n - \{0\}$, then*

$$|\widetilde{\nabla} f(z)| \leq \frac{(1 - |z|^2)|Rf(z)|}{|z|} + (1 - |z|^2)^{1/2}|\nabla_T f(z)| \leq 2|\widetilde{\nabla} f(z)|$$

for all f holomorphic in $\mathbb{B}_n$.

Proof. Set $w = z$ in the definition of $Q_f(z)$ (see Section 3.1) and recall from Proposition 1.18 that z is an eigenvector of the Bergman matrix $B(z)$ corresponding to the eigenvalue $(1 - |z|^2)^{-2}$. We obtain

$$|\widetilde{\nabla} f(z)| = Q_f(z) \geq \frac{(1 - |z|^2)|Rf(z)|}{|z|}.$$

Similarly, by restricting $w \in [z]^{\perp}$ in the definition of $Q_f(z)$ and using the fact that such a w is an eigenvector of $B(z)$ corresponding to the eigenvalue $(1 - |z|^2)^{-1}$ (see Proposition 1.18 again), we obtain

$$Q_f(z) \geq (1 - |z|^2)^{1/2}|\nabla_T f(z)|.$$

It follows that

$$\frac{(1 - |z|^2)|Rf(z)|}{|z|} + (1 - |z|^2)^{1/2}|\nabla_T f(z)| \leq 2|\widetilde{\nabla} f(z)|.$$

On the other hand, for any nonzero vector $w \in \mathbb{C}^n$, we can write

$$w = u + v, \qquad u \in [z], \quad v \in [z]^{\perp}.$$

By part (e) of Proposition 1.18, we have

$$\langle B(z)w, w \rangle = (1 - |z|^2)^{-2}|u|^2 + (1 - |z|^2)^{-1}|v|^2.$$

It follows that

$$\frac{|\langle \nabla f(z), \overline{w} \rangle|}{\sqrt{\langle B(z)w, w \rangle}} \leq \frac{|\langle \nabla f(z), \overline{u} \rangle| + |\langle \nabla f(z), \overline{v} \rangle|}{\sqrt{(1 - |z|^2)^{-2}|u|^2 + (1 - |z|^2)^{-1}|v|^2}}$$

$$\leq \frac{|\langle \nabla f(z), \overline{u} \rangle|}{\sqrt{(1 - |z|^2)^{-2}|u|^2}} + \frac{|\langle \nabla f(z), \overline{v} \rangle|}{\sqrt{(1 - |z|^2)^{-1}|v|^2}}$$

$$= (1 - |z|^2) \left| \left\langle \nabla f(z), \frac{\overline{u}}{|u|} \right\rangle \right| + (1 - |z|^2)^{\frac{1}{2}} \left| \left\langle \nabla f(z), \frac{\overline{v}}{|v|} \right\rangle \right|.$$

Since

$$(1 - |z|^2) \left| \left\langle \nabla f(z), \frac{\overline{u}}{|u|} \right\rangle \right| = \frac{(1 - |z|^2)|Rf(z)|}{|z|},$$

and

$$(1 - |z|^2)^{\frac{1}{2}} \left| \left\langle \nabla f(z), \frac{\overline{v}}{|v|} \right\rangle \right| \leq (1 - |z|^2)^{\frac{1}{2}}|\nabla_T f(z)|,$$

taking the supremum over all nonzero vectors $w \in \mathbb{C}^n$ gives

$$Q_f(z) \leq \frac{1}{|z|}(1 - |z|^2)|Rf(z)| + (1 - |z|^2)^{1/2}|\nabla_T f(z)|.$$

This completes the proof of the theorem. $\square$

We now consider the smoothness of a function $f \in \Lambda_\alpha, 0 < \alpha < 1$, in tangential directions. First recall that the radial derivative of a function $f \in \Lambda_\alpha$ can grow at most at the rate of $(1 - |z|^2)^{\alpha-1}$ as $|z| \to 1^-$; see Theorem 7.9. The following result demonstrates the maximum growth rate for the tangential and invariant gradients of $f \in \Lambda_\alpha$.

Theorem 7.23. *Suppose $n > 1$ and $f \in \Lambda_\alpha$.*

(a) If $0 < \alpha < \frac{1}{2}$, then there exists a constant $C > 0$ such that

$$(1 - |z|^2)^{-\alpha}|\widetilde{\nabla}f(z)| \le C, \qquad (1 - |z|^2)^{\frac{1}{2}-\alpha}|\nabla_T f(z)| \le C,$$

for all z in $\mathbb{B}_n$.
(b) If $\alpha = \frac{1}{2}$, then there exists a constant $C > 0$ such that

$$(1 - |z|^2)^{-\frac{1}{2}}|\widetilde{\nabla}f(z)| \le C \log \frac{2}{1 - |z|^2}, \qquad |\nabla_T f(z)| \le C \log \frac{2}{1 - |z|^2},$$

for all $z \in \mathbb{B}_n$.
(c) If $\frac{1}{2} < \alpha < 1$, then there exists a constant $C > 0$ such that

$$(1 - |z|^2)^{-\frac{1}{2}}|\widetilde{\nabla}f(z)| \le C, \qquad |\nabla_T f(z)| \le C,$$

for all $z \in \mathbb{B}_n$.

Proof. Since $\Lambda_\alpha = \mathcal{B}_{1-\alpha}$ for $0 < \alpha < 1$, the estimates for $|\widetilde{\nabla}f(z)|$ follow from Theorem 7.2. The estimates for $|\nabla_T f(z)|$ then follow from the corresponding ones for $|\widetilde{\nabla}f(z)|$ and the inequality

$$|\nabla_T f(z)| \le (1 - |z|^2)^{-1/2}|\widetilde{\nabla}f(z)|,$$

which can be found in the proof of Theorem 7.22. $\square$

We are going to consider smooth curves $\gamma : [a, b] \to \mathbb{S}_n$. Such a curve γ will be called a complex tangential curve if $\gamma'(t)$ is bounded on $[a, b]$ and $\langle \gamma(t), \gamma'(t) \rangle = 0$ for every $t \in [a, b]$.

A smooth function $h : [a, b] \to \mathbb{C}$ is Lipschitz of order α, where $0 < \alpha < 1$, if there exists a constant $C > 0$ such that

$$|h(s) - h(t)| \le C|s - t|^\alpha \tag{7.13}$$

for all s and t in $[a, b]$. In this case, we also write $h \in \Lambda_\alpha$.

Theorem 7.24. *Suppose $0 < \alpha < \frac{1}{2}$ and f is holomorphic in $\mathbb{B}_n$. If $f \in \Lambda_\alpha$ and $\gamma : [a, b] \to \mathbb{S}_n$ is a complex tangential curve, then the function $g = f \circ \gamma : [a, b] \to \mathbb{C}$ is Lipschitz of order 2α.*

Proof. For $0 < r < 1$ let

$$g_r(t) = f_r(\gamma(t)) = f(r\gamma(t)), \qquad t \in [a, b].$$

By the chain rule,

$$g_r'(t) = r \sum_{k=1}^{n} \frac{\partial f}{\partial z_k}(r\gamma(t))\gamma_k'(t),$$

where

$$\gamma(t) = (\gamma_1(t), \cdots, \gamma_n(t)).$$

By the definition of complex directional derivatives,

$$g_r'(t) = r|\gamma'(t)|\frac{\partial f}{\partial v}(r\gamma(t)),$$

where $v = \gamma'(t)/|\gamma'(t)|$ is a tangential direction. Applying Theorem 7.23, we obtain a constant $C > 0$ (depending on f, but independent of t and r) such that

$$|g_r'(t)| \leq C|\gamma'(t)|(1 - r^2)^{\alpha - \frac{1}{2}}$$

for all $t \in [a, b]$ and $0 < r < 1$. Since $\gamma'(t)$ is bounded, we find a constant $C' > 0$ (depending on f and γ) such that

$$|g_r'(t)| \leq C'(1 - r^2)^{\alpha - \frac{1}{2}}$$

for all $t \in [a, b]$ and $0 < r < 1$. It follows easily that

$$|g_r(t_1) - g_r(t_2)| \leq C'|t_1 - t_2|(1 - r^2)^{\alpha - \frac{1}{2}}$$

for all t_1 and t_2 in $[a, b]$ and all $0 < r < 1$.

Let $r = 1 - |t_1 - t_2|^2$, where we assume that $|t_1 - t_2| < 1$. Then

$$|g_r(t_1) - g_r(t_2)| \leq C'|t_1 - t_2|^{2\alpha}.$$

On the other hand, under the same assumptions on t_1, t_2, and r, we have

$$|g(t_k) - g_r(t_k)| = |f(\gamma(t_k)) - f(r\gamma(t_k))| \leq C''|\gamma(t_k) - r\gamma(t_k)|^{\alpha}$$
$$= C''(1 - r)^{\alpha} = C''|t_1 - t_2|^{2\alpha},$$

where $k = 1, 2$, and C'' is a positive constant dependent on f only. An application of the triangle inequality now gives

$$|g(t_1) - g(t_2)| \leq C'''|t_1 - t_2|^{2\alpha}$$

for all t_1 and t_2 in $[a, b]$ with $|t_1 - t_2| < 1$, where C''' is a positive constant dependent on g, but not on t_1 and t_2. This clearly shows that $g \in \Lambda_{2\alpha}$. $\qquad\square$

7.7 Duality

In Chapter 4 we realized the dual of H^p, $0 < p < 1$, as the Bloch space using an integral pairing that involves certain weighted volume measures on the ball. In this section we show that the dual space of H^p, when $0 < p < 1$, can also be identified with the Lipschitz space Λ_α for some $\alpha > 0$, using the more natural H^2 integral pairing with the surface measure on $\mathbb{S}_n$.

Theorem 7.25. *Suppose $0 < p < 1$ and $\alpha = n(p^{-1} - 1)$. Then the dual space of H^p can be identified with Λ_α under the integral pairing*

$$\langle f, g \rangle = \lim_{r \to 1^-} \int_{\mathbb{S}_n} f_r \bar{g} \, d\sigma, \qquad f \in H^p, g \in \Lambda_\alpha.$$

In particular, the above limit always exists.

Proof. Let $\beta = (n/p) - (n+1)$. Then a computation using Lemma 1.11 shows that

$$\int_{\mathbb{S}_n} f(\zeta)\overline{g(\zeta)} \, d\sigma(\zeta) = \int_{\mathbb{B}_n} f(z)\overline{R^{-1,\alpha}g(z)} \, dv_\beta(z),$$

where f and g are bounded holomorphic functions in $\mathbb{B}_n$. The desired result then follows from Theorems 4.51 and 7.19. $\qquad\qquad\qquad\qquad\square$

Another type of duality theorems can be proved for Λ_α using weighted volume integral pairings. For example, if $0 < \alpha < 1$, then $\Lambda_\alpha = \mathcal{B}_{1-\alpha}$, so Theorems 7.5 and 7.6 can be restated for Lipschitz spaces. Next we show that this can actually be done for Λ_α for any $\alpha > 0$.

Theorem 7.26. *Suppose $\alpha > 0$, $\beta > -1$, and $s = \beta - \alpha$. If $s > -1$, then $(\Lambda_{\alpha,0})^* = A^1_\beta$ (with equivalent norms) under the integral pairing*

$$\langle f, g \rangle_s = \lim_{r \to 1^-} \int_{\mathbb{B}_n} f(rz)\overline{g(rz)} \, dv_s(z).$$

Proof. Let t be any real parameter such that the fractional differential operator $R^{t,\alpha}$ is a bounded invertible operator from Λ_α onto $\mathcal{B}_0$; see Theorem 7.20. Then F is a bounded linear functional on $\Lambda_{\alpha,0}$ if and only if $F \circ R_{t,\alpha}$ is a bounded linear functional on $\mathcal{B}_0$. Combining this with Theorem 3.16, we see that $F \in (\Lambda_{\alpha,0})^*$ if and only if there exists a function $g \in A^1_s$ such that

$$F \circ R_{t,\alpha}(f) = \langle f, g \rangle_s, \qquad f \in \mathcal{B}_0.$$

It follows that

$$F(f) = \langle R^{t,\alpha} f, g \rangle_s, \qquad f \in \Lambda_{\alpha,0}.$$

It is easy to see from the Taylor series representation of $\langle \, , \, \rangle_s$ that

$$\langle R^{t,\alpha} f, g \rangle_s = \langle f, R^{t,\alpha} g \rangle_s.$$

Thus $F \in (\Lambda_{\alpha,0})^*$ if and only if there exists $g \in A_s^1$ such that

$$F(f) = \langle f, R^{t,\alpha} g \rangle_s, \qquad f \in \Lambda_{\alpha,0}.$$

Let $h = R^{t,\alpha} g$. Then by Theorem 2.19, $g \in A_s^1$ if and only if $h \in A_\beta^1$, and

$$F(f) = \langle f, h \rangle_s, \qquad f \in \Lambda_{\alpha,0}.$$

This completes the proof of the theorem. $\qquad\square$

Theorem 7.27. *Suppose $\alpha > 0$, $\beta > -1$, $0 < p \le 1$, and*

$$s = \frac{n+1+\beta}{p} - (n+1+\alpha).$$

If $s > -1$, then $(A_\beta^p)^* = \Lambda_\alpha$ under the integral pairing $\langle\ ,\ \rangle_s$, where the equality holds with equivalent norms.

Proof. If α and β satisfy $\beta - p\alpha > -1$, which is true when $p = 1$, then we can argue as follows. $F \in (A_\beta^p)^*$ if and only if $F \circ R^{t,\alpha} \in (A_{\beta-p\alpha}^p)^*$, because Theorem 2.19 tells us that $R^{t,\alpha}$ is a bounded invertible operator from $A_{\beta-p\alpha}^p$ onto A_β^p. It follows from this and Theorem 3.17 that $F \in (A_\beta^p)^*$ if and only if there exists a function $g \in \mathcal{B}$ such that

$$F \circ R^{t,\alpha}(f) = \langle f, g \rangle_s, \qquad f \in A_{\beta-p\alpha}^p,$$

where

$$s = \frac{n+1+\beta-p\alpha}{p} - (n+1) = \frac{n+1+\beta}{p} - (n+1+\alpha).$$

Equivalently, for $f \in A_\beta^p$, we have

$$F(f) = \langle R_{t,\alpha} f, g \rangle_s = \langle f, R_{t,\alpha} g \rangle_s = \langle f, h \rangle_s,$$

where $h = R_{t,\alpha} g$ belongs to Λ_α; see Theorem 7.19.

Also note that the case $0 < \alpha < 1$ follows from Theorem 7.6. In the more general case, we argue directly using the integral representations of Λ_α.

First assume that $f \in A_\beta^p$ and $g \in \Lambda_\alpha$. By Theorem 7.16, there exists a function $h \in L^\infty(\mathbb{B}_n)$ such that

$$g(z) = \int_{\mathbb{B}_n} \frac{h(w)\,dv_{\alpha+s}(w)}{(1 - \langle z, w \rangle)^{n+1+s}}.$$

By Fubini's theorem and the reproducing formula in Theorem 2.2,

$$\langle f_r, g \rangle_s = \int_{\mathbb{B}_n} \overline{h(w)}\,dv_{\alpha+s}(w) \int_{\mathbb{B}_n} \frac{f_r(z)\,dv_s(z)}{(1 - \langle w, z \rangle)^{n+1+s}}$$

$$= \int_{\mathbb{B}_n} f_r(w)\overline{h(w)}\,dv_{\alpha+s}(w).$$

Let $r \to 1$ and use Lemma 2.15. We find a constant $C > 0$ such that

$$|\langle f, g \rangle_s| \le C \|h\|_\infty \|f\|_{\beta,p}.$$

This shows that every function $g \in \Lambda_\alpha$ induces a bounded linear functional on A_β^p via the integral pairing $\langle \ , \ \rangle_s$.

Conversely, if $F \in (A_\beta^p)^*$, and if $f \in A_\beta^p$, then we deduce from the reproducing formula

$$f_r(z) = \int_{\mathbb{B}_n} \frac{f_r(w) \, dv_s(w)}{(1 - \langle z, w \rangle)^{n+1+s}}$$

that

$$F(f_r) = \int_{\mathbb{B}_n} f(w) F_z \left[\frac{1}{(1 - \langle z, w \rangle)^{n+1+s}} \right] dv_s(w),$$

where $0 < r < 1$. Define a holomorphic function g in $\mathbb{B}_n$ by

$$\overline{g(w)} = F_z \left[\frac{1}{(1 - \langle z, w \rangle)^{n+1+s}} \right].$$

Then

$$F(f_r) = \langle f_r, g \rangle_s,$$

and it follows from the homogeneous expansion of the kernel function that

$$\overline{R^{s,\alpha} g(w)} = F_z \left[\frac{1}{(1 - \langle z, w \rangle)^{n+1+s+\alpha}} \right].$$

Similarly,

$$\overline{RR^{s,\alpha} g(w)} = (n+1+s+\alpha) F_z \left[\frac{\langle z, w \rangle}{(1 - \langle z, w \rangle)^{n+2+s+\alpha}} \right].$$

By the boundedness of F on A_β^p and Theorem 1.12, we can find a constant $C > 0$ such that

$$|RR^{s,\alpha} g(w)| \le (n+1+s+\alpha) \|F\| \left[\int_{\mathbb{B}_n} \frac{dv_\beta(z)}{|1 - \langle z, w \rangle|^{p(n+2+s+\alpha)}} \right]^{\frac{1}{p}} \le \frac{C\|F\|}{1 - |w|^2}.$$

This shows that $R^{s,\alpha} g \in \mathcal{B}$, and hence $g \in \Lambda_\alpha$ by Theorem 7.16. The proof of the theorem is complete.   □

Notes

This chapter advocates the approach to Lispchitz spaces using Bergman-type kernels. In fact, for any $\alpha > 0$ the Lipschitz space Λ_α is simply the image of the Bloch space under a fractional integral operator of order α. This identification shows that the Lipschitz spaces are to the Bloch space just the same as the Besov spaces are to the Bergman spaces. In particular, atomic decomposition and duality using volume-integral pairings follow immediately from the corresponding results for the Bloch space.

The Lipschitz spaces play a special role in duality issues, namely, the dual space of H^p, $0 < p < 1$, can be identified with a Lipschitz space under the more natural integral pairing on the unit sphere. This was first done in [34] for the unit disk and in [133] for the unit ball.

Sections 7.2 and 7.6 are basically from [94]. For $\alpha \geq 1$, the Lipschitz spaces Λ_α can also be characterized by higher order differences, but we chose to omit this result because it seems to be more real variable in nature.

Exercises

7.1. Prove the theorems of this chapter whose proofs were left out.

7.2. Formulate and prove an atomic decomposition for $\mathcal{B}_{\alpha,0}$ and $\Lambda_{\alpha,0}$.

7.3. For $n = 1$ characterize lacunary series in Λ_α in terms of the coefficients.

7.4. Find a suitable integral pairing under which the dual space of $\Lambda_{\alpha,0}$ can be identified with B_1. Similarly, find an integral pairing under which the dual space of B_p, $0 < p \leq 1$, can be identified with Λ_α.

7.5. Suppose $\alpha > 0$ and

$$f(z) = \sum_m a_m z^m$$

is holomorphic in $\mathbb{B}_n$. Show that $f \in \Lambda_\alpha$ if and only if the function

$$g(z) = \sum_m |m|^\alpha z^m$$

is in the Bloch space.

7.6. Show that every polynomial is a pointwise multiplier of Λ_α, where $\alpha > 0$.

7.7. Suppose $\alpha > 1$ and f is holomorphic in $\mathbb{B}_n$. Show that $f \in \mathcal{B}_\alpha$ if and only if the function $(1 - |z|^2)^{\alpha-1}|f(z)|$ is bounded in $\mathbb{B}_n$.

7.8. Suppose $\alpha > 0$ and f is holomorphic in $\mathbb{B}_n$. If the functions

$$(1 - |z|^2)^\alpha \frac{\partial^m f}{\partial z^m}(z)$$

are bounded for each multi-index m with $|m| = k$, where k is a positive integer. Then the functions

$$(1 - |z|^2)^\alpha \frac{\partial^m f}{\partial z^m}(z)$$

are bounded for for each multi-index m with $|m| \leq k$.

7.9. Suppose $\alpha > 0$ and $f \in \Lambda_\alpha$. If $f(0) = 0$, show that there exist functions $f_k \in \Lambda_\alpha, 1 \le k \le n$, such that

$$f(z) = \sum_{k=1}^{n} z_k f_k(z)$$

for $z \in \mathbb{B}_n$.

7.10. Show that the condition $\alpha > \frac{1}{2}$ in part (a) of Theorem 7.2 is necessary.

7.11. Suppose $0 < \alpha < \frac{1}{2}$ and f is holomorphic in $\mathbb{B}_n$. Show that $f \in \Lambda_\alpha$ if and only if the function

$$h(z) = (1 - |z|^2)^{-\alpha} |\widetilde{\nabla} f(z)|$$

is bounded in $\mathbb{B}_n$, and $f \in \Lambda_{\alpha,0}$ if and only if h is in $C_0(\mathbb{B}_n)$ if and only if h is in $C(\overline{\mathbb{B}}_n)$.

7.12. Suppose $\alpha > 1, t > 0$, and s is a real parameter such that the fractional differential operator $R^{s,t}$ is well defined. Then a holomorphic function f in $\mathbb{B}_n$ belongs to $\mathcal{B}_\alpha$ if and only if the function

$$h(z) = (1 - |z|^2)^{\alpha+t-1} R^{s,t} f(z)$$

is bounded in $\mathbb{B}_n$, and $f \in \mathcal{B}_{\alpha,0}$ if and only if h is in $C_0(\mathbb{B}_n)$ if and only if h is in $C(\overline{\mathbb{B}}_n)$.

7.13. Suppose $\alpha > 1$ and s is a real parameter such that $R^{s,\alpha-1}$ is well defined. Show that a holomorphic function f in $\mathbb{B}_n$ belongs to the Bloch space $\mathcal{B}$ if and only if $R^{s,\alpha-1} f$ belongs to $\mathcal{B}_\alpha$. The little oh version of this result also holds.

7.14. Formulate and prove the little oh versions of Theorems 7.16.

7.15. Suppose f is holomorphic in $\mathbb{B}_n$. Then the following conditions are equivalent:

(a) $f \in \Lambda_1$.
(b) There exists a constant $C > 0$ such that

$$|f(z + h) + f(z - h) - 2f(z)| \le C|h|$$

holds whenever $z \in \mathbb{B}_n$ and $z \pm h \in \mathbb{B}_n$.
(c) f is in the ball algebra and there exists a constant $C > 0$ such that

$$|f(\zeta + h) + f(\zeta - h) - 2f(\zeta)| \le C|h|$$

for all $\zeta \in \mathbb{S}_n$ and $\zeta \pm h \in \mathbb{S}_n$.

7.16. Suppose $n > 1, \frac{1}{2} < \alpha < 1$, and $f \in \Lambda_\alpha$. If $\gamma : [a,b] \to \overline{\mathbb{B}}_n$ is a complex tangential curve, then the derivative of the function $f \circ \gamma : [a,b] \to \mathbb{C}$ is Lipschitz of order $2\alpha - 1$. See [94].

7.17. Suppose $n > 1$, $f \in \Lambda_{\frac{1}{2}}$, and $\gamma : [a, b] \to \overline{\mathbb{B}_n}$ is a complex tangential curve. Then the function

$$h(t) = f \circ \gamma(t), \qquad t \in [a, b],$$

satisfies

$$|h(t + h) + h(t - h) - 2h(t)| \leq Ch$$

for some constant $C > 0$ and all $t \in (a, b)$ and $h > 0$ with $t \pm h \in [a, b]$. See [94].

7.18. For $\alpha > 0$ define an operator T_α by

$$T_\alpha f(z) = (1 - |z|^2)^{n+1} \int_{\mathbb{B}_n} \frac{f(w)}{(1 - \langle z, w \rangle)^{2(n+1)}} (1 - |w|^2)^{\alpha-1} \, dv(w).$$

Show that T_α is an emedding of $\mathcal{B}_\alpha$ into $L^\infty(\mathbb{B}_n)$, and that T_α maps $\mathcal{B}_{\alpha,0}$ into $\mathbb{C}_0(\mathbb{B}_n)$.

7.19. For any $\alpha > 0$ define a function

$$d_\alpha : \mathbb{B}_n \times \mathbb{B}_n \to [0, \infty)$$

by

$$d_\alpha(z, w) = \sup\{|f(z) - f(w)| : \|f\|_\alpha \leq 1\},$$

where $\|\ \|_\alpha$ is the norm in $\mathcal{B}_\alpha$. Show that d_α is a distance in $\mathbb{B}_n$ and that

$$\lim_{w \to z} \frac{d_\alpha(z, w)}{|z - w|} = \sup\{|\nabla f(z)| : \|f\|_\alpha \leq 1\}$$

for every $z \in \mathbb{B}_n$. See [132] and [134].

7.20. With notation from the previous exercise, show that a holomorphic function f in $\mathbb{B}_n$ belongs to $\mathcal{B}_\alpha$ if and only if there exists a positive constant $C = C_f$ such that

$$|f(z) - f(w)| \leq Cd_\alpha(z, w)$$

for all z and w in $\mathbb{B}_n$. See [132] and [134].

7.21. Show that each space $\mathcal{B}_\alpha$ is isomorphic to $\mathcal{B}$ via an appropriate fractional differential (or integral) operator.

7.22. Suppose a is any real parameter such that the operator $R^{a,1}$ is well defined. Show that a holomorphic function f in $\mathbb{B}_n$ belongs to $\mathcal{B}_\alpha$ if and only if the function

$$(1 - |z|^2)^\alpha R^{a,1} f(z)$$

is bounded in $\mathbb{B}_n$. A little oh version of this result also holds.

References

1. P. Ahern and J. Bruna, Maximal and area integral characterizations of Hardy-Sobolev spaces in the unit ball of $\mathbb{C}^n$, *Rev. Mat. Iberoamericana* **4** (1988), 123-153.
2. P. Ahern and W. Cohn, Besov spaces, Sobolev spaces, and Cauchy integrals, *Michigan Math. J.* **39** (1972), 239-261.
3. P. Ahern, M. Flores, and W. Rudin, An invariant volume mean value property, *J. Funct. Anal.* **111** (1993), 380-397.
4. P. Ahern and W. Rudin, Bloch functions, BMO, and boundary zeros, *Indiana Univ. Math. J.* **36** (1987), 131-148.
5. P. Ahern and R. Schneider, Holomorphic Lipschitz functions in pseudo-convex domains, *Amer. J. Math.* **101** (1979), 543-565.
6. A.B. Aleksandrov, Function Theory in the Ball, in *Several Complex Variables II* (G.M. Khenkin and A.G. Vitushkin, editors), Springer-Verlag, Berlin, 1994.
7. J.M. Anderson, Bloch functions: The basic theory, in *Operators and Function Theory* (S.C. Power, editor), D. Reidel, 1985, 1-17.
8. J.M. Anderson, J. Clunie, and Ch. Pommerenke, On Bloch functions and normal functions, *J. Reine Angew. Math.* **270** (1974), 12-37.
9. J. Arazy, Multipliers of Bloch functions, *University of Haifa Mathematics Publications* **54**, 1982.
10. J. Arazy and S. Fisher, Some aspects of the minimal, Möbius invariant space of analytic functions on the unit disk, *Springer Lecture Notes in Math.* **1070**, 24-44.
11. J. Arazy and S. Fisher, The uniqueness of the Dirichlet space among Möbius invariant Hilbert spaces, *Illinois J. Math.* **29** (1985), 449-462.
12. J. Arazy, S. Fisher, and S. Janson, Membership of Hankel operators on the ball in unitary ideals, *J. London Math. Soc.* **43** (1991), 485-508.
13. J. Arazy, S. Fisher, and J. Peetre, Möbius invariant function spaces, *J. Reine Angew. Math.* **363** (1985), 110-145.
14. S. Axler, The Bergman space, the Bloch space, and commutators of multiplication operators, *Duke Math. J.* **53** (1986), 315-332.
15. S. Axler, Bergman spaces and their operators, in *Surveys of Some Recent Results in Operator Theory* (Volume 1, J.B. Conway and B.B. Morrel, editors), Pitman Research Notes in Math. **171**, 1988, 1-50.
16. S. Axler, J. McCarthy, and D. Sarason (editors), *Holomorphic Spaces*, Mathematical Sciences Research Institute Publications **33**, Cambridge University Press, 1998.

17. F. Beatrous and J. Burbea, Characterizations of spaces of holomorphic functions in the ball, *Kodai Math. J.* **8** (1985), 36-51.

18. F. Beatrous and J. Burbea, Holomorphic Sobolev spaces on the ball, *Dissertationes Mathematicae* **276**, Warszawa, 1989.

19. D. Békollé, C. Berger, L. Coburn, and K. Zhu, BMO in the Bergman metric on bounded symmetric domains, *J. Funct. Anal.* **93** (1990), 310-350.

20. J. Bennet, D. Stegenga, and R. Timoney, Coefficients of Bloch and Lipschitz functions, *Illinois J. Math.* **25** (1981), 520-531.

21. C. Berger, L. Coburn, and K. Zhu, Function theory on Cartan domains and the Berezin-Toeplitz symbol calculus, *Amer. J. Math.* **110** (1988), 921-953.

22. J. Bergh and J. Löfström, *Interpolation Spaces–An Introduction*, Springer-Verlag, Berlin, 1976.

23. J.S. Choa and B.R. Choe, A Littlewood-Paley type identity and a characterization of BMOA, *Complex Variables* **17** (1991), 15-23.

24. J.S. Choa, H.O. Kim, and Y.Y. Park, A Bergman-Carleson measure characterization of Bloch functions in the unit ball of $\mathbb{C}^n$, *Bull. Korean Math. Soc.* **29** (1992), 285-293.

25. B.R. Choe, Projections, the weighted Bergman spaces, and the Bloch space, *Proc. Amer. Math. Soc.* **108** (1990), 127-136.

26. B.R. Choe, W. Ramey, and D. Ullrich, Bloch-to-BMOA pullbacks on the disk, *Proc. Amer. Math. Soc.* **125** (1997), 2987-2996.

27. J. Cima and W. Wogen, Extreme points of the unit ball of the Bloch space, *Michigan Math. J.* **25** (1978), 213-222.

28. J. Cima and W. Wogen, On isometries of the Bloch space, *Illinois J. Math.* **24** (1980), 313-316.

29. J. Cima and W. Wogen, A Carleson measure theorem for the Bergman space of the ball, *J. Operator Theory* **7** (1982), 157-165.

30. R. Coifman and R. Rochberg, Representation theorems for holomorphic and harmonic functions in L^p, *Asterisque* **77** (1980), 11-66.

31. R. Coifman, R. Rochberg, and G. Weiss, Factorization theorems for Hardy spaces of several variables, *Ann. Math.* **103** (1976), 611-635.

32. A.E. Djrbashian and F.A. Shamoian, *Theory of A_α^p Spaces*, B.G. Tuebner Verlagsgesellschasft, Leipzig, 1988.

33. P. Duren, *Theory of H^p Spaces*, Academic Press, New York, 1970.

34. P. Duren, B. Romberg, and A. Shields, Linear functionals on H^p spaces with $0 < p < 1$, *J. Reine Angew. Math.* **238** (1969), 32-60.

35. C. Fefferman, Characterizations of bounded mean oscillation, *Bull. Amer. Math. Soc.* **77** (1971), 587-588.

36. C. Fefferman and E. Stein, H^p spaces of several variables, *Acta Math.* **129** (1972), 137-193.

37. T. Flett, On the rate of growth of mean values of holomorphic and harmonic functions, *Proc. London. Math. Soc.* **20** (1970), 749-768.

38. F. Forelli, The isometries of H^p, *Canadian J. Math.* **16** (1964), 721-728.

39. F. Forelli and W. Rudin, Projections on spaces of holomorphic functions in balls, *Indiana Univ. Math. J.* **24** (1974), 593-602.

40. M. Frazier and B. Jawerth, Decomposition of Besov spaces, *Indiana Univ. Math. J.* **34** (1985), 777-799.

41. S. Gallot, D. Hulin, and J. Lafontaine, *Riemannian Geometry*, Springer-Verlag, New York, 1993.

42. J. Garnett, *Bounded Analytic Functions*, Academic Press, New York, 1982.

43. D. Girela, Analytic functions of bounded mean oscillation, 61-170 in *Complex Function Spaces* (edited by R. Aulaskari), Joensuu, 2001.

44. M. Gowda, Nonfactorization theorems in weighted Bergman and Hardy spaces on the unit ball of $\mathbb{C}^n$, *Trans. Amer. Math. Soc.* **277** (1983), 203-212.

45. I. Graham, The radial derivative, fractional integrals, and the comparative growth of means of holomorphic functions on the unit ball in $\mathbb{C}^n$, in *Recent Developments in Several Complex Variables, Ann. Math. Studies* **100** (1981), 171-178.

46. K.T. Hahn and E.H. Youssfi, Möbius invariant Besov p-spaces and Hankel operators in the Bergman space on the unit ball, *Complex Variables* **17** (1991), 89-104.

47. K.T. Hahn and E.H. Youssfi, M-harmonic Besov p-spaces and Hankel operators in the Bergman space on the ball in $\mathbb{C}^n$, *Manuscripta Math.* **71** (1991), 67-81.

48. K.T. Hahn and E.H. Youssfi, Tangential boundary behavior of M-harmonic Besov functions, *J. Math. Anal. Appl.* **175** (1993), 206-221.

49. G.H. Hardy and J.E. Littlewood, Some properties of fractional integrals II, *Math. Z.* **34** (1932), 403-439.

50. W.W. Hastings, A Carleson measure theorem for Bergman spaces, *Proc. Amer. Math. Soc.* **52** (1975), 237-241.

51. H. Hedenmalm, B. Korenblum, and K. Zhu, *Theory of Bergman Spaces*, Springer-Verlag, New York, 2000.

52. K. Hoffman, *Banach Spaces of Analytic Functions*, Prentice Hall, Englewood Cliffs, New Jersey, 1962.

53. F. Holland and D. Walsh, Criteria for membership of Bloch space and its subspace, BMOA, *Math. Ann.* **273** (1986), 317-335.

54. L. Hörmander, L^p estimates for (pluri-)subharmonic functions, *Math. Scand.* **20** (1967), 65-78.

55. L.K. Hua, *Harmonic Analysis of Functions in Several Complex Variables in the Classical Domains*, Amer. Math. Soc., Providence, R.I., 1963.

56. K. Jarosz (editor), *Function Spaces*, Contemporary Mathematics **232**, Amer. Math. Soc., Providence, 1999.

57. M. Jevtic, A note on the Carleson measure characterization of BMOA functions in the unit ball, *Complex Variables* **17** (1992), 189-194.

58. F. John and L. Nirenberg, On functions of bounded mean oscillation, *Comm. Pure Appl. Math.* **14** (1961), 415-426.

59. C. Kolaski, Isometries of weighted Bergman spaces, *Canadian J. Math.* **34** (1982), 910-915.

60. A. Koranyi and S. Vagi, Singular integrals on homogeneous spaces and some problems of classical analysis, *Ann. Scuola Norm. Sup. Pisa Cl. Sci.(3)* **25** (1971), 575-648.

61. S. Krantz, *Function Theory of Several Complex Variables*, Wiley, New York, 1982.

62. S. Krantz and D. Ma, On isometric isomorphisms of the Bloch space on the unit ball of $\mathbb{C}^n$, *Michigan Math. J.* **36** (1989), 173-180.

63. O. Kures and K. Zhu, On the boundedness of a class of integral operators, preprint, 2004.

64. Z.J. Lou, Characterizations of Bloch functions in the unit ball, *Kodai Math. J.* **16** (1993), 74-78.

65. D. Luecking, A technique for characterizing Carleson measures on Bergman spaces, *Proc. Amer. Math. Soc.* **87** (1983), 656-660.

66. D. Luecking, Forward and reverse Carleson inequalities for functions in Bergman spaces and their derivatives, *Amer. J. Math.* **107** (1985), 85-111.

67. D. Luecking, Representation and duality in weighted spaces of analytic functions, *Indiana Univ. Math. J.* **34** (1985), 319-336.

68. T. Mazur, Canonical isometry on weighted Bergman spaces, *Pacific J. Math.* **136** (1989), 303-310.

69. J. Mitchell and K.T. Hahn, Representation of linear functionals in H^p spaces over bounded symmetric domains in $\mathbb{C}^n$, *J. Math. Anal. Appl.* **56** (1976), 379-396.

70. A. Nagel and W. Rudin, Möbius invariant function spaces on balls and spheres, *Duke Math. J.* **43** (1976), 841-865.

71. M. Nowark, On the Bloch space in the unit ball of $\mathbb{C}^n$, *Ann. Acad. Sci. Fenn.* **23** (1998), 461-473.

72. M. Nowark, Bloch and Möbius invariant Besov spaces on the unit ball of $\mathbb{C}^n$, *Complex Variables* **44** (2001), 1-12.

73. M. Nowark, Function spaces in the unit ball of $\mathbb{C}^n$, 171-197 in *Complex Function Spaces* (edited by R. Aulaskari), Joensuu, 2001.

74. V.L. Oleinik, Embedding theorems for weighted classes of harmonic and analytic functions, *J. Soviet Math.* **9** (1978), 228-243.

75. V.L. Oleinik and B.S. Pavlov, Embedding theorems for weighted classes of harmonic and analytic functions, *J. Soviet Math.* **2** (1974), 135-142.

76. C. Ouyang, Some classes of functions with exponential decay in the unit ball of $\mathbb{C}^n$, *Publ. Res. Inst. Math. Sci.* **25** (1989), 263-277.

77. C. Ouyang, An extension theorem for Bloch functions in the unit ball, *Acta Math. Sci.* **10** (1990), 455-461.

78. C. Ouyang, W. Yang, and R. Zhao, Characterizations of Bergman spaces and the Bloch space in the unit ball of $\mathbb{C}^n$, *Trans. Amer. Math. Soc.* **374** (1995), 4301-4312.

79. C. Ouyang, W. Yang, and R. Zhao, Möbius invariant Q_p spaces associated with the Green function on the unit ball, *Pacific J. Math.* **182** (1998), 69-99.

80. M. Pavlovic, Inequalities for the gradient of eigenfunctions of the invariant Laplacian in the unit ball, *Indag. Mathem.* **2** (1991), 89-98.

81. J. Peetre, *New Thoughts on Besov Spaces*, Duke Univ. Math. Series **1**, Durham, 1976.

82. J. Peetre, Invariant function spaces and Hankel operators–A rapid survey, *Exposition Math.* **5** (1986), 3-16.

83. M. Peloso, Möbius invariant spaces on the unit ball, *Michigan Math. J.* **39** (1992), 509-536.

84. Ch. Pommerenke, On Bloch functions, *J. London Math. Soc.* **2** (1970), 689-695.

85. Ch. Pommerenke, On univalent functions, Bloch functions, and VMOA, *Math. Ann.* **236** (1978), 199-208.

86. S. Power, Vanishing Carleson measures, *Bull. London Math. Soc.* **12** (1980), 207-210.

87. S. Power, Hörmander's Carleson theorem for the ball, *Glasg. Math. J.* **26** (1985), 13-17.

88. W. Ramey and D. Ullrich, Bounded mean oscillation of Bloch pullbacks, *Math. Ann.* **291** (1991), 591-606.

89. R.M. Range, *Holomorphic Functions and Integral Representation in Several Complex Variables*, Springer-Verlag, New York, 1986.

90. R. Rochberg, Decomposition theorems for Bergman spaces and their applications, in *Operators and Function Theory* (S.C. Power, editor), D. Reidel, 1985, 225-277.

91. R. Rochberg and S. Semmes, A decomposition theorem for BMO and applications, *J. Funct. Anal.* **67** (1986), 228-263.

92. L. Rubel and A. Shields, The second duals of certain spaces of analytic functions, *J. Australian Math. Soc.* **11** (1970), 276-280.

93. L. Rubel and R. Timoney, An extremal property of the Bloch space, *Proc. Amer. Math. Soc.* **75** (1979), 45-49.

94. W. Rudin, *Function Theory in the Unit Ball of* $\mathbb{C}^n$, Springer-Verlag, New York, 1980.

95. W. Rudin, *New Constructions of Functions Holomorphic in the Unit Ball of* $\mathbb{C}^n$, Amer. Math. Soc., Providence, 1986.

96. D. Sarason, Functions of vanishing mean oscillation, *Trans. Amer. Math. Soc.* **207** (1975), 391-405.

97. J. Shapiro, Macey topologies, reproduceing kernels, and diagonal maps on the Hardy and Bergman spaces, *Duke Math. J.* **43** (1976), 187-202.

98. J. Shi, Inequalities for the integral means of holomorphic functions and their derivatives in the unit ball of $\mathbb{C}^n$, *Trans. Amer. Math. Soc.* **328** (1991), 619-637.

99. A. Shields and D. Williams, Bounded projections, duality, and multipliers in spaces of analytic functions, *Trans. Amer. Math. Soc.* **162** (1971), 287-302.

100. A. Shields and D. Williams, Bounded projections, duality, and multipliers in spaces of analytic functions, *J. Reine Angew. Math.* **299/300** (1978), 256-279.

101. A. Shields and D. Williams, Bounded projections and the growth of harmonic conjugates in the unit disk, *Michigan Math. J.* **29** (1982), 3-26.

102. M. Spivak, *Differential Geometry*, Volume IV, Publish or Perish, Berkeley, 1979.

103. D. Stegenga, Bounded Toeplitz operators on H^1 and applications of the duality between H^1 and the functions of bounded mean oscillation, *Amer. J. Math.* **98** (1976), 573-598.

104. E. Stein, *Singular Integrals and Differentiability Properties of Functions*, Princeton Univ. Press, 1971.

105. E. Stein, *Boundary Behavior of Holomorphic Functions of Several Complex Variables*, Princeton Univ. Press, 1972.

106. M. Stoll, Rate of growth of p'th means of invariant potentials in the unit ball of $\mathbb{C}^n$, *J. Math. Anal. Appl.* **143** (1989), 480-499.

107. M. Stoll, Rate of growth of p'th means of invariant potentials in the unit ball of $\mathbb{C}^n$, II, *J. Math. Anal. Appl.* **165** (1992), 374-398.

108. M. Stoll, A characterization of Hardy spaces on the unit ball of $\mathbb{C}^n$, *J. London Math. Soc.* **48** (1993), 126-136.

109. M. Stoll, *Invariant Potential Theory in the Unit Ball of* $\mathbb{C}^n$, Cambridge University Press, London, 1994.

110. K. Stroethoff, The Bloch space and Besov spaces analytic functions, *Bull. Australian Math. Soc.* **54** (1996), 211-219.

111. R. Timoney, Bloch functions in several complex variables I, *Bull. London Math. Soc.* **12** (1980), 241-267.

112. R. Timoney, Bloch functions in several complex variables II, *J. Reine Angew. Math.* **319** (1980), 1-22.

113. R. Timoney, Maximal invariant spaces of analytic functions, *Indiana Univ. Math. J.* **31** (1982), 651-663.

114. D. Ullrich, Radial limits of M-subharmonic functions, *Trans. Amer. Math. Soc.* **292** (1985), 501-518.

115. D. Vukotić, A sharp estimate for A_α^p functions in $\mathbb{C}^n$, *Proc. Amer. Math. Soc.* **117** (1993), 753-756.

116. P. Wojtaszczyk, On functions in the ball algebra, *Proc. Amer. Math. Soc.* **85** (1982), 184-186.

117. P. Wojtaszczyk, Hardy spaces on the complex ball are isomorphic to Hardy spaces on the disk, $1 \le p < \infty$, *Ann. Math.* **118** (1983), 21-34.

118. S. Yamashita, Criteria for functions to be of Hardy class H^p, *Proc. Amer. Math. Soc.* **75** (1979), 69-72.

119. S. Yamashita, Holomorphic functions and area integrals, *Boll. Un. Mat. Ital. A* **6** (1982), 115-120.

120. R. Zhao, private communication.

121. K. Zhu, VMO, ESV, and Toeplitz operators on the Bergman space, *Trans. Amer. Math. Soc.* **302** (1987), 617-646.

122. K. Zhu, Positive Toeplitz operators on weighted Bergman spaces of bounded symmetric domains, *J. Operator Theory* **20** (1988), 329-357.

123. K. Zhu, Multipliers of BMO in the Bergman metric with applications to Toeplitz operators, *J. Funct. Anal.* **87** (1989), 31-50.

124. K. Zhu, *Operator Theory in Function Spaces*, Marcel Dekker, New York, 1990.

125. K. Zhu, On certain composition operators and unitary operators, *Proc. Symp. Pure Math.* **51** , Part II, 371-385, Amer. Math. Soc., Providence, RI, 1990.

126. K. Zhu, Möbius invariant Hilbert spaces of holomorphic functions in the unit ball of $\mathbb{C}^n$, *Trans. Amer. Math. Soc.* **323** (1991), 823-842.

127. K. Zhu, Analytic Besov spaces, *J. Math. Anal. Appl.* **157** (1991), 318-336.

128. K. Zhu, Duality of Bloch spaces and norm convergence of Taylor series, *Michigan Math. J.* **38** (1991), 89-101.

129. K. Zhu, Schatten class Hankel operators on the Bergman space of the unit ball, *Amer. J. Math.* **113** (1991), 147-167.

130. K. Zhu, A Forelli-Rudin type theorem, *Complex Variables* **16** (1991), 107-113.

131. K. Zhu, BMO and Hankel operators on Bergman spaces, *Pacific J. Math.* **155** (1992), 377-395.

132. K. Zhu, Bloch type spaces of analytic functions, *Rocky Mountain J. Math.* **23** (1993), 1143-1177.

133. K. Zhu, Bergman and Hardy spaces with small exponents, *Pacific J. Math.* **162** (1994), 189-199.

134. K. Zhu, Distances and Banach spaces of holomorphic functions on complex domains, *J. London Math. Soc.* **49** (1994), 163-182.

135. K. Zhu, Holomorphic Besov spaces on bounded symmetric domains, *Quart. J. Math. Oxford.(2)* **46** (1995), 239-256.

136. K. Zhu, Holomorphic Besov spaces on bounded symmetric domains II, *Indiana Univ. Math. J.* **44** (1995), 1017-1031.

137. K. Zhu, A sharp norm estimate of the Bergman projection on L^p spaces, preprint, 2003.

Index

(continued from page ii)